S0-ATH-795

Layered Materials for Structural Applications

MATERIALS RESEARCH SOCIETY
SYMPOSIUM PROCEEDINGS VOLUME 434

Layered Materials for Structural Applications

Symposium held April 8-11, 1996, San Francisco, California, U.S.A.

EDITORS:

J.J. Lewandowski
Case Western Reserve University
Cleveland, Ohio, U.S.A.

C.H. Ward
Air Force Office of Scientific Research
Bolling Air Force Base, D.C., U.S.A.

M.R. Jackson
General Electric Corporate R&D
Schenectady, New York, U.S.A.

W.H. Hunt, Jr.
Alcoa Technical Center
Alcoa Center, Pennsylvania, U.S.A.

PITTSBURGH, PENNSYLVANIA

Effort sponsored by the Air Force Office of Scientific Research, Air Force Material Command, USAF, under F49620-96-1-0179. The U.S. Government is authorized to reproduce and distribute reprints for Governmental purposes notwithstanding any copyright notation thereon. The views and conclusions herein are those of the authors and should not be interpreted as necessarily representing the official policies or endorsements, either expressed or implied, of the Air Force Office of Scientific Research or the U.S. Government.

This work was supported in part by the Office of Naval Research under Grant Number N00014-96-1-0741. The United States Government has a royalty-free license throughout the world in all copyrightable material contained herein.

Single article reprints from this publication are available through University Microfilms Inc., 300 North Zeeb Road, Ann Arbor, Michigan 48106

CODEN: MRSPDH

Published by:

Materials Research Society
9800 McKnight Road
Pittsburgh, Pennsylvania 15237
Telephone (412) 367-3003
Fax (412) 367-4373
Website: http://www.mrs.org/

Library of Congress Cataloging in Publication Data

Layered materials for structural applications : symposium held April 8–11, 1996, San Francisco, California, U.S.A. / editors, J.J. Lewandowski, C.H. Ward, M.R. Jackson, W.H. Hunt, Jr.
p. cm—(Materials Research Society symposium proceedings ; v. 434)
Includes bibliographical references and indexes.
ISBN 1-55899-337-1
1. Laminated materials—Congresses. 2. Coatings—Congresses.
I. Lewandowski, J.J. II. Ward, C.H. III. Jackson, M.R. IV. Hunt, W.H., Jr.
V. Series: Materials Research Society symposium proceedings ; v. 434.
TA418.9.L3L38 1996 96-31606
620.1'1—dc20 CIP

Manufactured in the United States of America

TABLE OF CONTENTS

PART I: APPLICATIONS

PART II: PROCESSING

*Invited Paper

PART III: STABILITY ISSUES

PART IV: MECHANICAL BEHAVIOR

*Invited Paper

*Invited Paper

PREFACE

Layered materials and systems based on metallic, intermetallic, polymeric and ceramic constituents are becoming increasingly important to meet the structural requirements of current and future high-performance products. In response to various research and development activities in these areas, Symposium U was organized to cover a range of topics dealing with layered materials for structural applications and was supported by contributions from The Air Force Office of Scientific Research and Office of Naval Research. The support of these organizations is gratefully acknowledged. This proceedings volume is based on the first MRS symposium dedicated to current research and development of layered materials which are being considered for a range of structural applications.

The meeting began with overviews on structural applications of layered systems and highlighted applications such as thermal barrier coatings, aircraft structural components, and wear-resistant coatings for a variety of applications. Processing techniques such as EB deposition processing, reactive sputter deposition, sedimentation processing, pressureless co-sintering, and rapid prototyping via laminated object manufacturing were subsequently covered in a following session. Microstructural stability issues were additionally covered and highlighted as a critical area requiring further investigation. The largest number of papers presented focused on the mechanical behavior and modeling of layered systems and revealed significant effects of layer thickness, spacing, and constituent properties on the fracture and fatigue behavior of such systems. While considerable work has investigated the issues of strength and toughness, less effort has been focused on the behavior of such systems under either cyclic loading or high-temperature conditions.

The symposium was well attended and attracted attendees from the academic community as well as from various industrial and government laboratories. The organizers would like to express their appreciation for the contributions of the session chairs and the individuals who served as reviewers for the manuscripts. In addition, the able editorial assistance of Jacqueline Blackburn at the Alcoa Technical Center is gratefully acknowledged. All of their efforts were vital to the successful conduct of the symposium and the rapid publication of these proceedings.

J.J. Lewandowski
C.H. Ward
M.R. Jackson
W.H. Hunt, Jr.

June, 1996

MATERIALS RESEARCH SOCIETY SYMPOSIUM PROCEEDINGS

Volume 395— Gallium Nitride and Related Materials, F.A. Ponce, R.D. Dupuis, S.J. Nakamura, J.A. Edmond, 1996, ISBN: 1-55899-298-7

Volume 396— Ion-Solid Interactions for Materials Modification and Processing, D.B. Poker, D. Ila, Y-T. Cheng, L.R. Harriott, T.W. Sigmon, 1996, ISBN: 1-55899-299-5

Volume 397— Advanced Laser Processing of Materials—Fundamentals and Applications, R. Singh, D. Norton, L. Laude, J. Narayan, J. Cheung, 1996, ISBN: 1-55899-300-2

Volume 398— Thermodynamics and Kinetics of Phase Transformations, J.S. Im, B. Park, A.L. Greer, G.B. Stephenson, 1996, ISBN: 1-55899-301-0

Volume 399— Evolution of Epitaxial Structure and Morphology, A. Zangwill, D. Jesson, D. Chambliss, R. Clarke, 1996, ISBN: 1-55899-302-9

Volume 400— Metastable Phases and Microstructures, R. Bormann, G. Mazzone, R.D. Shull, R.S. Averback, R.F. Ziolo, 1996 ISBN: 1-55899-303-7

Volume 401— Epitaxial Oxide Thin Films II, J.S. Speck, D.K. Fork, R.M. Wolf, T. Shiosaki, 1996, ISBN: 1-55899-304-5

Volume 402— Silicide Thin Films—Fabrication, Properties, and Applications, R. Tung, K. Maex, P.W. Pellegrini, L.H. Allen, 1996, ISBN: 1-55899-305-3

Volume 403— Polycrystalline Thin Films: Structure, Texture, Properties, and Applications II, H.J. Frost, M.A. Parker, C.A. Ross, E.A. Holm, 1996, ISBN: 1-55899-306-1

Volume 404— *In Situ* Electron and Tunneling Microscopy of Dynamic Processes, R. Sharma, P.L. Gai, M. Gajdardziska-Josifovska, R. Sinclair, L.J. Whitman, 1996, ISBN: 1-55899-307-X

Volume 405— Surface/Interface and Stress Effects in Electronic Materials Nanostructures, S.M. Prokes, R.C. Cammarata, K.L. Wang, A. Christou, 1996, ISBN: 1-55899-308-8

Volume 406— Diagnostic Techniques for Semiconductor Materials Processing II, S.W. Pang, O.J. Glembocki, F.H. Pollack, F.G. Celii, C.M. Sotomayor Torres, 1996, ISBN 1-55899-309-6

Volume 407— Disordered Materials and Interfaces, H.Z. Cummins, D.J. Durian, D.L. Johnson, H.E. Stanley, 1996, ISBN: 1-55899-310-X

Volume 408— Materials Theory, Simulations, and Parallel Algorithms, E. Kaxiras, J. Joannopoulos, P. Vashishta, R.K. Kalia, 1996, ISBN: 1-55899-311-8

Volume 409— Fracture—Instability Dynamics, Scaling, and Ductile/Brittle Behavior, R.L. Blumberg Selinger, J.J. Mecholsky, A.E. Carlsson, E.R. Fuller, Jr., 1996, ISBN: 1-55899-312-6

Volume 410— Covalent Ceramics III—Science and Technology of Non-Oxides, A.F. Hepp, P.N. Kumta, J.J. Sullivan, G.S. Fischman, A.E. Kaloyeros, 1996, ISBN: 1-55899-313-4

Volume 411— Electrically Based Microstructural Characterization, R.A. Gerhardt, S.R. Taylor, E.J. Garboczi, 1996, ISBN: 155899-314-2

Volume 412— Scientific Basis for Nuclear Waste Management XIX, W.M. Murphy, D.A. Knecht, 1996, ISBN: 1-55899-315-0

Volume 413— Electrical, Optical, and Magnetic Properties of Organic Solid State Materials III, A.K-Y. Jen, C.Y-C. Lee, L.R. Dalton, M.F. Rubner, G.E. Wnek, L.Y. Chiang, 1996, ISBN: 1-55899-316-9

Volume 414— Thin Films and Surfaces for Bioactivity and Biomedical Applications, C.M. Cotell, A.E. Meyer, S.M. Gorbatkin, G.L. Grobe, III, 1996, ISBN: 1-55899-317-7

Volume 415— Metal-Organic Chemical Vapor Deposition of Electronic Ceramics II, S.B. Desu, D.B. Beach, P.C. Van Buskirk, 1996, ISBN: 1-55899-318-5

MATERIALS RESEARCH SOCIETY SYMPOSIUM PROCEEDINGS

Volume 416— Diamond for Electronic Applications, D. Dreifus, A. Collins, T. Humphreys, K. Das, P. Pehrsson, 1996, ISBN: 1-55899-319-3

Volume 417— Optoelectronic Materials: Ordering, Composition Modulation, and Self-Assembled Structures, E.D. Jones, A. Mascarenhas, P. Petroff, R. Bhat, 1996, ISBN: 1-55899-320-7

Volume 418— Decomposition, Combustion, and Detonation Chemistry of Energetic Materials, T.B. Brill, T.P. Russell, W.C. Tao, R.B. Wardle, 1996 ISBN: 1-55899-321-5

Volume 420— Amorphous Silicon Technology—1996, M. Hack, E.A. Schiff, S. Wagner, R. Schropp, M. Matsuda, 1996, ISBN: 1-55899-323-1

Volume 421— Compound Semiconductor Electronics and Photonics, R.J. Shul, S.J. Pearton, F. Ren, C.-S. Wu, 1996, ISBN: 1-55899-324-X

Volume 422— Rare Earth Doped Semiconductors II, S. Coffa, A. Polman, R.N. Schwartz, 1996, ISBN: 1-55899-325-8

Volume 423— III-Nitride, SiC, and Diamond Materials for Electronic Devices, D.K. Gaskill, C. Brandt, R.J. Nemanich, 1996, ISBN: 1-55899-326-6

Volume 424— Flat Panel Display Materials II, M. Hatalis, J. Kanicki, C.J. Summers, F. Funada, 1996, ISBN: 1-55899-327-4

Volume 425— Liquid Crystals for Advanced Technologies, T.J. Bunning, S.H. Chen, W. Hawthorne, N. Koide, T. Kajiyama, 1996, ISBN: 1-55899-328-2

Volume 426— Thin Films for Photovoltaic and Related Device Applications, D. Ginley, A. Catalano, H.W. Schock, C. Eberspacher, T.M. Peterson, T. Wada, 1996, ISBN: 1-55899-329-0

Volume 427— Advanced Metallization for Future ULSI, K.N. Tu, J.W. Mayer, J.M. Poate, L.J. Chen, 1996, ISBN: 1-5899-330-4

Volume 428— Materials Reliability in Microelectronics VI, W.F. Filter, J.J. Clement, A.S. Oates, R. Rosenberg, P.M. Lenahan, 1996, ISBN: 1-55899-331-2

Volume 429— Rapid Thermal and Integrated Processing V, J.C. Gelpey, M. Öztürk, R.P.S. Thakur, A.T. Fiory, F. Roozeboom, 1996, ISBN: 1-55899-332-0

Volume 430— Microwave Processing of Materials V, M.F. Iskander, J.O. Kiggans, E.R. Peterson, J.Ch. Bolomey, 1996, ISBN: 1-55899-333-9

Volume 431— Microporous and Macroporous Materials, R.F. Lobo, J.S. Beck, S. Suib, D.R. Corbin, M.E. Davis, L.E. Iton, S.I. Zones, 1996, ISBN: 1-55899-334-7

Volume 432— Aqueous Chemistry and Geochemistry of Oxides, Oxyhydroxides, and Related Materials J.A. Voight, B.C. Bunker, W.H. Casey, T.E. Wood, L.J. Crossey, 1996, ISBN: 1-55899-335-5

Volume 433— Ferroelectric Thin Films V, S.B. Desu, R. Ramesh, B.A. Tuttle, R.E. Jones, I.K. Yoo, 1996, ISBN: 1-55899-336-3

Volume 434— Layered Materials for Structural Applications, J.J. Lewandowski, C.H. Ward, W.H. Hunt, Jr., M.R. Jackson, 1996, ISBN: 1-55899-337-1

Volume 435— Better Ceramics Through Chemistry VII—Organic/Inorganic Hybrid Materials, B. Coltrain, C. Sanchez, D.W. Schaefer, G.L. Wilkes, 1996, ISBN: 1-55899-338-X

Volume 436— Thin Films: Stresses and Mechanical Properties VI, W.W. Gerberich, H. Gao, J-E. Sundgren, S.P. Baker 1996, ISBN: 1-55899-339-8

Volume 437— Applications of Synchrotron Radiation to Materials Science III, L. Terminello, S. Mini, D.L. Perry, H. Ade, 1996, ISBN: 1-55899-340-1

Prior Materials Research Society Symposium Proceedings available by contacting Materials Research Society

Part I

Applications

ADVANCED AIRCRAFT ENGINE MICROLAMINATED INTERMETALLIC COMPOSITE TURBINE TECHNOLOGY

R. G. ROWE, D. W. SKELLY, M. R. JACKSON, M. LARSEN AND D. LACHAPELLE†

GE Corporate Research and Development, Schenectady, NY 12309
† - GE Aircraft Engines, Cincinnati, OH 45215

ABSTRACT

Higher gas path temperatures for greater aircraft engine thrust and efficiency will require both higher temperature gas turbine airfoil materials and optimization of internal cooling technology. Microlaminated composites consisting of very high temperature intermetallic compounds and ductile refractory metals offer a means of achieving higher temperature turbine airfoil capability without sacrificing low temperature fracture resistance. Physical vapor deposition, used to synthesize microlaminated composites, also offers a means of fabricating advanced turbine blade internal cooling designs. The low temperature fracture resistance of microlaminated Nb(Cr)-Cr_2Nb microlaminated composites approached 20 MPa$\sqrt{m}$ in fracture resistance curves, but the fine grain size of vapor deposited intermetallics indicates a need to develop creep resistant microstructures.

HIGH TEMPERATURE INTERMETALLIC COMPOSITE TURBINE AIRFOIL CONCEPTS

Requirements for higher aircraft engine performance, higher thrust to weight ratio and greater fuel efficiency have pushed turbine gas path temperatures higher and higher. The materials and designs of turbine blade airfoils have evolved to meet these requirements. Figure 1 highlights the two significant evolutionary changes that have occurred in the HP turbine blade's configuration. Cooled HP turbine blade design configurations of the 1960's used equiaxed nickel based alloys with radial cooling channels and film holes exiting the leading and trailing airfoil edges. This technology step from uncooled turbine blades into cooled turbine blade designs was substantial for the time because it allowed the engine's turbine rotor inlet temperatures to operate well above 1100°C (~2000°F) and provide a high standard of durability.

Over the last several of decades, many advances occurred in the cooling design and materials of HP turbine blades. The developments were made by improving and using complex multi-pass serpentine internal cooling circuits with fully film cooled airfoil exteriors, investment casting techniques, and advanced single crystal nickel based superalloys. These advances allowed aircraft engines of the 1980's and 1990's to increase the turbine rotor inlet temperatures an additional 300°C (~500°F) with improved turbine blade durability and reduced cooling flows.

Figure 2 shows the maximum temperature capability of superalloys as a function of their approximate year of introduction [1]. It can be seen that the rate

---

This work was partially supported by U.S. Air Force contracts F33615-91-C-5613 and F33615-92-C-5977, WRDC, Wright Patterson AFB, OH

of increase of superalloy temperature capability has slowed in recent years. The reason for this is that although single crystal processing increased the fraction of the melting point for operation, superalloy maximum use temperatures are ultimately defined by their melting point.

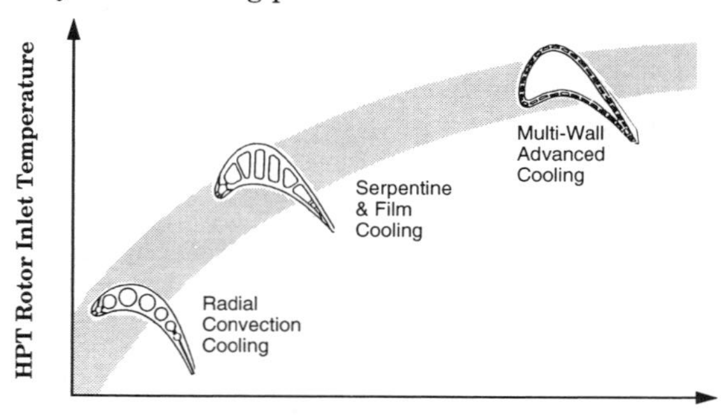

Figure 1. HP turbine blade cooling design and materials evolution

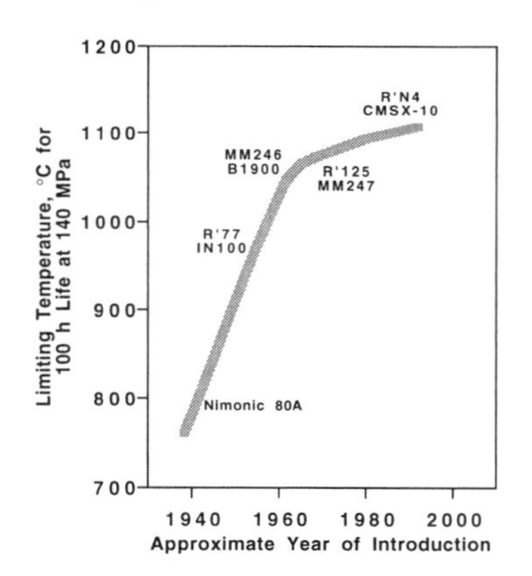

Figure 2. The evolution of superalloy temperature capability.

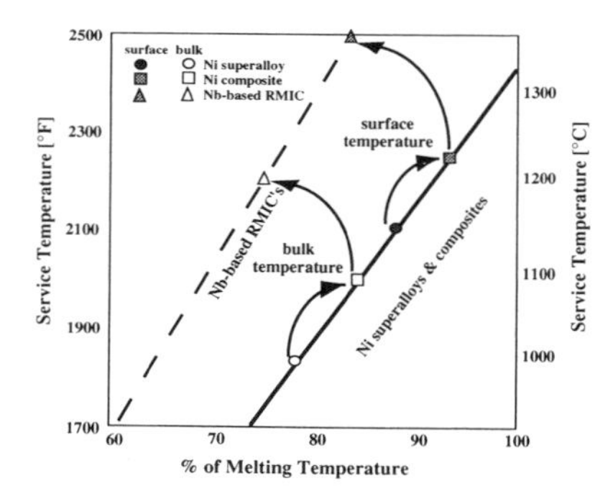

Figure 3. Service temperature vs $\%T_m$ for superalloy and advanced Nb-based composites.

High temperature intermetallic compounds such as Cr_2Nb, Nb_5Si_3, Nb_2Al, or Cr_3Si retain strength (as measured by microhardness indentation) to temperatures above 1200°C and in some cases, above 1400°C. The disadvantage is that they are very brittle and are difficult to fabricate as monolithic components [1]. One approach to overcome the limitation of brittleness in these intermetallic compounds is to design composite materials utilizing high temperature metals as a toughening component but relying upon the high temperature intermetallic phase for elevated temperature strength and creep resistance. Figure 3 shows approximate service

[1] The high temperature ductile to brittle transition temperature of intermetallic compounds such as Cr2Nb and Ti5Si3 often leads to fracture during cool-down due to thermal gradient stresses.

temperature as a function of the percent of the melting temperature of high performance alloys [2]. The circle and filled circle points show estimates of the bulk and surface temperature conditions for monolithic single crystal turbine blades. The increase in temperature capability that is made possible by the introduction of an intermetallic composite with superalloy toughening is shown by the open and filled square symbols. Because a superalloy is used as the toughening component, composites of this type follow the same temperature - fractional melting temperature curve as for superalloys. The potential surface temperature in intermetallic composite blade applications may extend to 90% of the melting point of the superalloy.

The potential temperature that may result from the use of niobium-based alloy toughening in a high temperature intermetallic composite is shown by the open and filled triangular points. The advantage of a niobium-based toughening component is that the blade can operate at a lower percentage of the melting point and still can achieve higher temperature capability.

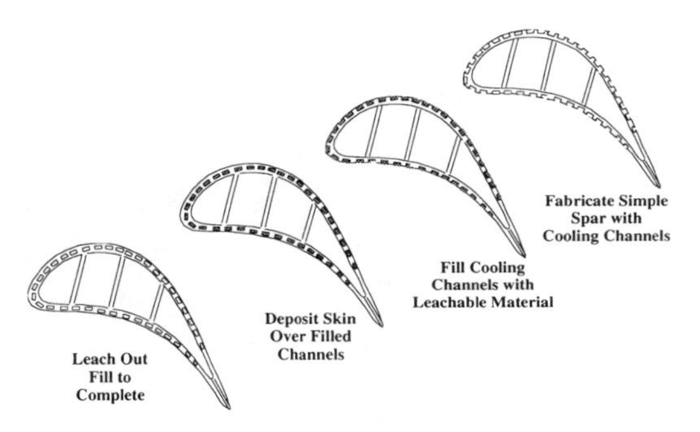

Figure 4. PVD synthesis of complex near-surface-cooled airfoil designs utilizing a PVD composite skin over a fabricated simple spar.

While serpentine and film cooling superalloy turbine airfoil designs can be manufactured by casting and directional solidification, the most complex multi-wall cooling design shown in Figure 1 may require new techniques. One technique that has promise for economical manufacturing of thin skinned near-surface cooling designs is high rate physical vapor deposition (PVD). Figure 4 schematically illustrates a method of PVD turbine blade fabrication. Because precise internal cooling cavities do not have to be cast into the spar, lower cost, higher yield techniques can be used to fabricate the internal spar structure. Cooling channels can either be cast in from the shell mold of can be machined into the surface of the spar and filled with a temporary filler. The filled surface is then finished and the high temperature intermetallic composite surface skin is deposited by PVD. Chemical leaching of the temporary cooling passage filler completes the process of forming a blade with a high performance near-surface cooling design. Higher temperature composite materials combined with innovative manufacturing processing has the potential to increase the effectiveness of the cooling designs and yield another major step in engine performance. It may be possible to allow turbine rotor inlet temperatures to increase an additional 300°C (~500°F) while maintaining turbine blade durability and cooling flow savings.

Airfoil Materials Performance

There are a number of requirements that high temperature materials for turbine airfoil applications must meet. These are: strength and creep resistance at elevated temperatures, fracture strength and fracture toughness at low temperatures, oxidation resistance, resistance to thermal cycling and shock, microstructural stability at elevated temperatures and manufacturability at competitive costs. Intermetallic composites for turbine blade applications must be chosen and processed to meet all of these requirements. It is very early in the development cycle for high temperature intermetallic composites but promising high temperature strength, oxidation resistance and low temperature toughening have been observed in independent experiments on different composite systems.

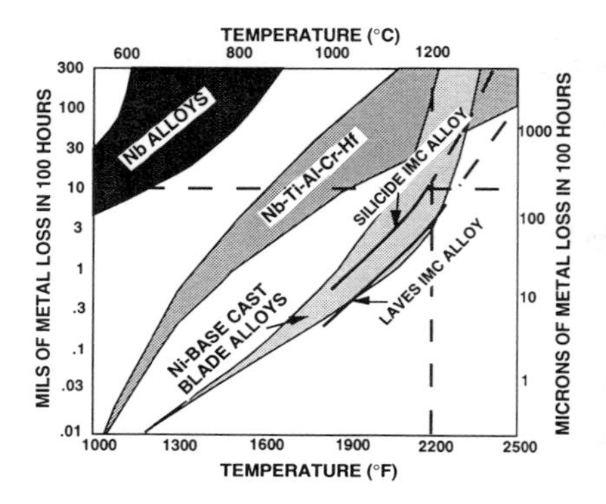

Figure 5. Relative oxidation behavior of high temperature materials

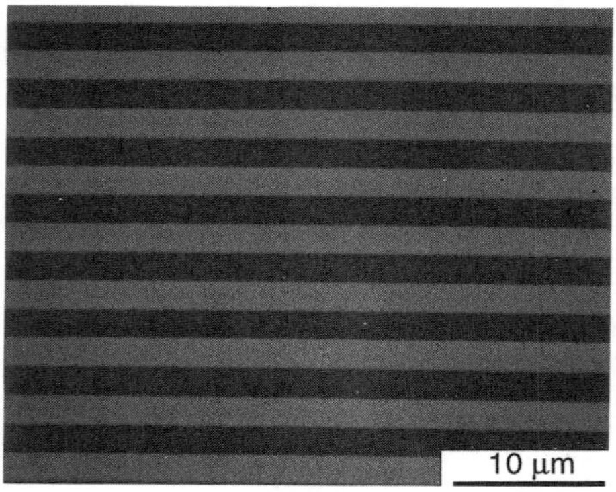

Figure 6. Backscattered electron micrograph of an as-deposited Cr_2Nb-Nb microlaminated composite with equal 2 μm thick metal and intermetallic layer thicknesses.

Oxidation resistance will be one of the most critical requirements for intermetallic composite applications in aircraft turbine components at temperatures above 1200°C. All composites will require coating for long term oxidation resistance but the intermetallic composite system must also have sufficient intrinsic oxidation resistance to prevent catastrophic oxidation in the event of a coating failure. Poor oxidation resistance has always limited the application of niobium-based alloys in aircraft engine turbine components. Figure 5 is a logarithmic plot of the 100 h metal loss vs temperature for superalloys and niobium based alloys which shows the relative difference in oxidation resistance between nickel and niobium-based alloys [2]. New Ti, Hf, Al, Cr alloyed niobium alloys have recently been developed with improved oxidation resistance compared to commercial niobium alloys and which approach the oxidation behavior of older nickel-based superalloys at temperatures of from 1150-1200°C and higher [3-8]. In addition, the oxidation resistances of high temperature Nb(Si)-Nb_5Si_3 and Nb(Cr)-Cr_2Nb composites are close to that of the most oxidation resistant nickel based superalloys over the same temperature range. It is expected that intermetallic composite compositions with even greater oxidation resistance will be developed.

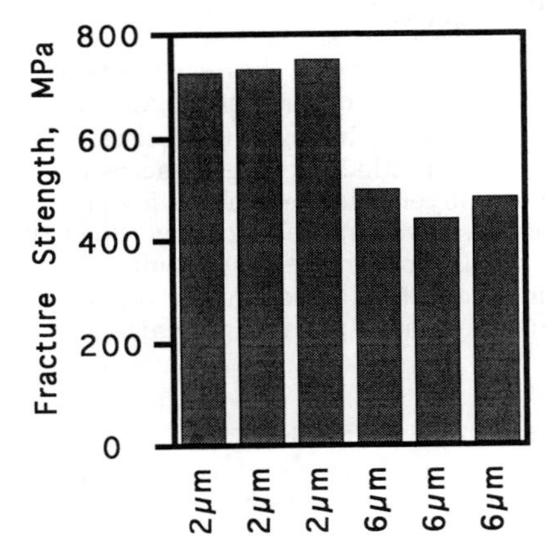

Figure 7. Room temperature tensile fracture strengths of multiple samples of 1200°C/2h aged Cr_2Nb-Nb microlaminated composites.

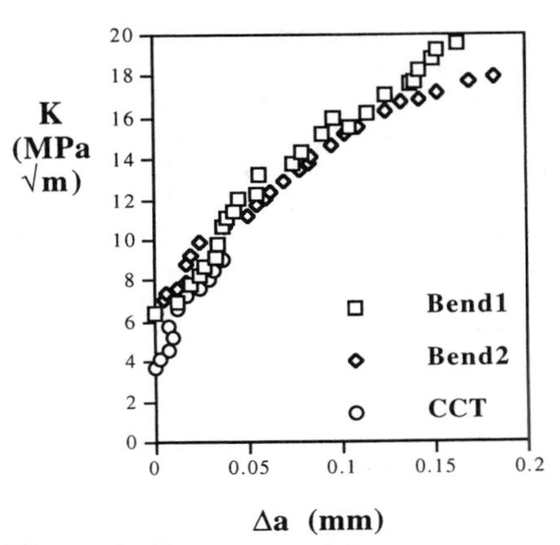

Figure 8. Fracture resistance curves (R-curves) for bend and center cracked tensile samples of 1200°C/2h aged Cr_2Nb-Nb microlaminated composites with 6 μm metal and intermetallic layer thicknesses.

Synthesis of Tough Intermetallic Matrix Composites

A process of high temperature intermetallic composite synthesis which permits the fabrication of advanced turbine blade designs is also a critical requirement. Microlaminated high temperature intermetallic composites with low interstitial contamination and with good structural integrity have been synthesized by PVD [9,10]. The process has the capability for the synthesis of a wide range of intermetallic matrix composites with independently tailored layer thicknesses, volume fraction and layer composition for optimized properties. Composites of Nb-Cr_2Nb with different layer compositions and layer thicknesses were also synthesized as shown in Table 1. Figure 6 shows a metallographic section of as-deposited PVD Nb-Cr_2Nb microlaminated intermetallic composite L17 which had equal 2 μm thick layers of Nb(Cr) and Cr_2Nb.

Table 1. Microlaminated Cr_2Nb-Nb Composite Compositions (at.%)

ID No.	Intermetallic Composition	Metal Composition	Layer Thickness	No. Int.	No. Met.
L17	Cr-41Nb	Nb-4.7Cr	2 μm	32	33
L60	Cr-42Nb	Nb-3.3Cr	6 μm	11	11
L72	Cr-38Nb	Nb-7.9Cr	2 μm	34	35

After deposition, the microlaminates were crystallized at 1200°C and tested at room temperature. The fracture strengths of the 2μm and 6 μm layer thickness Nb-Cr_2Nb microlaminated composites L17 and L72 are shown in Figure 7 [11]. High room temperature fracture strengths infer that metal toughening overcomes the inherent brittleness of the intermetallic layer. Fracture resistance curves established toughening by direct measurement. Figure 8 shows the resistance of a 6μm layer thickness Nb-Cr_2Nb microlaminated composite to crack propagation [11,12]. Analyses of the room temperature fracture strengths and the fracture resistance curves are consistent with toughening by ductile Nb bridging across cracks in the intermetallic layer [12,13]. The 6 μm layer thickness had a lower fracture strength because its fracture resistance curve was not as steep as the curve for the 2 μm layer thickness composite, but its maximum toughness was much higher. Optimization of fracture toughness and strength will require the balancing of these two characteristics relative to the requirements of specific applications.

Elevated Temperature Strength

The microstructure of Nb-Cr_2Nb microlaminate L17 after 1200°C crystallization annealing is shown in Figure 9. The metal layer had coarsened bcc grain structure but the intermetallic layer had a very fine Cr_2Nb+Nb microstructure. Although the room temperature tensile and bend strength was high, three point bend tests of the Nb-Cr_2Nb microlaminates at elevated temperatures showed that these ultra fine grained intermetallic layers had little strength. Figure 10 is a plot of load vs displacement for three point bending of 0.75 mm thick microlaminated composite bars that were fabricated by HIP bonding five 0.15 mm thick sheets of 2 μm layer thickness Nb-Cr_2Nb sheet. The microlaminated Nb-Cr_2Nb composites had little strength at 1000°C and 1200°C. Post test metallography revealed that the outer fiber intermetallic layers had deformed rather than fractured during the tests.

Figure 9. TEM micrograph of a Cr_2Nb-Nb microlaminate with 2 μm metal and intermetallic layers after a heat treatment of 2h at 1200°C.

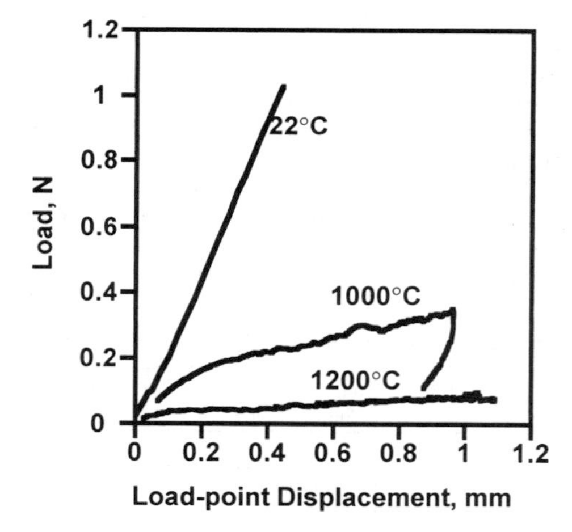

Figure 10. Load-displacement curves for 0.75 mm thick 1200°C/3h HIP bonded three point bend specimens of Cr_2Nb-Nb microlaminate L17. Samples were tested at RT, 1000°C and 1200°C.

Good high temperature strength and creep resistance have been demonstrated in other high temperature intermetallic composite systems. Directionally solidified Nb-Cr_2Nb discontinuous intermetallic composites with ~50 vol.% Cr_2Nb have been produced and tested. Tensile strengths of 250 MPa at 1000°C and 150 MPa at 1200°C have been reported [2]. Directionally solidified composites of Nb-Nb_5Si_3 have been reported to have a tensile strength of 370 MPa at 1200°C [2]. The creep resistance of both DS and extrusion-aligned discontinuous Nb-Nb_5Si_3 intermetallic composites have been shown to be better than that of high performance single crystal superalloys [2]. These results demonstrate that there is a significant elevated temperature strength payoff to be realized using high temperature intermetallic composites.

Microstructural Evolution

Because surface deposition technology may permit a wider selection of composite compositions and more advanced designs to be considered, it is important to develop techniques to produce coarse grained creep resistant intermetallic composite microstructures by PVD. This may be accomplished by increasing the deposition temperature to allow crystallization during deposition or by selecting an intermetallic composite system that has a wider temperature window between the onset of grain boundary migration and the upper intermetallic stability limit.

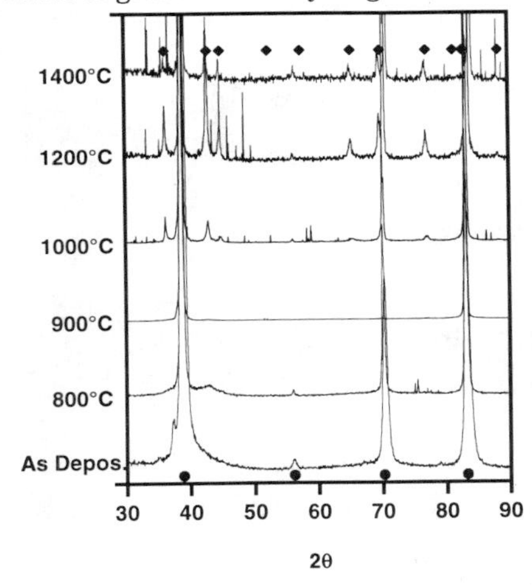

Figure 11. X-ray diffraction spectra of Cr_2Nb-Nb microlaminate L17 after heat treatment for 2h at various temperatures.

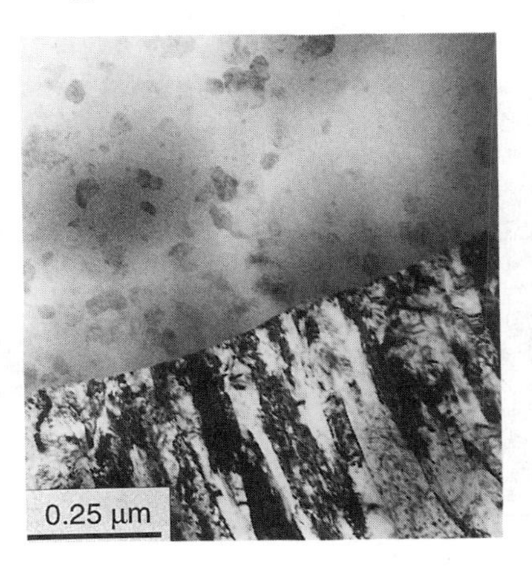

Figure 12. TEM micrograph of as-deposited microlaminate L17. the upper portion is the layer with the intermetallic composition and the lower layer is the bcc Nb(Cr) metal layer.

The Nb-Cr_2Nb microlaminate composite L17 that was shown in Figure 6 was deposited by PVD at an estimated substrate temperature of from 300°C to 500°C. X-ray diffraction of the as-deposited microlaminated composite showed that only bcc or B2 structure was present, Figure 11. Figure 12 is a TEM micrograph of an as-deposited microlaminate which shows both the intermetallic layer (top portion of micrograph) and the metal layer (bottom portion of the micrograph). The metal Nb(Cr) layer had a columnar bcc grain structure but the intermetallic layer consisted of metastable microcrystalline bcc/B2 grains which probably grew from an amorphous Cr_2Nb precursor.

The microlaminated composites were annealed for two hours at temperatures from 800°C to 1400°C and examined by x-ray diffraction to follow crystallization of the Cr_2Nb phase. Cr_2Nb crystallization did not occur at temperatures below 1000°C and full transformation of Cr_2Nb did not occur until 1200°C. Binary Nb-Cr phase diagrams indicate that the Cr_2Nb phase should be stable at low temperatures so the failure of Cr_2Nb to completely transform until 1200°C, indicates that the kinetics of Cr_2Nb phase growth are very sluggish.

Figure 13(a) shows the microstructure of the Nb-Cr_2Nb microlaminate after cycling 100 times from 700°C to 1200°C. The same microstructure was produced by isothermal annealing at 1200°C for 24 h and the two tests show that the composite is relatively stable at 1200°C. X-ray intensity measurements as well as metallographic comparisons, Figures 13(b) and 13(c), showed that the fraction of the Cr_2Nb phase decreased after annealing at 1400°C due to increased solubility of Cr in the Nb metallic phase. Because of this, 1400°C appears to be an effective upper temperature limit of stability for the composition of this microlaminated composite. Therefore it was not possible to coarsen the Cr_2Nb intermetallic layer grain size by heat treatment.

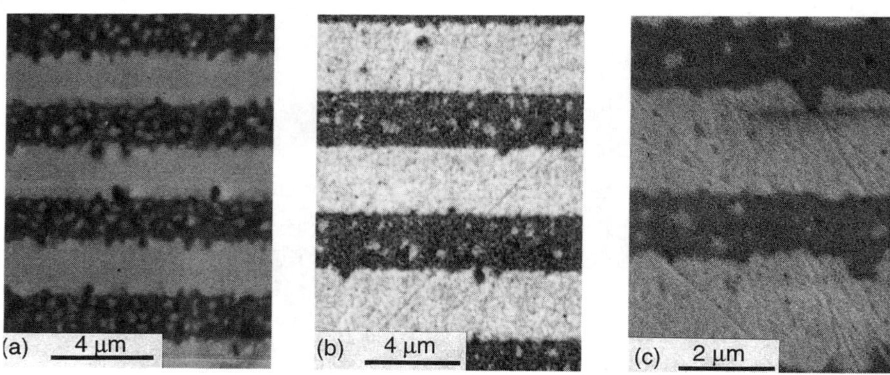

Figure 13. Backscattered electron micrographs of Cr_2Nb-Nb microlaminate L17 after heat treatments of: (a) 100 cycles from RT to 1200°C. (b) 2h at 1200°C and (c) 2h at 1400°C. The intermetllic layers imaged dark.

Microstructural Stability

Figure 14 shows the microstructure of the Nb-Cr_2Nb microlaminate after cycling 100 times from 700°C to 1200°C. A similar microstructure was obtained after isothermal annealing at 1200°C for 24 h, demonstrating that at 1200°C, microlaminate L17 was morphologically stable. The composition of the Nb(Cr) layer of microlaminate L17 shifted and a lower intermetallic layer thickness was observed after annealing at 1400°C, however. Microlaminated Nb-Cr_2Nb composite L72 was synthesized with a nearly single phase Cr_2Nb intermetallic layer to improve elevated temperature stability. Its microstructures after heat treatment at 1200°C and 1400°C are shown in Figures 14. It exhibited less reduction in the intermetallic layer thickness after annealing at 1400°C than microlaminate L17 but exaggerated protrusions of the intermetallic phase into the metal layer were observed which suggested less microstructural stability.

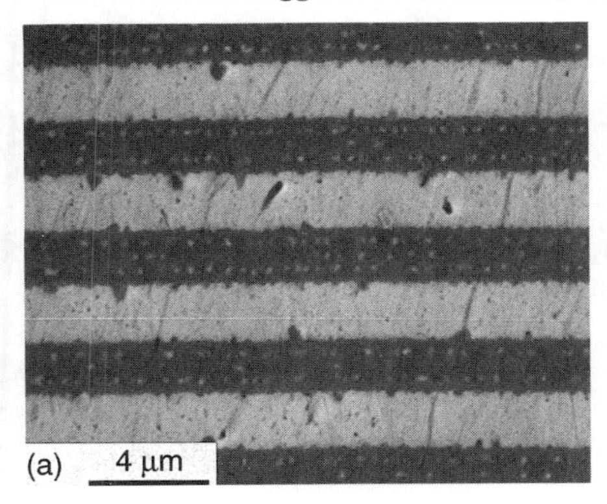

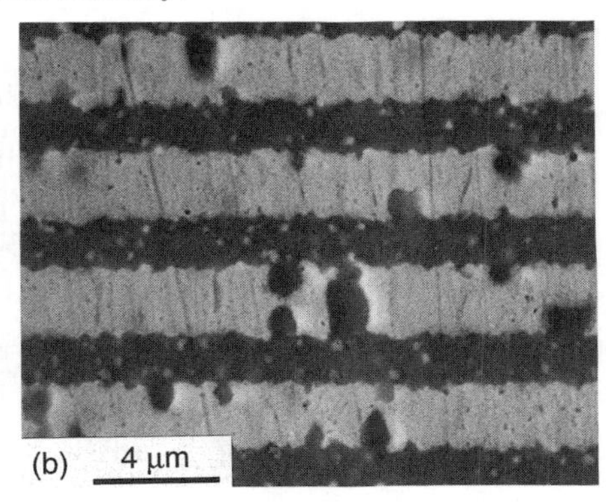

Figure 14. Backscattered electron micrographs of high Cr Cr_2Nb-Nb microlaminate L72 after heat treatments of: (a) 2h at 1200°C and (b) 2h at 1400°C.

Small protrusions of intermetallic layer into the metal layer were seen in both L17 and L72 after 1200°C annealing, Figures 13(a) and 14(a). Many of these coincided with columnar grain boundaries in the metal layer and may be grain boundary grooving of the Nb(Cr) layer to equilibrate boundary energies. The deep protrusions at 1400°C, Figure 14(b), were also associated with grain boundaries in the metal layer and may have grown from grooved grain boundaries. Since they extend into and in some cases almost all of the way through the metal layer, they cannot be simple interfacial scalloping due to interdiffusion [14].

In the most recent study of the Nb-Cr binary alloy system, Thoma and Perepezko, reported a Nb composition of 9.1 at.% Cr for the metal and 64.5 at.% Cr for the Cr_2Nb phase at 1200°C [15]. They showed that the metal composition shifts to 14.4 at.% Cr and the intermetallic to 63.6 at.% Cr at 1400°C. During heat treatment at 1400°C, we should therefore observe little change in the composition of the Cr_2Nb layer but a large change in the composition of the metal layer. This can only be accomplished by diffusion of Cr from the surface of the intermetallic layer into the metal layer until the metal layer composition reaches 14.4 at.% Cr. Diffusion in the Nb-Cr system is slow, [15], so that metal layer grain boundary diffusion may dominate bulk diffusion.

Groove instability and deep protrusions may have occurred because of the effect of grain boundary diffusion on the relative surface energies of the Nb grain boundaries and Nb-Cr_2Nb interphase boundaries at the root of grain boundary grooves. Vogel and Ratke observed such an influence in the liquid-solid Bi-Cu alloy system [16]. Their model predicts that grain boundary diffusion can lead to instability of grain boundary grooves at a solid-liquid interface. Since there may be a high grain boundary diffusive flux in the Nb-Cr_2Nb microlaminated composite systems of this study, a solid state counterpart to their observations may be responsible for these deep intermetallic protrusions. Protrusion formation may also increase the rate of transport of Cr into the interior of the Nb layer because of an increase in the surface area of the Nb-Cr_2Nb interface. Further work will be necessary to confirm this hypothesis.

SUMMARY

Significant advances in aircraft engine turbine performance has been projected by incorporating advanced designs with new higher temperature materials. Further increases in temperature capability of superalloys is ultimately limited by the melting point of the superalloys. Higher temperature intermetallic materials will require a composite design because of their inherent brittleness and difficulty of fabrication as monolithic components. The method of composite synthesis will have to be compatible with manufacture of advanced cooling designs at an affordable cost.

We have demonstrated that high temperature intermetallic composites can be synthesized by PVD with little interstitial pickup and with a good balance of low temperature fracture strength and fracture resistance. The PVD process appears to have inherent advantages in the fabrication of the most advanced turbine blade cooling designs.

The high temperature strength of the initial PVD microlaminated intermetallic composites was inadequate because of an ultrafine grained intermetallic layer microstructure. However, high strength and creep resistance have been observed in aligned discontinuous composite systems. This suggests that if the microstructure of the intermetallic layer can be coarsened that high temperature strength goals may be met.

Improvements of the oxidation resistance of both the Nb-based metal toughening layer and the composite system show that at high temperatures, achieving oxidation resistance goals may also be possible.

The stability of microlaminated composites at temperatures of 1400°C has been shown to be affected by non-equilibrium penetration of deep protrusions of intermetallic phase into the metal layer. The mechanism of protrusion formation may be related to grain boundary diffusion so that achievement of composite systems with little change in equilibrium metal and intermetallic composition over the temperature range of operation may eliminate this problem. The mechanism of protrusion formation must be firmly established to be able to predict the conditions under which protrusions occur, however.

ACKNOWLEDGMENTS

The authors would like to acknowledge the assistance of Mssrs. C. Canestraro and R. A. Nardi and Ms. K. K. Denike for assistance in carrying out the synthesis and testing of microlaminated composite materials. This work was partially supported by USAF Contract No. F33615-91-C-5613, Materials Directorate, Wright-Patterson AFB, OH.

REFERENCES

1. C. T. Sims and W. C. Hagel, The Superalloys, John Wiley, NY 1972.
2. M.R. Jackson, B.P. Bewlay, R.G. Rowe, D.W. Skelly and H.A. Lipsitt, "High Temperature Refractory Metal-Intermetallic Composites", Journal of Metals, 46 (1), pp 39-44, 1996.
3. M. R. Jackson, "Ductile Low Density Alloys Based on Niobium", Proc Intl. Conf on Tungsten and Refr. Metals 2, A. Bose and R. J. Dowding, eds, MPIF, Princeton, 1995, pp. 657-664.
4. M.R. Jackson, K.D. Jones, S.C. Huang and L.A. Peluso, "Response of Nb-Ti Alloys to High Temperature Air Exposure", Refractory Metals Extraction, Processing and Applications, K.C. Liddell, D.R. Sadoway and R.G. Bautista, eds., TMS, Warrendale, PA, 1991, pp. 335-346.
5. M. R. Jackson, "Nb-Ti-Al-Hf-Cr Alloy", U.S. Patent No. 4,931,254, June 5, 1990.
6. M. R. Jackson, "Hf Containing High Temperature Nb-Ti-Al Alloy", U.S. Patent No. 4,956,144, September 11, 1990.
7. M. R. Jackson, "Chromium Containing High Temperature Alloy (Nb-Ti-Al-Cr)", U.S. Patent No. 4,990,308, February 9, 1991.
8. M. R. Jackson, "Hafnium Containing Niobium, Titanium, Aluminum High Temperature Alloy", U.S. Patent No. 5,006,307, April 9, 1991.
9. R. G. Rowe and D. W. Skelly, "The Synthesis and Evaluation of Nb_3Al-Nb Laminated Composites", Mat Res. Soc. Symp. Proc., 273, Materials Research Society, 1992, pp. 411-415.
10. R. G. Rowe, D. W. Skelly, M. Larsen, J. Heathcote, G.E. Lucas and G. R. Odette, "Properties of Microlaminated Intermetallic-Refractory Metal Composites", Mat Res. Soc. Symp. Proc., 322, Materials Research Society, 1994, pp. 461-472.
11. R. G. Rowe, D. W. Skelly, M. Larsen, J. Heathcote, G. R. Odette and G.E. Lucas, "Microlaminated High Temperature Intermetallic Composites", Scripta Met. et Mater., 11, pp. 1487-1492, (1994).
12. J. Heathcote, G. R. Odette, G.E. Lucas and R. G. Rowe, "Mechanical Properties of Metal-Intermetallic Microlaminate Composites", This Proceedings, Materials Research Society.
13. R. G. Rowe, D. W. Skelly, M. Larsen, G.E. Lucas, G. R. Odette, J. Heathcote, H.-C. Cao and A. G. Evans, Final Report, Contract Number F33615-91- C-5613, and , "In-Situ Synthesis of Intermetallic Matrix Composites", Materials Directorate, Wright-Patterson AFB, OH, to be published.
14. J.C. Malzahn Kampe, T.H. Courtney and Y. Leng, "Shape Instabilities of Plate-Like Structures -- I. Experimental Observations in Heavily Cold-Worked In Situ Composites", Acta Metall, 37, pp 1735-1745 (1989),
15. D.J. Thoma and J.H. Perepezko, "An Experimental Evaluation of the Phase Relationships and Solubilities in the Nb-Cr System", Mater. Sci and Engrg., A156, pp 97-108, (1992).
16. H.J. Vogel and L. Ratke, "Instability of Grain Boundary Grooves Due to Equilibrium Grain Boundary Diffusion", Acta Met. et Mater., 39, pp 641-649 (1991).

MECHANICS OF METAL MATRIX LAMINATES

J.L. TEPLY
Alcoa Technical Center, 100 Technical Drive, Alcoa Center, PA 15068, teply_jl@atc.alcoa.com

ABSTRACT

The mechanics of fracture and fatigue crack propagation of ARALL Laminates and Discontinuously Reinforced SiC Aluminum (DRA) laminates is studied. In these two, supposedly different, material systems the fracture and fatigue performance is closely related to their capability to delaminate. The delamination shape, length, rate of growth, effect on fatigue and fracture toughness are analyzed in the both lamination systems. The similarities between these two lamination systems fatigue and fracture failure modes are determined and summarized. Design for compression is also mantioned.

INTRODUCTION

The idea of lamination has been practiced by metal makers for centuries [1]. The unifying force which drives engineers to use laminates is the unique opportunity to use the excellent performance of simple materials in multi-functional applications. For example, cross-lamination is used to resist multidirectional loads by superior strength of unidirectionally reinforced fiber composites. Similarly, lamination is used to combine thermal or sound insulation performance of foams with structural performance of fibrous composites to create structural sandwiches - a special form of laminates.

In this paper, we focus on design and analysis of metal laminates for fracture toughness and fatigue crack growth. More specifically, the two lamination systems of ARALL Laminates and Discontinuously Reinforced SiC Aluminum (DRA) laminates are analyzed. Not only are these systems tougher and fatigue resistant, but both products also offer improvements in other properties such as high strength, Young's modulus, strain hardening rate, etc. For example, see [2], [3], [4] and [5].

The fracture and fatigue performance of these supposedly different laminates is closely related to their capability to delaminate. In these systems, delamination is followed by crack bridging which in turn is accompanied by an increase in fracture toughness and fatigue crack growth resistance. Although the laminates are noticeably different in their material selection, the mechanics behind laminate toughening is strikingly similar in both systems. In this paper we study the similarities of fracture and fatigue performance with the firm belief that the understanding of these similarities will lead to the new, improved metal laminates.

The effects of delamination, such as compression after impact, is generally considered detrimental to the performance of laminates. However, if a laminate is designed for delamination , the detrimental effects of delamination can be virtually eliminated. This is especially true for metal laminates. In metal laminates the number of layers and functional interfaces can be optimized because the nearly isotropic properties of metals ask for no or a minimum number of cross-layers. It means that the individual layers can be thick enough to effectively resist compressive loads. Additionally, plasticity of the metal layers can be employed to design laminates in which damage and failures are preceded by visually detectable dents and permanent deformations, respectively.

In this paper, we will focus on the benefits of controlled delamination in the aforementioned lamination systems. For the brevity of this paper, we focus on delamination effects and assume that the detrimental effects of delamination were minimize below the critical level for the entire life time.

First, we review the mechanics of ARALL Laminates. From there, we introduce some initial models and experiments describing mechanics of DRA laminates. Finally, by comparing these two different systems, we conclude with some recommendation for future metal laminates.

Fatigue Crack Growth and Fracture Toughness of ARALL Laminates.

ARALL Laminates were primarily developed in cooperation between TU Delft and Fokker Aircraft. Five years ago, Alcoa and AKZO created a joint venture to commercialized ARALL Laminates and other laminate alternatives such as GLARE.

ARALL Laminates consist of aluminum sheet layers which are interspersed with layers of aramid fiber/epoxy composite. An illustrative example of the 3/2 lay-up is shown in Figure 1. When an optimized ARALL laminate is subjected to a tension-tension cyclic load, a fatigue crack initiates and propagates in the aluminum layers. Stresses from the aluminum layers are gradually transferred to the aramid layers, which posses tough aramid fibers that stay intact, via shear forces in a compliant, resin rich layer that lays between the aluminum and aramid layers, see Figure 1.

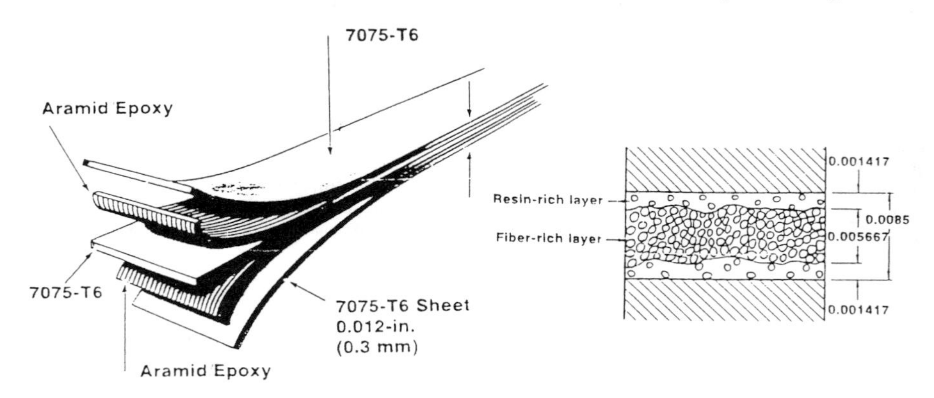

Figure 1. Schematic of ARALL Laminate and location of resin rich layer.

Using Figure 1, one can describe the mechanisms of the delamination. At the moment when fatigue crack elongates at the aluminum layer, a local single lap shear element is created along the newly created crack tip edges. Although shear stresses and geometry are slightly different from that of a single lap shear joint, it is possible to approximate the shear stresses in the resin rich layer by a modified single lap shear solution

$$\tau_i(x,y) = \tau_{oi}(x,o) * e^{-g(x,y)} \qquad (1)$$

where x is from the small crack elongation interval (a,a+da), $\tau_{oi}(x,o)$ is the extreme shear stress along the newly created crack edge, and g(x,y) is the function for the exponentially decaying shear stress. To determine $\tau_{oi}(x,o)$ and g(x,y), one can simply use tests combined with finite element analysis.

According to (1), the highest shear stresses concentration is expected along the new crack edge at the crack tip. Due to this high shear stress concentration, the failure of the fiber/matrix interface along the resin rich adhesive layer is almost simultaneous with the fatigue crack elongation. This delamination provides a finite bridging length to the fibers in the crack tip elongation interval (a,a+da). This sequence of events give rise to fiber strains below the failure strain. Without delamination, very high local strain, in excess of the failure strain, would immediately brake the fibers in the wake of the fatigue crack tip elongation.

Details of the mechanics of fiber bridging and additional factors affecting the fatigue crack propagation in ARALL Laminates such as residual stresses, shear deformation of the resin rich layer, crack closure, crack length, delamination length are well described in work of Marrissen [6], [7].

One of the important parameters affecting fatigue crack growth as well as residual strength of ARALL Laminates is the shape of delamination zone. Marissen [6] assumed an elliptical shape using the following argument: The shear force which causes delamination along the fiber/matrix interface are proportional to the fiber bridging force. Therefore, the fiber bridging force is proportional to the fiber strain which, subsequently, is equal to the crack tip displacement divided by the delamination length. Putting it together leaves one with

$$\varepsilon = v(x/a)/d(x/a)$$
$$F_b = E\Pi r^2 \varepsilon \qquad (2)$$

where 2a is the crack length, E is Young's modulus of fiber, and r is fiber radius. Since the crack opening is elliptical, delamination should also be elliptical. If delamination is not elliptical, closing forces are not constant. At the location with shorter delamination d(x/a), bridging forces F_b and shear stress $\tau_{oi}(x,o)$ are higher forcing delamination to propagate faster. Hence, during subsequent fatigue cycles, the delamination contour tends to assume an elliptical shape because, under such a condition, the delamination rate is uniform along the entire bridging length.

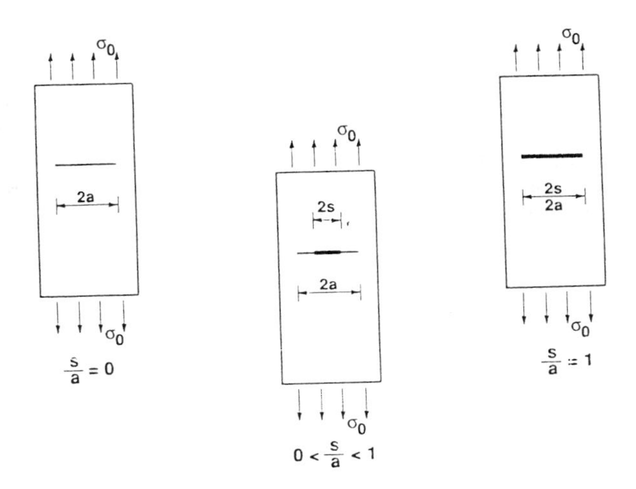

Figure 2. ARALL Panels with saw cuts.

This logic of elliptical delamination applies only to the panel in which the fatigue crack emanates from small holes. However, in the case of fiber breakage described by Roebroeks in [8], the delamination shape changes. The effects of different delamination shapes were studied by Teply and DiPaolo [9], and Macheret and Teply [10]. Fiber breakage was simulated by a saw cut, see Figure 2. A typical delamination for laminates with an initial saw cut was triangular as shown in Figure 3. Nonetheless, in some panels a concave delamination shape was observed. Using the

Westergaard [11] solution for a center crack panel, Macheret estimated the effect of delamination shape on the reduction of the stress intensity factor, $\Delta K = (K_{max}-K_{min})$, in the aluminum layers.

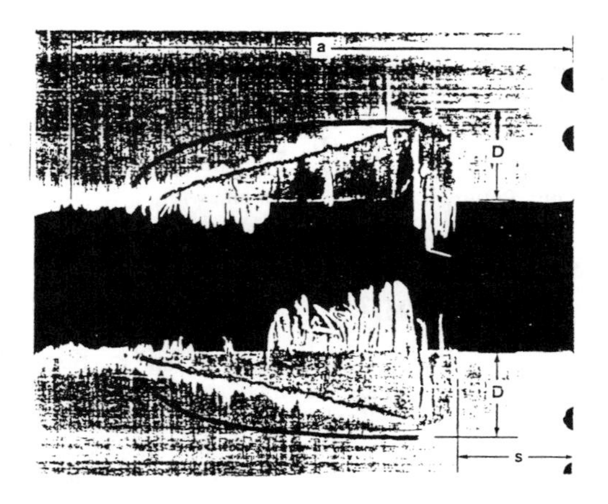

Figure 3. Triangular (linear) shape of delamination in the panels with an initial saw cuts. (Elliptical delamination is shown for contrast.)

The results are briefly summarized in Figure 4. The effect of delamination shape on fatigue crack growth is pronounced. As shown in Figure 4, elliptical delamination is less efficient than triangular one. The latter is about 40 percent more efficient for short fatigue cracks (s/a greater than 0.7) and produces crack tip stress intensity reduction factor up to 3 times larger for long fatigue cracks (s/a less than 0.3). The importance of crack delamination length, D, is also depicted in Figure 4. According to Figure 4 and (2), a longer delamination produces a lower bridging force resulting in less effective crack arresting mechanism. The length of D depends on the rate of delamination growth. Delamination growth rate for ARALL Laminates is captured in two Alcoa internal publications [12] and [13].

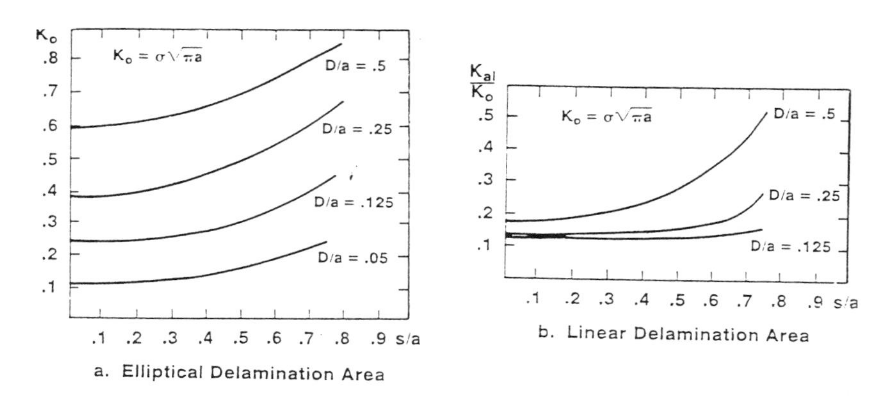

Figure 4. Effect of triangular and ellictical delamination on DK (Kmax - Kmin)

As mentioned, DiPaolo and Teply [9] extended the effect of fiber bridging and delamination to the residual strength of center crack panels. The conclusions in [9] are typical for many fiber reinforced metal laminates. It was observed that the high strength fiber layer breaks abruptly when a first fiber bundle breaks. Due to a large amount of energy stored in the unbroken fibers, it is conceivable that the location of the first fiber bundle failure is random and is most likely determined by the presence of a fiber defect. This is true for the laminates with an elliptical delamination because the fiber bridging forces are uniform. As shown in [9] that a fiber defect also dominates the failure of laminates with non-elliptical delamination. Although stress is non-uniform and stress concentrations are present in the bridging fibers, the location of fiber failure is still random and strongly affected by the presence of a fiber defect rather than the stress concentrations. Thus, using the ultimate strain failure criterion, the fiber bridging length (a-s), not the delamination shape, is only significant factor to determine the residual strength of ARALL type laminates.

The redistribution of forces to the unbroken fibers is very rapid at the time of the final failure. The postmortem analysis of fractured panels [9] showed only very limited delamination at the crack tip during the steady crack growth prior to the final failure. This means that the individual layers of ARALL Laminates react as a single laminate and are constrained by surrounding materials.

This is in sharp contrast with the behavior of DRA laminates which strongly depends on delamination and the subsequent increase in the number of thin layers resisting the progress of the final fracture tear. This effect of tougher, thinner layers resisting laminate tear will be studied in more detail in the next section.

The final note on ARALL Laminates deals with the balancing of their tensile and compressive performance. The alternation of compliant resin layers with stiffer aluminum and aramid layers has some detrimental effect on short column buckling of these laminates. Using layer-wise theory of plates [14],[15], Teply, Reddy and Barbero [16] captured the effect of resin rich layers on buckling and vibration of ARALL Laminates. To fully understand compressive behavior of ARALL Laminates, buckling of aramid fibers has to be accounted for in determining the buckling limits of aramid layers [17].

Fatigue Crack Growth and Fracture Toughness of DRA Laminates

DRA laminates consist of two material types, SiC reinforced aluminum layers and unreinforced layers of a compatible aluminum alloy. Using hot press bonding [18], a stack of alternating unreinforced and reinforced blanks is heated and pressed to required final thickness and size. During this quite rapid horizontal expansion, surface oxides are broken and an adequate metallurgical bond is established. The strength of the bond, affected by press and blank temperature, forming rate, reduction ratio, and by alloy selection, is obviously the important factor for the performance of DRA laminates.

Spray forming is considered to be a future manufacturing alternative that offers economical solution to the cost of DRA laminates. One of the several possible spray manufacturing arrangement is shown in Figure 5. Linear spray nozzles are arranged is series; unreinforced and reinforced alloys are alternatively sprayed on a moving substrate to create the layers of a future laminate. The rollers at the end of this manufacturing unit are used to consolidate the laminate and heal the porosity which is inherent to this spray process.

Until now, all laminates presented in this paper have been manufactured using press bonding method. Although the microstructure of individual layers may be different for press bonded and spray formed laminates, it is believed that the major mechanisms of delaminations, layer bridging, fatigue and fracture, will remain the same for both manufacturing processes. Since microstructural differences in layers will affect interface properties, it is also reasonable to assume that the relative interaction between the major failure mechanisms will be different for press bonded and spray formed laminates. Hence, certain tests will have to be performed to adjust the press formed model to spray formed laminates.

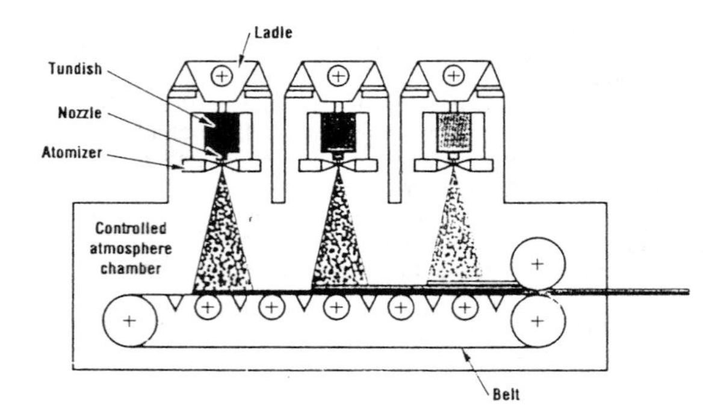

Figure 5. Spray forming of DRA laminates

Experiments were performed in Alcoa Technical Center (ATC) to study, understand and model failure mechanisms of DRA laminates [19]. Compact test (CT) specimens were machined from DRA samples. The samples were manufactured at Lawrence Livermore National Laboratories under CRADA agreement between Alcoa and LLNL.

Each compact test specimen was used for both fatigue and fracture failure testing. To accomplished this, the CT specimens were manufactured with a shorter initial notch length than a typical R-curve specimen should have. The fatigue precracking of the specimens was used to collect the data on fatigue crack growths and fatigue mechanisms in DRA laminates. When a recommended crack length was reached, the specimen was loaded to slightly beyond its maximum load and R-curve recorded. Additional fatigue testing followed; the details of the total test schedule are in [18].

Observations from DRA laminate fatigue and R-curve tests are opposite to those from ARALL Laminates. During fatigue crack propagation almost no or negligible delamination was observed. On the other hand, the delamination at the crack tip during the steady crack propagation before the final failure of a R-curve test specimen is the major contributor to DRA toughening mechanism. The failure of notched DRA laminates is dominated by the three following toughening mechanisms: crack tip delamination, the bridging of unbroken ductile layers over cracked brittle layers, and by changing fracture mode from the plain strain mode of partially constrained layers to the plane stress mode of thin, delaminated layers. The extent of the crack tip delamination depends the alloy selection and, of course, the laminate processing parameters.

To a design engineer DRA laminates basically behave like metals. First, the visible fatigue crack in the two outside layers is the actual fatigue crack length in the laminate. Due to negligible fatigue delamination, no estimate or expensive non-destructive testing is needed to determine the amount of layer bridging to calculate the residual strength or life of a DRA panel. On the other end, DRA laminates experience a considerable delamination during the final failure stages. Such delamination promotes longer steady crack growth before the final failure.

Using design for delamination control, DRA laminates offer a wide range of material properties that would be difficult, almost impossible to achieve by a single alloy. For example, we can decouple strength and fracture toughness in heat treatable alloys, we can also design laminates with high specific modulus and acceptable fracture toughness, etc.

Since design for delamination is the focal points of DRA laminates performance, the rest of this section deals with delamination only. The effect of delamination on the fracture toughness, fatigue, modulus, strength, etc., of DRA laminates is distinctively different in three directions. These directions are shown in Figure 6. but in this paper, we will deal only with fatigue and fracture performance in the crack divider direction.

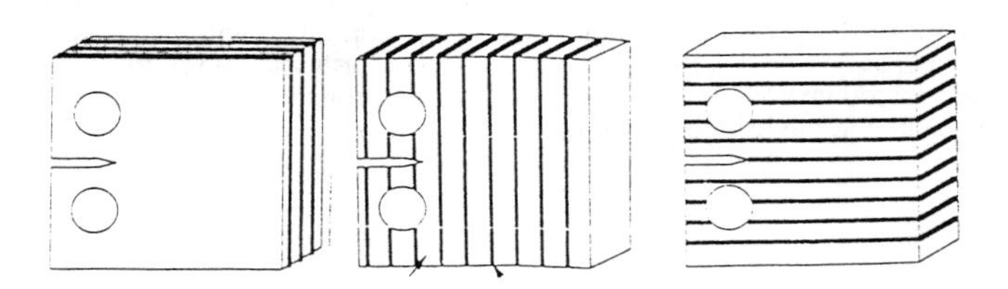

Figure 6. From left to right, crack divider, crack arrester and short transverse direction in DRA laminates

The sequence of failure and delamination mechanisms in the crack divider direction are depicted in Figure 7. However, before we describe them, let us analyze the fracture toughness of laminates with no delamination. Due to the low fracture toughness of the reinforced layers, a crack prefers to begin to propagate earlier in these layers than in the unreinforced ones. If the interface between these two layers does not delaminate, the steady crack growth would be dominated by the brittle fracture of the reinforced material. In this case, the fracture toughness of the laminate can be estimated as a weighted average of the fracture toughness of constitutive materials

$$K_I = w_u K_{u,IC}\,(1+B_u \exp^2(A_u t_{tot}/t_{ou})) + w_r K_{r,IC}\,(1+B_r \exp^2(A_r t_{tot}/t_{or})) \tag{3}$$

where w_u and w_r are the weight factor associated with unreinforced and reinforced layers, respectively, t_{tot} is the total thickness of the laminate, t_{ou} and t_{or} are the thickness associated with the plain strain failure mode of the unreinforced and reinforced material, respectively, and A_u,A_r,B_u and B_r are material constants.

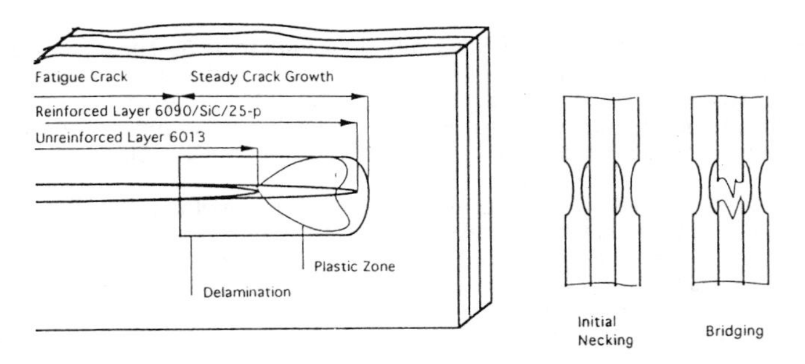

Figure 7. Fracture failure of DRA laminates.

Provided appropriate materials and processing parameters are selected, the layers will delaminate at the same time as the steady crack growth begins in the reinforced layers. Hence, the unreinforced layers will bridge the crack tip of the reinforced layers. The bridging distance and force are functions of the failure strain and hardening rate of the unreinforced alloy. Large failure strain promotes longer bridging distance and higher hardening give rise to higher bridging/closing forces.

The onset of delamination can be associated with the necking ability of the unreinforced alloy. As it is shown in Figure 7, the unreinforced material begins to neck in the crack tip area. This necking gives rise to peeling forces and subsequent delamination in the crack tip zone allowing for the crack growth in the reinforced layers. When peeling forces are combined with the shear forces from the uneven crack propagation, the delamination can propagate further allowing for additional crack growth in the reinforced layers. This uneven steady crack growth continues till the critical fracture conditions are reached in the bridging ligaments.

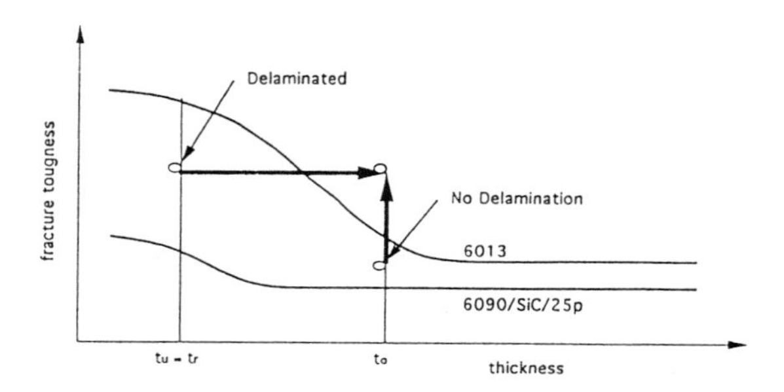

Figure 8. Effect of thickness change in delaminated DRA laminates

In addition to layer bridging, the delamination also gives rise to another important toughening factor. The fracture mode of the individual layers changes from the constrained mode dominated by the laminate thickness to the unconstrained plane stress mode of individual layers. It means that the thicknesses of unreinforced and reinforced layers, t_u and t_r, should be substituted for the total thickness, t_{tot}, in. (3), and the individual contributions summed up

$$K_I = \Sigma(w_u K_{u,IC} (1+B_u \exp^2(A_u t_u/t_o)) + w_r K_{r,IC} (1+B_r \exp^2(A_r t_r/t_o))) \tag{4}$$

The impact of thickness change is schematically shown in Figure 8. The top curve is the fracture toughness of unreinforced 6013 alloy. The bottom curve is the fracture toughness of 6090/SiC/25p (reinforced 6090 aluminum by 25 percent of SiC). Depending on the relative thickness of the reinforced and unreinforced layers to the thickness of the laminate, the fracture toughness of the laminate can increase by factor of 2 if the laminate delaminates.

Assuming weights, w_u and w_r, are equal to the volume fraction of unreinforced and reinforced layers, respectively, the fracture toughness of 50/50 DRA laminates can be calculated from (3) and (4). (A 50/50 DRA laminate consists of 50 percent of reinforced and 50 percent of unreinforced layers.) The calculation for a non-delaminated and delaminated 50/50 6090/SiC/25p/6013 DRA laminate is shown in Figure 9. The first preliminary test data from delaminated specimens are included. Additional verification testing in under way in Alcoa.

As mentioned, delamination and thickness effect are combined with layer bridging and crack tip closing forces. The model which accounts for the bridging forces is complex and its introduction is beyond the scope of this paper. Basically, the modeling approaches which were used in [9] and [10] are modified and applied to DRA. This work will be reported separately.

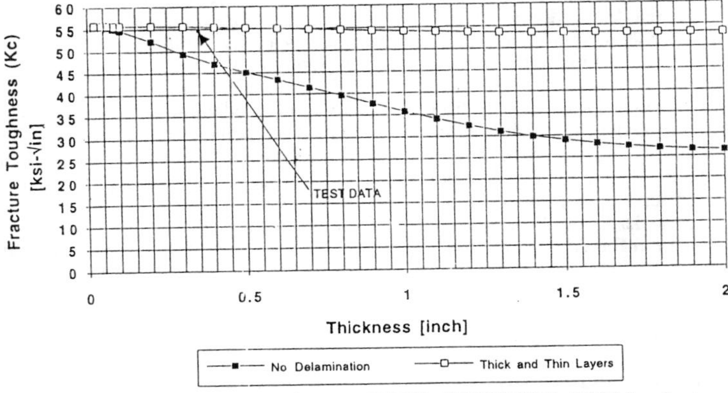

Figure 9. Fracture toughness of 50/50 6090/SiC/25p/6013 laminates

One of the many simplified ways on how to consider the effect of the bridging forces is to adjust the weight factors, w_u and w_r, in (4). In general, the weight factor for the delaminated reinforced layers will be higher than their volume fraction. Conversely, the weight factor for the reinforced layers will be smaller than their volume fracture

$$w_u = v_u + b_u$$

$$w_r = v_r - b_r. \qquad (5)$$

Therefore, when w_u is equal to v_u and w_r is equal to v_r , Equation (4) is a lower bound on the fracture toughness of the laminates in which delamination and bridging forces are significant. For example, the upper curve in Figure 9 is the lower bound for 50/50 6090/SiC/25p/6013 laminates.

CONCLUSIONS

Wide variety of metal laminate performance can be achieved when delamination is understood and accounted for in the early stages of laminate design. Two different laminate systems were reviewed in this paper to demonstrate this statement. The first was a metal, polymer, fiber system (ARALL Laminates). Here delamination occured during service fatigue loads. Fiber bridging forces were effective and crack arrest was possible. Depending on fiber type, bridging fibers could break under more general service loads. It was shown that fiber breakage affected both delamination shape and crack propagation rate.

The residual strength and the final fracture are dominated by the brittle fiber layers. The reason is that only negligible delamination accompanies this failure mode. Moreover, it was shown that the residual strength of an ARALL panel strongly depended on the fiber bridging length.

Such behavior poses a challenge for design engineers. To account for the bridging effect of unbroken fibers , the length of this zone must be known. However, under general service load, the bridging fibers at the crack wake can be severed. Hence, the bridging length (a-s) is unknown if visual inspection is used. Using visual inspection, the fatigue crack length in the aluminum layers, a, can be determined by a visual inspection of the two most outside aluminum layers. It is impossible to determine the crack length, s, of the fiber reinforced layers. An expensive non-destructive inspection have to be applied to determine bridging length in order to reliably account for the effect of unbroken fibers on the residual strength.

In most practical applications of metal, polymer, fiber laminates, design engineers consider the crack in the fiber and aluminum layers to be the same length as the crack detected in the two outside layers, which means that the bridging length is not considered in the calculation of residual strength. This conservative approach, dictated by today's inspection methods, does not allow the use of all the advantages of fatigue delamination in these laminate systems.

The second laminate system consisted of discontinuously reinforced aluminum layers combined with tougher unreinforced alloys - DRA laminates. This system behaves opposite to the metal, polymer, fiber systems. Fatigue crack propagates in all layers simultaneously. Also, delamination is small and can be neglected. Under such conditions, the fatigue crack propagation rate is dominated by the tortures path in the reinforced layers. No delamination means that (3) can be applied to estimate the da/dN of DRA laminates when stress intensity factors in (3) are replaced by da/dN curves. Since reinforced layers dominate fatigue crack growth, it is expected that the weight factors will be in favor of reinforced layers; i.e., for fatigue behavior, (4) would have opposite signs for b_u and b_r.

As mentioned in the previous section, delamination and significant crack bridging control the steady crack growth during the final stages of a panel failure. Hence, the R-curve of DRA laminates is significantly influenced by tougher, unreinforced layers.

In summation, the aforementioned two examples of metal laminates showed that fatigue and fracture performance can be significantly altered and required performance achieved by design for delamination.

REFERENCES

1. *Hierarchical Structures in Biology as a Guide for New Material Technology*, NMAB - 464, (National Academy Press, Washington, D. C. 1994).

2. R. J. Bucci, et al., in *ARALL Laminates Mechanical Behavior*, (Proceeding of ASM Material Week 1986, Orlando, Florida, 1986).

3. R. J. Bucci, et al., in *ARALL Laminates Results from a Cooperative Test Program,* (Proceedings of Advanced Material Technology 1987), (32nd International SAMPE Symposium and Exhibition, Anaheim, California, April 1987).

4. D.R. Laseur and C. K. Syn, in *Metallic Laminates for Engineering Applications*, (Proceedings of the 8th CIMTEC-World Ceramic Congress and Forum on New Material, 1994), Article SV-L08.

5. C. K. Syn, D. R. Laseur, and O. D. Sherby, in *Processing and Mechanical Properties of Laminated Metal Composites of Al/Al-25 vol% SiC and Ultrahigh Carbon_Steel/Brass*, (Int'l Conf. Advanced Synthesis of Engineered Structural Materials, San Francisco, 31 August - 2 September 1992).

6. R. Marissen, *Fatigue Crack Growth in ARALL A Hybrid Aluminum-Aramid Composite Material - Crack Growth Mechanisms and Quantitative Prediction of Crack Growth Rates*, Ph.D. Dissertation, TU Delft 1988.

7. R. Marissen, L. B. Vogelsang, in *Development of a New Hybrid Material: ARALL*, (Intercontinental SAMPE Meeting, Cannes, France, January 1981).

8. R. H. J. J. Roebroecks, *Constant Amplitude Fatigue of ARALL-2 Laminates*, Ph.D. Dissertation, TU Delft, October 1987.

9 . J. L. Teply, B. A. DiPaolo and R. J. Bucci, in *Residual Strength of ARALL Laminate Panels*, (19th International Sampe Technical Conference, Crystal City, VA, 13 -15 October, 1987).

10. Y. Macheret, J. L. Teply and E. F. M. Winter, in *Delamination Shape Effects in Aramid-Epoxy-Aluminum (ARALL) Laminates with Fatigue Cracks*, (Composites '88 International Conference, Boucherville, Quebec, 8 - 9 November, 1988).

11. H. M, Westergard, Am. Soc. Mechanical Engineers, Journal of Applied Mechanics, Series **a**, Vol. **66**, pp. 49 (1939).

12. J. Awerbuch, *Experimental Investigation of Delamination in ARALL Laminate*, Final Report, (Department of Mechanical Engineering, Drexel University, Philadelphia, PA 19101, May 1987).

13. R. S. Long, *Analytical Investigation of Delamination Progression in ARALL Laminates,* Product Design and Mechanics Division, Alcoa Technical Center, PA 15069, Report No. 57-88-24, August 1988 (unpublished).

14. J. N. Reddy, E. J. Barbero and J. L. Teply, *Plate Bending Element Based on a Generalized Laminate Plate Theory*, International Journal for Numerical Methods in Engineering, Vol. **28**, (October 1989).

15. E. J. Barbero, J. N. Reddy and J. L. Teply, *Accurate Determination of Stresses in ARALL Laminates Using a Generalized Laminate Plate Theory*, American Society of Mechanical Engineers, Applied Mechanics Division, Vol. **100**, (1989).

16. J. L. Teply, E. J. Barbero and J. N. Reddy, *Bending, Vibration and Stability of ARALL Laminates Using a Generalized Laminate Plate Theory*, International Journal of Solids and Structures, Vol. **27**, No.**5**, (1990), pp. 585-599.

17. J. R. Yeh and J. L. Teply, *Compressive Response of Kevlar/Epoxy Composites*, Journal of Composite Materials, Vol. **22**, (March 1988).

18. C. K. Syn, D. R. Laseur, K. L. Cadwell, O. D. Sherby, and K. R. Brown in *Laminated Metal Composites of Ultrahigh Carbon Steel/Brass and Al/Al-SiC: Processing and Properties*, Developments in Ceramic and Metal -Matrix Composites, Ed. K. Upadhya, (The Mineral, Metals & Material Society, 1991).

19. R. W. Bush, *Extrinsic Toughening of Discontinuously Reinforced Aluminum: An Experimental Study of DRA Laminates*, Engineering Design Center, Alcoa Technical Center, PA 15069, Report No. 57-95-08, August 1995 (unpublished).

MICROSTRUCTURE AND THERMAL CONDUCTIVITY OF THERMAL BARRIER COATINGS PROCESSED BY PLASMA SPRAY AND PHYSICAL VAPOR DEPOSITION TECHNIQUES

K. S. Ravichandran, R. E. Dutton*, S. L. Semiatin* and K. An
Department of Metallurgical Engineering, The University of Utah, Salt Lake City, UT 84112.
*Wright Laboratory, Materials Directorate, WL/MLLN, Wright Patterson AFB, OH 45433.

ABSTRACT

The temperature dependence of the thermal conductivity of multilayer coatings made by a plasma spray technique as well as some coatings made by physical vapor deposition (PVD) was investigated. The multilayer coatings consisted of a varying number of layers of Al_2O_3 and ZrO_2 stabilized by 8%Y_2O_3. Plasma sprayed coatings exhibited a large reduction in thermal conductivity at all temperatures when compared to the bulk monolithic materials. This reduction was found to be due to porosity as well as thermal resistance brought about by interfaces in the coatings. A comparable reduction in thermal conductivity was achieved in monolithic ZrO_2 as well as in a composite coating deposited by the PVD technique. Microstructural factors that may be responsible for this reduction are discussed.

INTRODUCTION

Improvements in the efficiency of gas turbine require the highest operating temperatures possible. Because the Ni-base superalloys used as turbine materials rapidly loose strength and oxidize above 1000°C, a reduction in service temperature is often accomplished by the use of thermal barrier coatings [1,2]. Traditionally, such coatings have been applied by plasma spray [1] or physical vapor deposition [2] onto turbine components with an intermediate NiCoCrAlY alloy bond coating to improve adherence and to reduce oxidation. The thermal conductivity of these coatings is sensitive to the deposition technique, microstructure, density, and interface thermal resistance between layers [3]. The general objective of this research was to examine the relationship between the coating microstructure and thermal conductivity. Specifically, the thermal conductivities of multilayer coatings involving alternating ZrO_2 and Al_2O_3 layers, deposited by plasma spray (PS) as well as the coatings deposited by physical vapor deposition (PVD) technique, were investigated. These configurations were expected to provide reduced thermal conductivity due to the interfaces present in these coatings.

EXPERIMENTAL PROCEDURE

Plasma sprayed coatings were obtained by spraying alternating layers of ZrO_2 and Al_2O_3 onto a 3mm thick superalloy substrate measuring 62.5mm X 12.5mm. Powders of ZrO_2 stabilized with 8%Y_2O_3 with an average particle size of about 10μm and Al_2O_3 with an average particle size of about 5μm were used. A Plasma Technik Spray system, with a single spray nozzle and dual powder feeder at the Thermal Spray Laboratory of the State University of New York, Stony Brook, NY, was used. Calibration sprays were performed to control the layer thickness during the actual multilayer spray deposition. The ZrO_2 and Al_2O_3 powders were alternately fed in to the spray gun, and deposition was carried out for a specified period, determined from the calibration trials and the required layer thickness. To determine if residual porosity could be closed by sintering, heat treatment of these coatings was performed at 1300°C for 50 hrs in a furnace under flowing argon. However, the coatings detached from the substrate as units after such a sintering heat treatment. Porosity levels were determined using measurements of coating mass and volume as well as by the point counting technique on micrographs.

PVD coatings were deposited using a proprietary PVD process. A monolithic ZrO_2 coating and a composite coating comprising of ZrO_2 and Al_2O_3 were deposited.

Mat. Res. Soc. Symp. Proc. Vol. 434 © 1996 Materials Research Society

Thermal conductivity (TC) measurements were made by the Thermophysical Properties Research Laboratory of Purdue University, West Lafayette, IN, using the laser flash method. Details of this technique are available elsewhere [4]. TC measurements at room temperature and at various temperatures to 1000°C were performed. Measurements were made on the coatings with substrate in the as-sprayed condition, but only on the coatings in the detached condition after heat treatment.

RESULTS AND DISCUSSION

(a) Plasma Sprayed Coatings

The single layer Al_2O_3 and ZrO_2 PS coatings were designated as A1 and Z2, respectively. The multilayer coatings were identified as AZ11, AZ21, AZ41 and AZ81 in which the first number after Z identifies the number of alternating layers each of Al_2O_3 and ZrO_2 (AZ41 consists of 4 layers each of Al_2O_3 and ZrO_2). Microstructures of some of the coatings are shown in Fig. 1 and 2, for the as-sprayed condition and after heat treatment, respectively. The porosity levels in the coatings were: A1: 18.8%; Z2: 11.9%; AZ11: 15.5%; AZ21: 17%; AZ41: 12.7%; AZ81: 13.7%. While there appears to be no change in porosity in Al_2O_3 layers, a slight reduction in porosity in ZrO_2 layers after heat treatment can be seen.

The thermal conductivity data as a function of temperature are presented in Figs. 3 and 4 for coatings in the as-sprayed and in the heat-treated conditions, respectively. In Fig. 3, it can be seen that the thermal conductivity levels of monolithic as well as multilayer coatings are lower than that of dense materials. At temperatures <200°C, the TC levels of coatings and the monolithic ZrO_2 are comparable. On the other hand, the TC values are at least a factor of two lower than that of monolithic ZrO_2 at temperatures >200°C. The TC of Al_2O_3 coating is slightly higher than that of the other coatings. The TC levels of all the multilayer coatings average around the monolithic ZrO_2 coating, but are slightly lower than the trend calculated from the monolithic TC values on the basis a of series arrangement of alternate layers (for a bi-layer, $\lambda = \lambda_{Al2O3}\lambda_{ZrO2} / (\lambda_{Al2O3}t_{ZrO2} + \lambda_{ZrO2}t_{Al2O3})$, in which λ is thermal conductivity and t is the thickness fraction of each layer). Also included in the figure is the TC for a bi-layer calculated using the dense TC values reported in the literature and accounting for the effect of porosity on thermal conductivity of each layer. The data for dense Al_2O_3 and ZrO_2 were taken from Refs. [5] and [6], respectively. The effect of porosity on thermal conductivity was incorporated using the relationship: $\lambda = \lambda_d [1\text{-}P^{2/3}]$ (λ_d is the TC of dense material and P is the volume fraction of porosity). It can be seen that the measured TC data for multilayers are significantly lower than this value at all temperatures. This suggests that the reduction in thermal conductivity of plasma sprayed coatings can be explained only partially on the basis of the porosity of coatings, and that other microstructural factors are equally important.

The TC levels of all the coatings after heat treatment were significantly higher than those in the as-sprayed condition as shown in Fig. 4. It is to be noted that while the data for all the as-sprayed coatings showed a significant change with temperature, the TC data after sintering were largely temperature-independent. The thermal conductivities of the multilayer coatings are in agreement with the estimated bi-layer thermal conductivity using the data of monolithic coatings as well as the predictions based on the dense materials after accounting for the effects of porosity on thermal conductivity. In the latter, the porosity data of as-sprayed coatings were used in the calculation because there was only a small change in porosity during heat treatment.

These results indicate that several microstructural factors should be considered in understanding thermal conductivity changes after heat treatment. Porosity and thermal resistance at interfaces can significantly influence the thermal conductivity in solids [7,8].

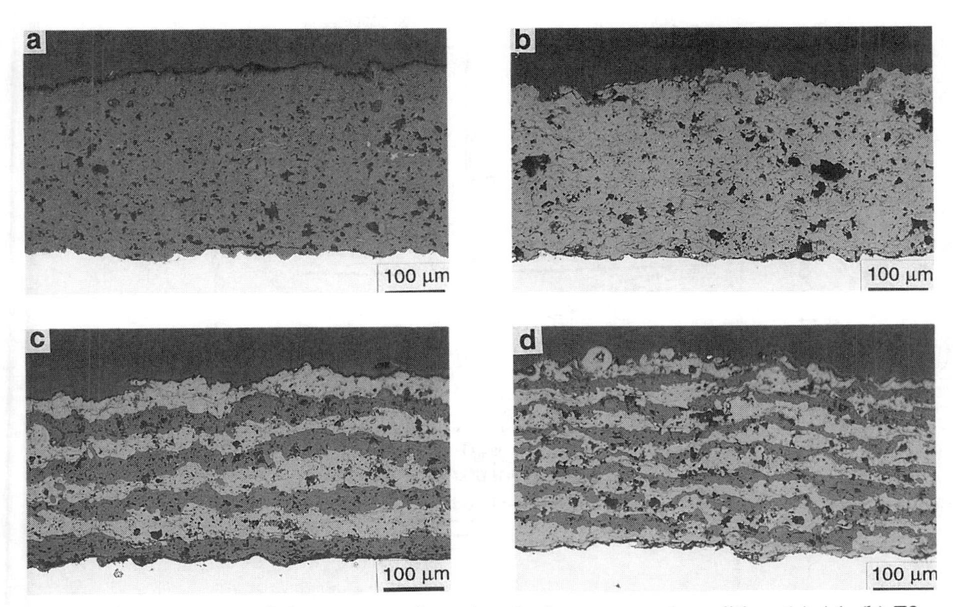

Fig. 1. Microstructures of plasma sprayed coatings in the as-sprayed condition. (a) A1, (b) Z2, (c) AZ41 and (d) AZ81.

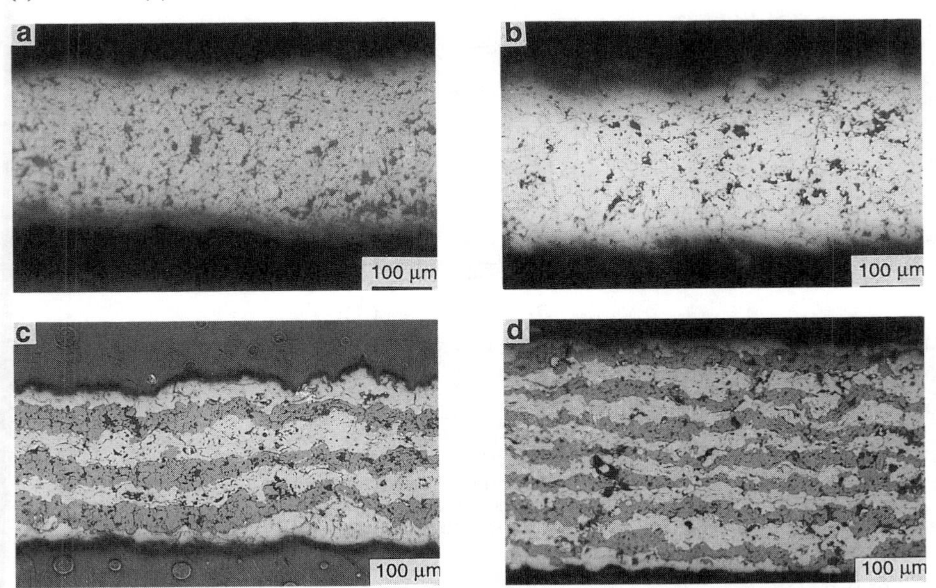

Fig. 2. Microstructures of plasma sprayed coatings after the heat treatment. (a) A1, (b) Z2, (c) AZ41 and (d) AZ81.

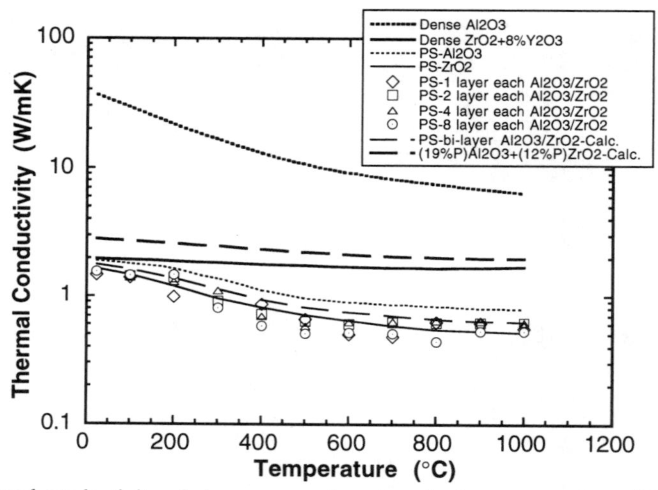

Fig. 3. Thermal conductivity of plasma sprayed coatings in the as-sprayed condition.

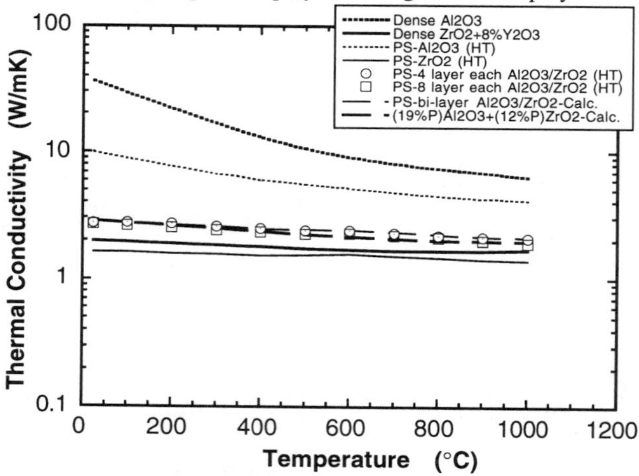

Fig. 4. Thermal conductivity of plasma sprayed coatings after heat treatment at 1300°C for 50 hrs. HT refers to the heat-treated condition.

Since porosity levels changed only a little, this is not a major factor. On the other hand, interfaces between the splats in thermal sprayed coatings have been suggested to contribute to reduced thermal conductivity due to the interface thermal resistance [3,8]. In addition, there are interlayer interfaces and the interface between the coating as a unit and the substrate. Thus, although the bulk of the reduction in thermal conductivity appears to be due to an interface-type effect, at this stage it is not possible to determine the specific contributions from the different interfaces.

(b) PVD Coatings

Microstructures of the PVD coatings were very different from those deposited by the plasma spray technique. Both coatings exhibited a columnar microstructure typical of the PVD process [9,10]. Fig. 5 shows the microstructures of monolithic ZrO_2 coating as seen in the optical and scanning electron microscopes. The microstructures of the composite coating are not shown here because of its proprietary nature.

Thermal conductivity data for the PVD coatings are presented in Fig. 6. At temperatures above 500°C, both coatings have similar thermal conductivity levels. However at lower temperatures, the composite coating is seen to have a lower thermal conductivity, the magnitude of this difference increasing at lower temperatures.

Several microstructural factors must be considered to interpret the thermal conductivities of the coatings with respect to that of the bulk monolithic ZrO_2 and to explain the differences between the coatings themselves. First, the differences between bulk material and the coatings need to be examined. A preliminary X-ray diffraction study indicated that the amount of monoclinic ZrO_2 (M-ZrO_2) in both the coatings was negligible and the coatings consisted entirely of tetragonal ZrO_2 (T-ZrO_2) phase. However, the reference bulk ZrO_2 consisted of a significant amount of M-ZrO_2 in addition to T-ZrO_2. Additionally, strong (200) and (111) textures were observed in the coatings. Hence, both the absence of M-ZrO_2 phase and the presence of anisotropy in coatings may explain part of the reduction in the thermal conductivities of both coatings compared to that of the bulk material. Since the effects of M-ZrO_2 versus T-ZrO_2 phase proportion and crystal orientation on thermal conductivity are not clear, these effects could not be quantitatively assessed at present.

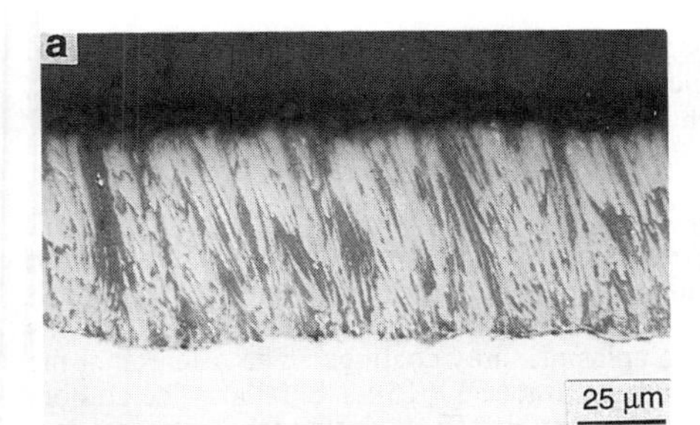

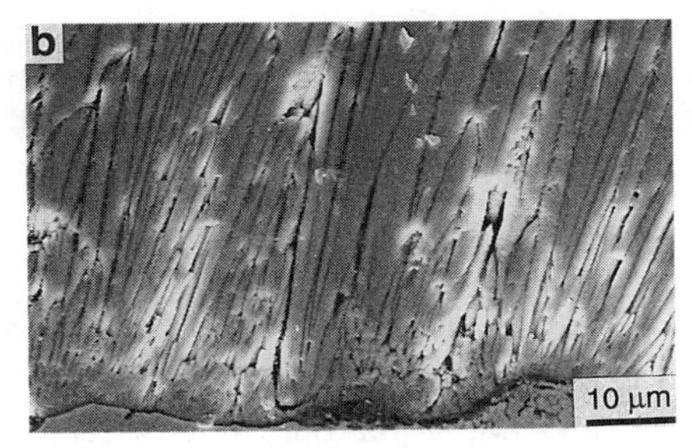

Fig. 5. Microstructure of ZrO_2 coating deposited by the PVD technique as seen in (a) optical microscope and (b) scanning electron microscope.

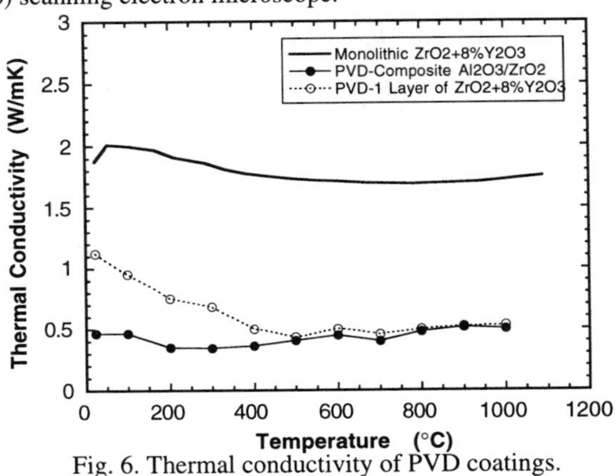

Fig. 6. Thermal conductivity of PVD coatings.

The microstructure of the composite coating differs from that of the monolithic coating in several aspects. First, the composite coating had a (111) texture compared to a (200) texture in the monolithic coating. Secondly, the composite had unique microstructural arrangement that differed from the monolithic coating. X-ray diffraction indicated that the Al_2O_3 was present in an amorphous form. Further research is necessary to understand the impact of these differences and other factors on the thermal conductivity of the composite coatings.

CONCLUSIONS

1. The thermal conductivity of the plasma sprayed multilayer coatings comprising alternating Al_2O_3/ZrO_2 layers was comparable to the monolithic ZrO_2 coating made by the same technique. This similarity appears to be due to porosity and the thermal resistance due to interfaces in the microstructure.

2. After heat treatment, the effect of interface thermal resistance on conductivity was absent. However, because the coatings detached after heat treatment, it is not clear whether the elimination of thermal resistance was due to the increased contact between internal interfaces or the elimination of the interface between the coatings as units and the substrate.

3. The thermal conductivity of the PVD coatings was comparable to that of the plasma sprayed coatings although their microstructures were entirely different. Such a large reduction in thermal conductivity may be due to the proportion of M-ZrO_2 versus T-ZrO_2 as well as the presence of anisotropy in the coatings.

4. The PVD composite Al_2O_3/ZrO_2 coating showed a significantly lower thermal conductivity compared to that of the monolithic ZrO_2 coating at temperatures below 500°C. This may be due to its unique microstructural characteristics, compared to monolithic coating.

ACKNOWLEDGMENTS

The authors thank Mr. T. Broderick, Processing Science Group, Wright Laboratory, Materials Directorate for his interest, support and encouragement throughout this study. The authors also thank Dr. S. Sampath, State University of New York, Stony Brook, NY for his help and discussions during the manufacture of the plasma spray coatings. The research at the University of Utah was supported through Air Force contract F33615-92-C-5900. The authors also acknowledge Mr. Douglas R. Barker of UES, Inc., for his efforts during this program.

REFERENCES

1. R. A. Miller, Surf. Coat. and Tech., **10**, 1 (1987).
2. H. Lammermann and G. Kienel, Adv. Mater. Processes, **140**, 18 (1991).
3. L. Pawlowski and P. Fauchais, Int. Metall. Rev., **31**, 271 (1992).
4. R. E. Taylor, J. Phy. E. Sci. Inst., **13**, 1193 (1980).
5. D. P. H. Hasselman, L. F. Johnson, L. D. Bentsen, R. Syed, H. M. Lee and M. V. Swain, Am. Ceram. Soc. Bull., **66**, 799 (1987).
6. W. N. D. Santos and R. Taylor, High Temp. - High Press., **25**, 89 (1993).
7. M. V. Roode and B. Beardsley, ASME Paper 88-GT-278, Presented at the Gas Turbine and Aeroengine Congress, Amsterdam, The Netherlands, June 6-9, (1988).
8. C. H. Liebert and R. E. Gaugler, Thin Solid Films, **73**, 471 (1980).
9. J. A. Thornton, Ann. Rev. Mater. Sci., **7**, 239 (1977).
10. B. E. Paton and B. A. Movchan, Thin Solid Films, **54,** 1 (1978).

MULTI-PHASE FUNCTIONALLY GRADED MATERIALS FOR THERMAL BARRIER SYSTEMS

M.R. JACKSON, A.M. RITTER, M.F. GIGLIOTTI, A.C. KAYA, J.P. GALLO, General Electric CRD, Schenectady, NY 12309 jacksmr@crd.ge.com

ABSTRACT

Metallic candidates for functionally graded material (FGM) coatings have been evaluated for potential use in bonding zirconia to a single crystal superalloy. Properties for four materials were studied for the low-expansion layer adjacent to the ceramic. Ingots were produced for these materials, and oxidation, expansion and modulus were determined. A finite element model was used to study effects of varying the FGM layers. Elastic modulus dominated stress generation, and a 20-25% reduction in thermal stress generated within the zirconia layer may be possible.

INTRODUCTION

Jet engine and gas turbine hot section components can be protected from the 1350-1650°C combustion gases by thermal barrier coatings (TBCs). TBC systems with insulating ceramics, oxidation resistant bond coats, and strong substrates, have been studied for 20 years [1,2]. Designed to reduce heat transfer to metal components, TBCs offer two forms of protection. At steady state, the temperature of the underlying metal is reduced, relative to the uncoated metal with identical cooling flow. During transient conditions, when flame temperatures increase or decrease rapidly, the rate of temperature change in the metal is reduced and thermomechanical shock is less severe.

Zirconia has the best success as a TBC, with low thermal conductivity and high thermal expansion, compared to other oxides. Its expansion behavior is still much lower than for the Ni and Co alloys it protects, creating thermal stresses during exposure that add to other stresses causing delamination and spallation. Stabilizing additions, such as the 6–8 wt% yttria-stabilized ZrO_2 (YSZ), avoid the volume change from crystallographic transformation which can cause cracking and spallation [3-5].

A metallic oxidation resistant bond coat deposited onto the metal substrate anchors the ceramic mechanically, and its roughness helps to accomodate thermal mismatch. The bond coat is usually an MCrAlY composition (where "M" is Ni, Co, or both) or a diffusion aluminide, and is more oxidation-resistant and corrosion-resistant than the underlying superalloy. Bond coat oxidation is a major factor in TBC failure [6,7] for bond coats processed by plasma-spraying [8] and by PVD [9]. MCrAlY oxidation resistance has also been substantially improved by aluminiding prior to deposition of the ceramic [8], to produce an increased Al content on the outer surface of the bond coat.

The driving force for eventual TBC spallation can be reduced if the TBC/substrate expansion differential is reduced [10]. TBCs deposited onto IN 718 and IN 909 were furnace cycled to evaluate the concept of matching thermal expansion of substrate and oxide. IN 909 is much lower in expansion than IN 718, and approaches the expansion of YSZ. On cycling to 900 and to 1015°C, the TBC on IN 718 failed early (360 and 72 cycles, respectively), while the TBC on IN 909 was unfailed when testing was terminated (4932 and 216 cycles, respectively). At 1100°C, TBC failure on IN 909 occurred, but at a significant gain in cyclic life over IN 718 (1026 and 178 cycles, respectively). These results indicate that a low-expansion substrate can offer extended TBC lifetime. However, the strong Ni and Co-base superalloys have expansion similar to IN 718 expansion behavior.

FGMs have demonstrated mechanical performance improvements as bond coats for TBC systems. The residual thermal stresses present in parts as a result of fabrication and service can be reduced by grading the structure [11]. The FGMs can be made of stepwise homogeneous-composition layers, or can be graded continuously through the thickness, to reduce stresses. Increased coating adhesion and TBC lives have been shown, and the thermal cycling/ thermal shock resistance of graded structures was superior to that of monolithic coatings [12-15].

Mat. Res. Soc. Symp. Proc. Vol. 434 © 1996 Materials Research Society

To the present, TBC systems have been used to extend airfoil life beyond the uncoated design life. Reliability of the TBC has been insufficient to allow incorporation into the design. Major improvements in turbine efficiency and performance will be achievable when the TBC can be relied upon to survive. These improvements result directly from designs allowing higher gas temperatures and/or reductions in cooling air, without a rise in metal temperatures. The concept of using FGMs to increase that reliability may be essential to achieving the full potential of TBCs. Clearly, there has been improvement in mechanical performance at the bond coat/TBC interface through the use of FGMs. However, incorporating low-conductivity ceramics in the graded region will lead to slightly increased temperatures in the metallic region closest to the TBC. TBC failure is related to oxidation of the bond coat in conventional TBCs, and FGMs with ceramic gradients have suffered from the same limitations. Our research is aimed at achieving good expansion matching using metallic bond coatings with low-expansion phases, to achieve the lowest bond coat temperatures.

EXPERIMENT

A total of 7 alloys were selected for evaluation (Table I), low-expansion alloys 1-4 to serve next to the TBC, intermediate layer alloys 5-6, and alloy 7 as the alloy next to the substrate. Only alloys 1-4 are described in this paper. Compositions are shown schematically in Figure 1. All alloys were directionally solidified at 20cm/h to produce sound material, and test pins 2.5 cm long and .25 cm in diameter were taken in a transverse orientation to produce a multi-grained, non-directional sample axis for dynamic modulus, dilatometric thermal expansion and oxidation. A pin of each was tested in one-hour cyclic oxidation to 1100°C for 525 hours. FGM temperatures and stresses were modeled in a finite element analysis using estimated properties.

RESULTS

Alloy Structures

Directional solidification was used to produce low-porosity alloys. The phases expected for alloy 1 from the NiAlCr phase diagram [16] were observed. For alloys 2-4 (microstructure of alloy 2 shown in Fig. 2), no detailed NiAlCrC diagram exists. Based on the NiCrC ternary, the carbide in equilibrium with Cr, NiAl and Ni (α, β γ, respectively) is expected to be Cr_7C_3; this was verified by X-ray diffraction.

Table I - Candidate Alloy Compositions for FGMs (phase v/o and alloy elemental a/o)

	α	β	γ	M_7C_3	**Ni**	**Cr**	**Al**	**C**
1	45	25	30	0	36.5	51.1	12.4	0
2	30	25	30	15	37.8	45.4	12.3	4.5
3	15	25	30	30	39.2	39.7	12.1	9.0
4	0	25	30	45	40.5	34.0	12.0	13.5
5	15	25	45	15	48.2	33.7	13.6	4.5
6	15	45	25	15	46.2	30.5	18.8	4.5
7	0	25	60	15	58.5	22.0	14.9	4.6

α-98Cr 1Ni 1Al; β-4Cr 60Ni 36Al; γ-20Cr 70Ni 9.9Al 0.1C; M_7C_3-60Cr 10Ni 30C (a/o)

[approximate phase chemistries at 1100 °C]

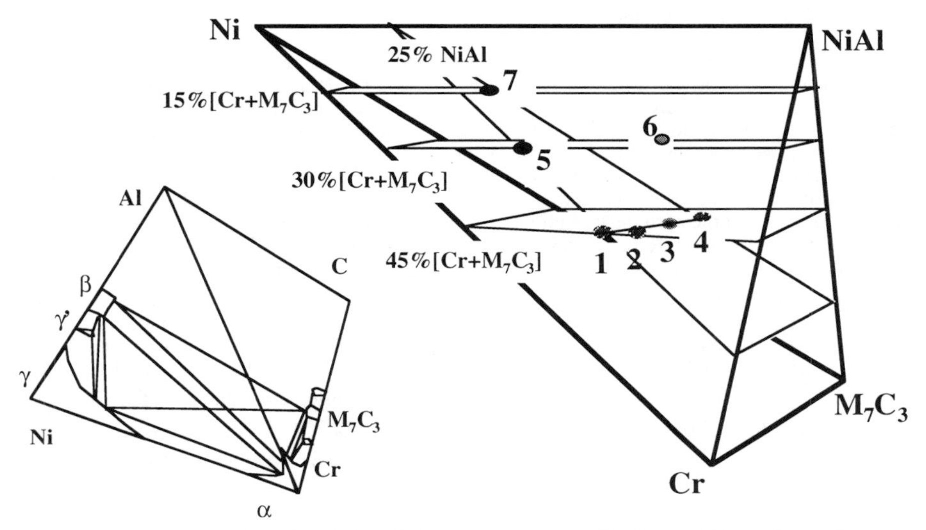

Figure 1 - Approximate locations in Ni-Cr-Al-C quaternary space for the candidate alloys. Low-expansion alloys 1-4 contain ~45volume percent of αCr plus M_7C_3.

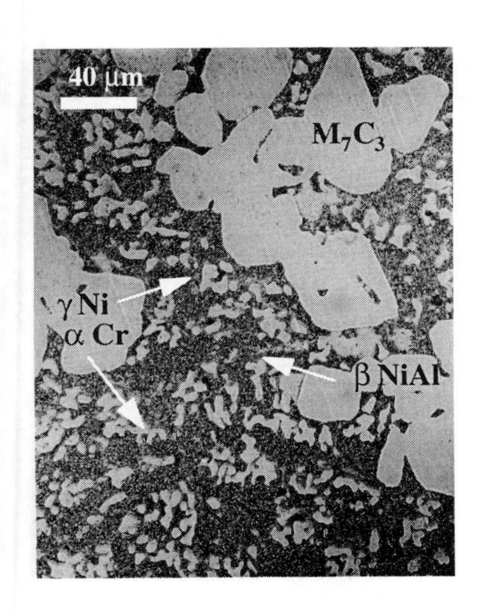

Figure 2-Microstructure of alloy 2 after equilibration at 1100°C. The large bright phase is M_7C_3, the small bright phase is αCr, the light gray phase is βNiAl, and the dark matrix is γNi solid solution.

Physical Properties

Pin samples were subjected to differential dilatometric evaluation over the range 23-1200°C, with a heating rate of 10°C/min, with an alumina standard. Rule-of-mixture estimates of thermal expansion were made before testing, by using known data for the three metal phases and their expected volume fractions, and using an assumed expansion behavior for the carbide that was somewhat lower than that of Cr (~0.85 x % expansion of Cr). These estimated values predicted expansion of ~1.4% (1.38% for alloy 4, 1.45% for alloy 1) from room temperature to 1100°C. These values are all greater than the 1.22% expansion of YSZ. However, the measured expansion showed the carbide-free alloy 1 to be lower in expansion than expected, and quite close to YSZ. The carbide-containing alloys 2-4 showed substantially greater expansion than expected, and greater than alloy 1 or YSZ (Figure 3). Small differences in the volume percentages of the low-expansion αCr and/or carbide phases from those assumed in Table I could account for the differences in expansion behavior. All four alloys exhibited markedly increased rates of expansion above 1000°C. At approximately this temperature, there is a four-point invariant reaction in the NiCrAl equilibrium diagram [16]. At temperatures greater than 1000°C, there is a γ-β phase field that isolates γ' from α, while below that temperature the α-γ' field replaces the γ-β field. As temperature rises above 1000°C, the volume fraction of α in the structure is reduced, and the suppression in expansion behavior of the alloys due to the αCr phase is lessened.

Elastic modulus measurements were made at room temperature using a dynamic ultrasonic method. Values of 241-248 GPa were estimated, based on values of 290-310 GPa for the Cr and carbide phases, and values of 193-207 GPa for γ and β. Measured values were 221-228 GPa for alloys 1-4, with the range too small to see any trend.

Oxidation Behavior

Pin samples .25cm in diameter and 2.5 cm in length were subjected to one-hour cyclic oxidation in a static air furnace, with periodic removal for visual evaluation and weight measurements. Weight gain per unit area is shown in Figure 4, and a typical microstructure is shown in Figure 5. Oxidation rate is relatively low for alloy 1, and rate of oxidative attack is

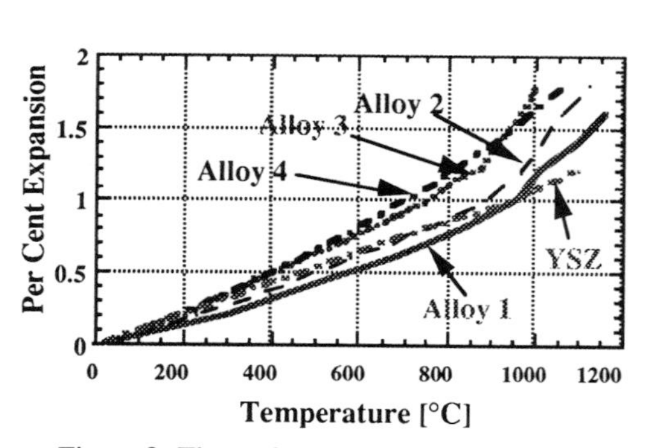

Figure 3- Thermal expansion behavior measured for alloys 1-4, compared to yttria-stabilized zirconia.

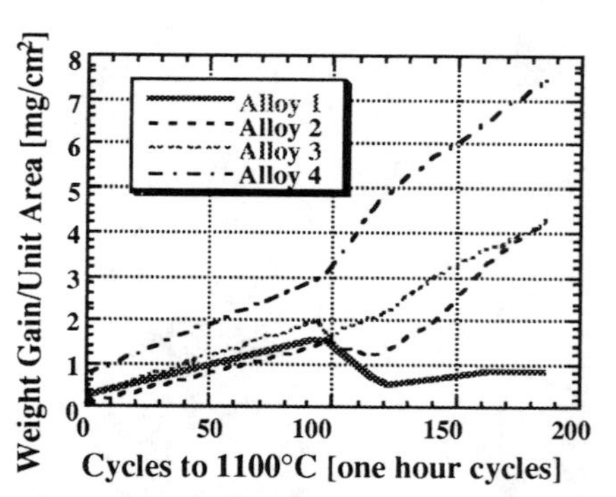

Figure 4-Weight change in cyclic oxidation for alloys 1-4: 50 minutes at 1100°C and 10 minutes for cooling cycle.

seen to increase with carbide volume fraction. Samples were continued through 525 hours of cyclic oxidation, with alloy 1 showing a gain of $14.5mg/cm^2$, alloy 2 a loss of 17.6, alloy 3 a gain of 16.3, and alloy 4 a gain of $72mg/cm^2$. For reference, if a weight gain reflected only growth of M_2O_3, with no simultaneous spallation, a weight gain of $\sim 2mg/cm^2$ would represent 10μm of oxide growth. If a weight loss reflected only spallation of oxide, with no simultaneous oxidation, a weight loss of $\sim 7mg/cm^2$ would represent 10μm of metal loss to oxide spallation.

Alumina formation resulted in substantial Al depletion (Figure 5). The alumina scale is not protective, and loss of β phase due to Al diffusion to the surface oxide growth front is evident, as is internal oxidation where the Al loss has been greatest. For reference, formation of 10μm of alumina consumes >10μm of NiAl, >27μm of a 20a/o Al alloy, and >50μm of a 10a/oAl alloy.

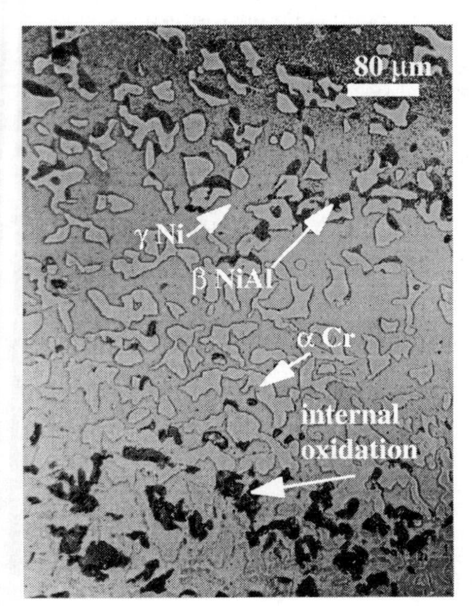

Figure 5-Microstructure of oxidized alloy 2 after 525 cycles to 1100°C. The outer oxidized surface is out of view at the bottom. The unaltered structure is at the extreme top of the micrograph. Substantial depletion of Al from the structure has eliminated the NiAl phase from the outer regions of the sample. Internal oxidation has led to attack of the Cr-rich phases, both α (shown here) and the carbide (not shown).

Model Predictions

Some 24 elastic cases were run in a finite element analysis of thermal stresses generated in a cooled structure in a hot gas (thermal stress assumed to be zero at room temperature). The bond coat was treated as either a monolithic structure or as a stepwise FGM with three layers. The different combinations of the layer next to the TBC as alloy 1, 2, 3 or 4; the mid layer as either alloy 5 or 6; and the layer adjacent to the substrate as alloy 7, were considered. Different thicknesses of the layers were assumed, but with a constant bond coat thickness of 100μm. The TBC surface temperature was 1200°C, the bond coat/TBC interface was 965°C, the bond coat/superalloy interface was 954°C, and the cold wall of the superalloy was 810°C. The range in elastic modulus and expansion behavior assumed for alloys 1-4 resulted in calculated elastic stresses varying from ~250MPa to ~400MPa. If actual values were used, based on measured expansion and room temperature modulus, the four alloys are expected to be closer in stress, and possibly to be less stressed due to their lower actual moduli. The actual expansions of alloys 2-3

The calculated stresses in the YSZ TBC, ~25MPa, are about 20% lower than the calculated value for the TBC on a monolithic coating on a single crystal superalloy. This amount of reduction may be significant in reducing the driving force for TBC loss due to alumina spallation or TBC delamination, but cases of stresses in the TBC with a more conventional bond coat or a metal/oxide FGM bond coat is yet to be calculated.

CONCLUSIONS

It appears possible to essentially match the expansion behavior of YSZ TBCs using alloys which are combinations of metal phases αCr, βNiAl, and γNi. Using the moduli of such alloys to predict elastic stresses in FGM bond coats results in reasonable stress levels, considering experience with calculations of more conventional bond coats. Building gentle gradient structures in all-metallic materials can involve simple variation of the proportion of α phase through the thickness of the FGM. It may be necessary to modify FGM microstructures and chemistries to promote better oxidation resistance (a richer Al source within the FGM), and to develop greater phase stability if TBC/FGM interface temperatures exceed 1000°C. The possibility of stress reduction in the TBC in such a material system is very promising.

ACKNOWLEDGEMENTS

The authors wish to acknowledge the financial support of the Air Force Office of Scientific Research, Captain Charles Ward, through contract F49620-95-C-0028. E.H. Hearn carried out the measurements on oxidation and expansion, and performed metallographic evaluations. R.J. Petterson performed directional solidification, and S.A. Weaver measured elastic modulus.

REFERENCES

[1] C.H. Leibert and S. Stecura, "Thermal Barrier Coating System," U.S. Patent 4,055,705, October 25,1977.
[2] R.D. Dowell, "Coating for Metal Surfaces," U.S. Patent 3,911,891, October 14, 1975.
[3] N. Iwamoto, Y. Makino and Y. Arata, *Proc. Int. Thermal Spraying Conf.*, The Hague, Netherlands, p. 267 (1980).
[4] G. Johner and K.K. Schweitzer, *Thin Solid Films119*, p. 301 (1984).
[5] S. Stecura, *Thin Solid Films 150*, p. 15 (1987).
[6] S. Stecura, *Thin Solid Films 136*, p. 241 (1986).
[7] P. Sahoo and R. Raghuraman, *Proc. 1993 National Thermal Spray Conf.*, Anaheim, CA, p. 369 (1993).
[8] D.J. Wortman, B.A Nagaraj and E.C. Duderstadt, *Mat. Sci. Eng. A121*, p. 433 (1989).
[9] E.Y. Lee and R.D. Sisson, Jr., *Proc. 7th Int. Thermal Spray Conf.*, Boston, MA, p. 55 (1994).
[10] G.D. Smith and J.A.E. Bell, *Physical Metallurgy of Controlled Expansion Invar-Type Alloys,* TMS, Warrendale, PA, p. 283 (1990).
[11] Y. Itoh and H. Kashiwaya, *J. Ceramic Society of Japan100*, pp. 476–481, (1992).
[12] J.R. Rairden, GE-CRD Report No. 90CRD236, Schenectady, NY (1991).
[13] H.E. Eaton and R.C. Novak, *Ceramic Eng. & Sci. Proc., Vol. 7*, p. 727 (1986).
[14] J. Musil, J. Filipensky, J. Ondracek and J. Fiala, *Proc.Int. Thermal Spray Conf.*, Orlando, FL, p. 525 (1992).
[15] L.S. Wen, K. Guan, S. W. Qian, C. -F. Lin, and L. S. Fu, "Ceramic Thermal Barrier Coating for Adiabatic Diesel Engine," *Acad Sinica Materials Protection 25*, pp. 14–17 (1992).
[16] A. Taylor and R.W. Floyd, "The Constitution of Ni-Rich Alloys of the Ni-Cr-Al System," *JIM 81*, pp. 451-464 (1953).

EXAMINATION OF IN-SERVICE COATING DEGRADATION IN GAS TURBINE BLADES USING A SMALL PUNCH TESTING METHOD

J. KAMEDA*, T. E. BLOOMER*, C. R. GOLD*, Y. SUGITA#, M. ITO# and S. SAKURAI+
*Center for Advanced Technology Development, Iowa State University, Ames, IA 50011.
#Electric Power R & D Center, Chubu Electric Power Co., Inc., Nagoya, 458, Japan.
+Mechanical Engineering Research Laboratory, Hitachi Ltd., Hitachi, 317, Japan.

ABSTRACT

This paper describes examination of in-service coating degradation in land based gas turbine blades by means of a small punch testing (SP) method and scanning Auger microprobe (SAM). SP tests on coated specimens with unpolished surfaces indicated large variations of the mechanical properties because of the surface roughness and curvature in gas turbine blades. SP tests on polished specimens better characterized the mechanical degradation of blade coatings. The coated specimens greatly softened and the room temperature ductility of the coatings and substrates tended to decrease with increasing operation time. The ductile-brittle transition temperature of the coatings shifted to higher temperatures during the blade operation. From SAM analyses on fracture surfaces of unused and used blades, it has been shown that oxidation and sulfidation near the coating surface, which control the fracture properties, result from high temperature environmental attack.

INTRODUCTION

Recently, land based gas turbines have been widely applied in a combined cycle of electric power stations to improve the fuel efficiency and environment. Advanced coating techniques enable one to enhance the performance of gas turbine blades that are operated under high thermal/applied stresses and high temperature aggressive environments. However, it is well recognized [1,2] that the mechanical degradation of coatings and substrates in gas turbine blades inevitably occurs as a result of microstructural/chemical evolution. Thus the degradation characterization of gas turbine blades is of importance to extend the remaining life and maintain safe operation.

Since gas turbine blades have a complex shape and coating degradation is highly localized near the blade surface, it is difficult to apply standard testing techniques to evaluate the mechanical properties of coatings. An attempt has been made to study the local mechanical properties in blade coatings by applying a miniaturized small punch (SP) testing method [3-5]. This paper summarizes recent results of the mechanical and microstructural/chemical degradation of gas turbine blades induced in-service. The relationship of the mechanical properties of the blade coatings to the microstructural/chemical evolution analyzed by scanning Auger microprobe (SAM) is presented in order to clarify the degradation mechanism of the blades.

EXPERIMENTAL METHOD

The materials used in this study were gas turbine blades made of a nickel base superalloy (René 80) substrate and CoNiCrAlY coating. The turbine blades were unused, operated for 8946 h and 21338 h using combined fuels of liquefied natural gas (LNG) and kerosene, and used for 22000 h mainly under LNG. The used blades are designated as 9Kh, 21Kh and L22Kh. The coating thickness varied from 150 to 250 μm.

As shown in Fig. 1, disk-shaped SP coating specimens (6 mm ϕ and 0.5 mm thick) were extracted from the near surface region of the various blades using an electrical discharging machine. SP substrate specimens were prepared by machining off the coating. The coating and substrate surface of SP specimens were mechanically polished using emery paper (1000 grit) to remove the surface roughness and curvature. The surface layer of the coated specimens was removed by 20-35 μm. Unpolished SP coating specimens and notched SAM specimens (2 mm x 3 mm x 10 mm) were also prepared.

Mat. Res. Soc. Symp. Proc. Vol. 434 © 1996 Materials Research Society

Figure 1. Extraction of disk-shaped small punch (SP) specimen from gas turbine blade.

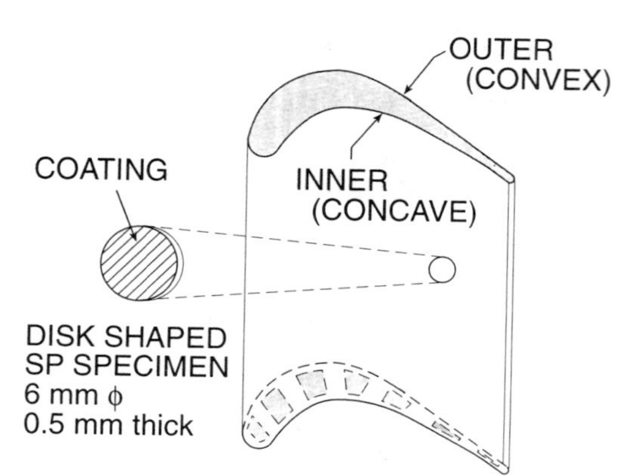

A puncher with a hemispherical tip (diameter of 2.4 mm) and specially designed specimen holders consisting of lower and upper dies and clamping screws were used for SP tests [3,6]. The details of a high temperature SP testing method are indicated in Ref. [3]. SP specimens were loaded in a screw-driven Instron testing machine with the cross head speeds of 2×10^{-5} and 8×10^{-6} m/s at 22 °C and elevated temperatures. SP tests were carried out in air in a temperature range from 22 to 950 °C

Hydrogenated SAM specimens of the unused and used blades were broken in an ultra high vacuum chamber (1.5×10^{-8} Pa). Mechanically polished cross sections of the SAM specimen were sputter-cleaned under Ar atmosphere (5×10^{-6} Pa) at 1.25 keV. The chemistry on the fracture surface and cross section of the blade specimens was examined using a cylindrical mirror analyzer (5 keV) of Physical Electronics Model 660.

RESULTS AND DISCUSSION

Some typical load vs. deflection curves obtained from SP tests on coating specimens with unpolished and polished surfaces are illustrated in Figs. 2 and 3. SP tests at 22 °C on the unpolished coating specimens indicated various deformation modes depending on the different surface roughness and curvature of gas turbine blades (Fig. 2). SP tests on the polished coating specimens consistently showed elastic and plastic bending behavior at various temperatures (Fig. 3). The polished SP specimen tests exhibited higher module with smaller deviations than the unpolished ones due to the absence of the surface roughness and curvature effects. Thus the coating degradation was examined using SP specimens with polished surfaces. At all the temperatures except 950 °C, the yield strength of the coated specimens, estimated from the yield load (P_y) [7], tended to decrease with increasing operating time [3,5]. Most of the coated specimens greatly softened and had almost the same yield strength at 950 °C though the 21Kh coating showed larger scattering. The substrate specimen did not show softening while in-service

Brittle coating cracks initiated at the critical defection (δ_f), represented by a decrease in the loading rate (Fig. 3). The crack formation occurred before the onset of membrane stretching deformation that can be observed in ductile materials during the punch loading [6]. The onset of ductile cracks occurred at higher temperatures without inducing a loading rate change unlike brittle cracking (Fig. 3). In such cases, load-interrupting SP tests were repeated at several deformation stages to determine the value of δ_f. The ductility (ε_f) of the coatings and substrates was defined at the initiation stage of cracks [3,4] and then estimated from δ_f [7,8]. Figure 4 indicates the temperature dependence of ε_f obtained from SP tests on the coating and substrate specimens. The ductility of the coatings and substrates at 22 °C decreased with increasing operation time. All the coatings and unused substrate exhibited lower values of ε_f at

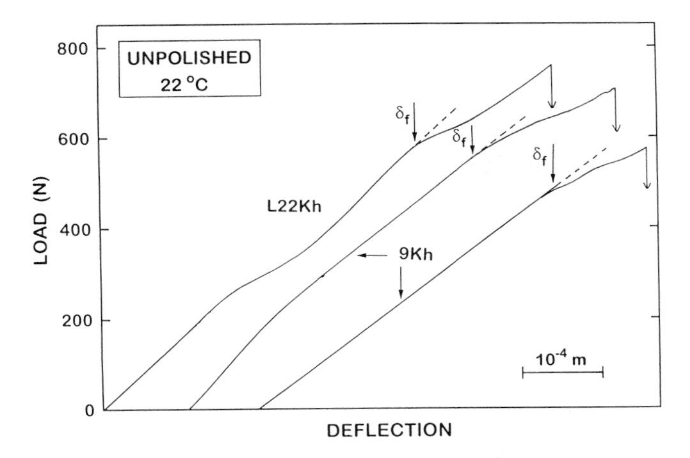

Figure 2. Three typical load vs. deflection curves obtained from SP tests on 9Kh and L22Kh coating specimens with unpolished surfaces at 22 °C. The SP tests were interrupted after the initiation of cracks at δ_f.

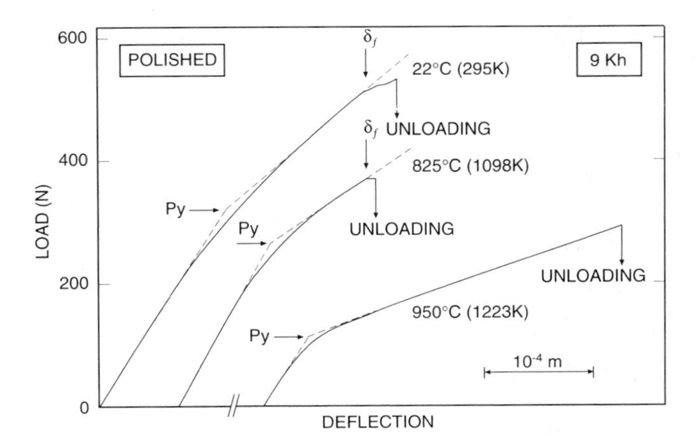

Figure 3. Typical load vs. deflection curves obtained from SP tests on polished 9Kh coating specimens at 22, 825 and 950 °C indicating yield load (P_y) and critical deflection to brittle cracking (δ_f).

825 °C, compared with those at 22 °C. As the testing temperature was further raised, the ductility of the coatings and substrate increased and the ductile-brittle transition behavior emerged. Both the 9Kh and 21Kh coatings depicted a higher ductile-brittle transition temperature (DBTT) by 90 °C than the unused one. The DBTT of the substrate was slightly lower than that of the unused coating and remained unchanged during the blade operation. It has been shown [5] that the L22Kh and thermally aged coatings possess different extents of the mechanical degradation. The SP test on the unused substrate indicated lower ductility than

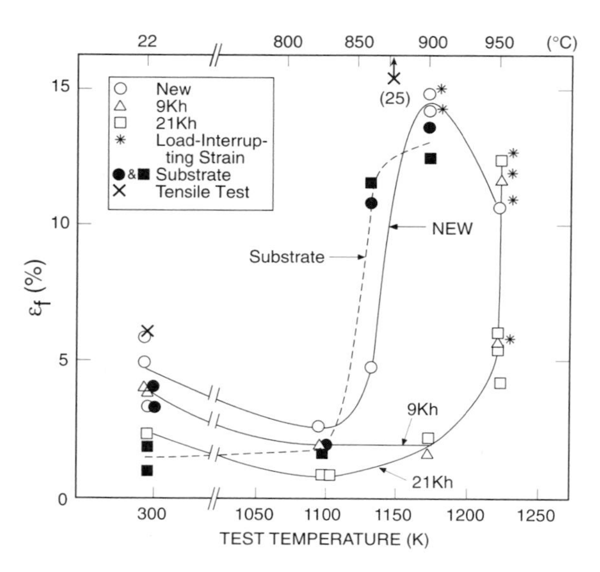

Figure 4. Temperature dependence of ductility (ε_f) obtained from SP tests on unused, 9Kh, and 21Kh coatings, and unused and 21Kh substrates. The result is compared with tensile test data of unused substrate.

tensile tests because the definition of ductility and the stress state are different. It should also be noted that the unpolished SP specimen tests provide higher ductility with larger standard deviations than the polished ones partly due the lower modulus [5].

Cracking morphologies observed in the deformed SP coating specimens can be summarized [3]. Brittle cracks initiated at the center of SP specimens and propagated along the radial direction in all the coatings at 22 °C. At elevated temperatures, many brittle cracks more discretely grew along random directions in the used blades. Ductile cracks were highly localized near the surface of the unused coating specimen and propagated in a zigzag mode. In the used coatings, cracks nucleated at oxides near the specimen surface, which had been formed due to the environmental attack. The nucleated cracks extended into the coating matrix accompanied by some crack tip opening.

In order to clarify the mechanical degradation of the coatings, the microstructural/chemical evolution of the blades was investigated using SAM. The unused blade indicated the formation of oxides, consisting of major alloying elements (Ni, Co, Cr and Al), and pores near the coating/substrate interface [3,5]. The blade operation led to an increase in the density and size of the oxides and pores near the interface. Figure 5 delineates a scanning electron micrograph indicating interfacial cracking, and O and S maps on fracture surfaces of the 9Kh coating. Oxidation substantially occurred near the coating surface and interface. Oxides near the coating surface predominantly had the form of Al_2O_3 unlike those observed near the interface [3,5]. S enrichment was significantly observed in all the used coatings but not in the unused and thermally aged ones, which is intensified near the surface region [3,5]. Thus the sulfidation of the coatings is believed to have occurred during the blade operation.

We now turn to discuss the relationship between the mechanical degradation and microstructural/chemical evolution in order to evaluate the degradation mechanism of the coatings. It was found [5] that the initiation of coating cracks predominantly takes place near the specimen surface but not near the interface. Thus it is evident that coupling oxidation and sulfidation near the coating surface produces in-service ductility loss at 22 °C and elevated

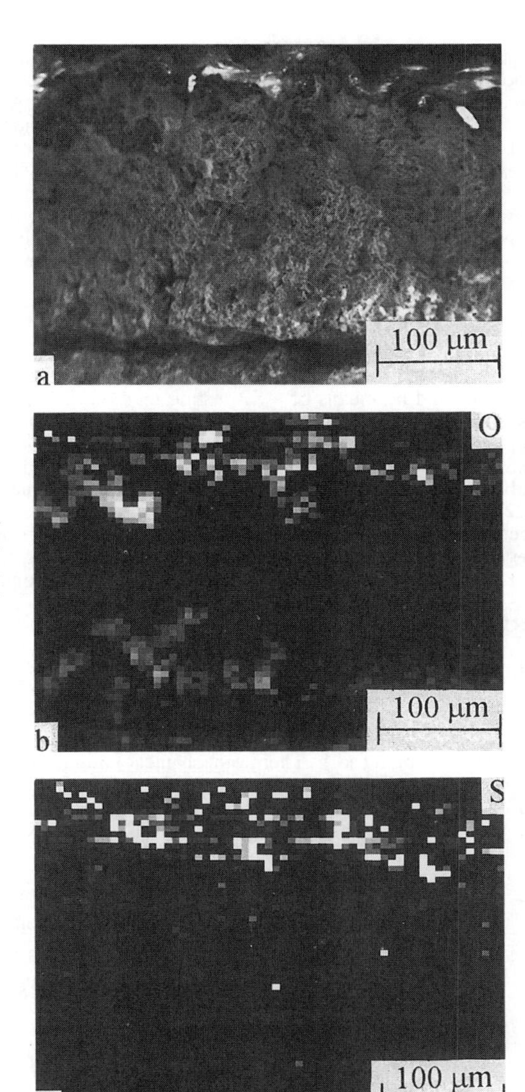

Figure 5. (a) SEM micrograph indicating interfacial cracking, and (b) O and (c) S maps on fracture surface of 9Kh coating and substrate near interface. The oxide observed near the coating surface had the form of Al_2O_3.

temperatures. Moreover, all of the coatings and unused substrate exhibited lower ductility at 825 °C in air than those at 22 °C (Fig. 4). Under Ar atmosphere, however, the 9Kh coating had higher ductility (7.5%) [5]. Hence the high temperature degradation of the blades is ascribed to environmental effects. In intermetallic coatings and nickel base superalloys, it is possible [9] that oxygen atoms absorbed from the specimen surface preferentially diffuse along stressed grain boundaries or interfaces. The formation of either oxides or segregated elemental oxygen, or both, exert dynamic embrittling effects. When increasing the testing temperature, the interaction of oxygen with dislocations in the grain matrix becomes stronger than with boundaries and the stress relaxation ahead of the crack tip dominates. In this way, the dynamic embrittling effect becomes weak so that the ductile-brittle transition behavior appears.

The low cycle fatigue (LCF) behavior of the unused and used coatings has been studied elsewhere [4]. The LCF life had a tendency to decrease during the blade operation. The LCF result is consistent with the ductility and DBTT change (Fig. 4). However, the LCF life showed larger data scattering than the fracture properties under monotonic loading. This is probably because the surface deformation controlling the LCF life is strongly influenced by the distribution of oxides formed near the coating surface. For the similar reason, we also can see some variations in the yield strength of the 21Kh coating and the ductility of the used coatings at 950 °C [3,5].

CONCLUSIONS

In-service mechanical degradation of blade coatings in land based gas turbines has been investigated by means of a SP testing method. SP tests on coated specimens with unpolished surfaces indicated various deformation modes and ductility variations depending on the surface conditions in used gas turbine blades. It is shown that SP tests on polished coating specimens are capable of characterizing well the mechanical degradation of gas turbine blades. During the blade operation, the coated specimens softened and the ductility of the coating and substrate at 22 °C decreased. The DBTT of the coatings increased in-service while that of the substrate remained the same. From scanning Auger microprobe analyses on fracture surfaces, it is evident that oxidation and sulfidation near the coating surface, which produce the ductility loss of the coatings, resulted from high temperature environmental attack.

REFERENCES

1. R. Viswanathan and J. M. Allen (eds.), Life Assessment and Repair Technology for Combustion Turbine Hot Section Components (ASM International, Materials Park, 1990).

2. H. Sehitoglu (ed.), Thermomechanical Fatigue Behavior of Materials, ASTM STP 1186, (American Society for Testing and Materials, Philadelphia, 1993).

3. Y. Sugita, M. Ito, N. Isobe, S. Sakurai, C. R. Gold, T. E. Bloomer and J. Kameda, Mater. Manuf. Proc. **10-5**, 987 (1995).

4. Y. Sugita, M. Ito, S. Sakurai, C. R. Gold, T. E. Bloomer and J. Kameda; in Materials Aging and Component Life Extension, edited by V. Bicego, A. Nitta and R. Viswanathan (EMAS, West Midland, 1995), p. 307.

5. J. Kameda, T. E. Bloomer, Y. Sugita, M. Ito and S. Sakurai, to be published.

6. J. M. Baik, J. Kameda and O. Buck, Scripta Metall. **17**, 1143 (1983); in The Use of Small-scale Specimen for Testing Irradiated Materials, ASTM STP 888, edited by W. R. Corwin and G. E. Lucas (American Society for Testing and Materials, Philadelphia, 1986), p. 92.

7. X. Mao and H. Takahashi, J. Nucl. Mater. **150**, 42 (1987).

8. J. Kameda and X. Mao, J. Mater. Sci. **27**, 983 (1992).

9. C. T. Liu and C. L. White, Acta Metall. **35**, 643 (1987).

Part II

Processing

REACTIVE SPUTTER DEPOSITION OF SUPERHARD POLYCRYSTALLINE NANOLAYERED COATINGS

William D. Sproul
BIRL, Northwestern University, 1801 Maple Avenue, Evanston, IL 60201 USA
e-mail: wsproul@nwu.edu

ABSTRACT

Nanometer-scale multilayer nitride coatings, also known as polycrystalline nitride superlattice coatings, such as TiN/NbN or TiN/VN with hardnesses exceeding 50 GPa, are deposited by high-rate reactive sputtering. The high hardness is achieved by carefully controlling several deposition parameters in an opposed cathode, unbalanced magnetron sputtering system: the target power, the reactive gas partial pressure, the substrate bias voltage and current density, and the substrate rotation speed. Target power controls the deposition rate and the thickness for each layer in conjunction with the reactive gas partial pressure at each target, which also affects the composition of each layer, and the substrate rotation speed. Split partial pressure control is necessary when each layer requires a different partial pressure to be stoichiometric. Fully dense, well-adhered coatings with the highest hardness are deposited when the negative substrate bias exceeds -130 V and the substrate ion current density is 4-5 mA cm^{-2}. The work on polycrystalline superlattice coatings is being extended into oxide systems. Oxide coatings can now be sputter deposited using pulsed dc power, which prevents arcing on both the target and the substrate. Pulsed dc power along with partial pressure control of the reactive gas leads to significantly higher deposition rates for the oxide films compared to sputtering with conventional rf power.

INTRODUCTION

There is much interest in the deposition of nanometer-scale, multilayer films, also known as superlattices, to achieve enhanced properties in the films. In the area of hard nitride films, it has been shown for single crystal nitride superlattice films composed of thin (2-4 nm) alternating layers of titanium nitride and vanadium nitride (TiN/VN) or titanium nitride and niobium nitride (TiN/NbN) that the hardness of the films can exceed 50 GPa when the superlattice period, which is the bilayer thickness, is in the range of 4-8 nm [1,2]. These same superlattice films have also been deposited in polycrystalline form on polycrystalline substrate materials by high-rate reactive sputtering with careful control of the deposition parameters, and they have achieved the same high hardness as the single crystal films [3-6].

Our understanding of the hardening mechanisms in the nitride superlattice films has been enlightened by the model by Chu and Barnett [7]. This model is based on restricted dislocation movement within a superlattice film in two different regimes. In the first regime when the superlattice period is less than the optimum value for peak hardness, it is more difficult to move a dislocation within a layer than it is to move it between layers. The force needed to move the dislocation across the boundary between the layers is related to the difference in dislocation line energies or shear modulus for the two layer materials and to the width of the interface between the two layers. If the difference in shear modulus is high, the force needed to move the dislocation will be high, and this force will increase as the superlattice period increases.

As the superlattice period and the force to move a dislocation increases, a point is reached when it is easier to move a dislocation within a layer than it is to move it across the interface

Mat. Res. Soc. Symp. Proc. Vol. 434 © 1996 Materials Research Society

between the layers. When dislocation movement is within the layer, the force to move the dislocation decreases as the superlattice period increases.

The agreement between the superlattice hardening model and the experimental results is excellent [7]. The model predicts the peak in hardness when there is a difference in shear modulus between the two layer materials; and when the interface width is taken into account, the experimental values for hardness are close to those predicted by the model. This model provides the basis for choosing materials to achieve an increase in hardness and strength for a superlattice system, and one of the key points is that there should be a difference in shear modulus for the two layer materials if an increase in hardness is to be expected. It also accurately predicts that there is very little hardness increase when there is little or no difference in shear modulus for the two materials as exits for the NbN/VN superlattice system. Experimentally [3] it has been verified that there is no increase in hardness for the NbN/VN polycrystalline superlattice system.

The materials used in the superlattice systems that have been used to verify the model of Chu and Barnett have the same crystal and slip systems. For example, TiN, NbN, and VN all have the B1 sodium chloride structure, and they all have the same {111} <110> slip system. The question has been asked if the model applies to other material systems with different slip systems, and work is now underway to determine if the model is applicable to oxide superlattice systems [8-10].

The reactive sputter deposition of oxide films has recently been made much easier with the use of pulsed direct current (dc) power combined with the use of partial pressure control of the reactive gas [11,12]. Partial pressure control is very important in the production of superhard polycrystalline superlattice films, and its use will be reviewed in the following sections. This paper will first review the deposition of the nitride superlattice films, and then it will show how the developments of the nitride work combined with recent advances in power supply technology make it possible to deposit multilayer oxide films at high deposition rates.

EXPERIMENTAL PROCEDURES

All of the work on both nitride and oxide superlattice systems was carried out in an opposed-cathode, unbalanced magnetron sputtering system [13], which has recently been modified to include pulsed dc power. Advanced Energy MDX 10 kW power supplies provide the basic dc power to each of the sputtering cathodes and to the substrate table, and this dc power is transformed into pulsed dc power with the insertion of variable frequency (up to 50 kHz) Advanced Energy Sparc-Le arc suppression units [11] between the power supplies and the cathodes or substrate table. In addition to pulsed dc power being available for the substrate bias, rf power can be used if required. When multiple Sparc-Le units are used together in the same system, they must be run in a master-slave set-up where one unit, the master, controls the pulsing in the other slave units. Independent Sparc-Le units running on the same process will not work since they will be out of phase with one another, causing the plasma to extinguish.

The Sparc-Le units are not needed for the deposition of the nitride superlattice coatings since the nitrides are conductive and do not build up a surface charge. Nitrides can be deposited with conventional dc power, but in the event that an arc does occur with the nitrides the Sparc-Le units will quench it faster or prevent it from occurring in the first place.

Partial pressure control of the reactive gas requires the use of an on-line sensor that can generate a signal quickly that is representative of the reactive gas partial pressure in the vicinity of the reaction at the substrate. A differentially pumped Leybold-Inficon Quadrex 100 quadrupole mass spectrometer was used to generate the required partial pressure signal, and this signal was used instead of the gas flow signal to control the partial pressure of the reactive gas.

In this opposed cathode sputtering system, the reactive gas is injected into the process through manifolds surrounding each cathode. When the two coating materials require different partial pressures to form the stoichiometric compound such as in the TiN/NbN superlattice system, split partial pressure control was implemented by using a master-slave partial pressure control system. A master controller, which received the feedback signal of the average partial pressure from the mass spectrometer, controlled two slave controllers. Different amounts of the reactive gas were injected into the chamber at each target when the slave controllers were set at different settings. MKS 260 mass flow controllers were used for this master-slave partial pressure control system.

Substrates were mounted on a rotating substrate holder, and the speed of rotation could be varied from 0 to 28 rpm. Individual layers of each material were deposited onto the substrates as they passed in front of each target, and the bilayer structure was built up into a thick coating (3-5 μm) by multiple rotations in front of the targets. To prevent cross contamination of one target material onto the other, the diameter and height of the substrate holder was such that it was greater than that of the sputtering targets. No other shielding was needed. The targets were of the Materials Research Corporation (MRC) Mu Inset design and were nominally 12.7 cm wide by 38.1 cm long. The substrate holder was 15.2 cm in diameter by 40 cm long.

A typical coating sequence was as follows. Once the substrates, which were pieces of high speed steel, cemented carbide, stainless steel, silicon, glass, or other materials, had been cleaned outside the chamber, they were mounted on the substrate holder, which was loaded into the chamber through a load lock. Once the main chamber reached a pressure of 1.3×10^{-4} Pa or better, the samples were given a dc sputter etch in an argon atmosphere at a pressure of 3.3 Pa for 15 minutes. When the etch was completed, the total pressure was reduced to 1 Pa, power was applied to the targets and the substrate holder, and the reactive gas was turned on. Typical target power was 5 kW to each target, and the partial pressure depended on the materials being deposited. Substrate bias was usually in the -100 to -150 V range, and the ion current density was 4 to 5 mA cm^{-2}. With the rotating substrate holder, the closest point of approach between the target and the substrate was 6.3 cm when the 15.3 cm diameter substrate holder was used.

The thickness of each layer as the substrate passed in front of each target was determined by the power to the target, by the reactive gas partial pressure, and by the rotation speed of the substrate. For the TiN/NbN system, the average partial pressure as measured by the mass spectrometer was 0.04 Pa. However, NbN requires a partial pressure of 0.053 Pa to form the cubic NbN structure, whereas TiN requires 0.027 Pa. With the split partial pressure control, twice as much reactive gas was delivered to the manifold next to the Nb target as was delivered to the one next to the Ti target. Equal amounts of reactive gas are supplied to each target for the TiN/VN system, and the average partial pressure during deposition was 0.027 Pa.

RESULTS AND DISCUSSION

Nitride Superlattice Films

The opposed-cathode, unbalanced magnetron sputtering system is a very effective system for depositing nanometer-scale, multilayer films. Individual layer thicknesses from 1 to 100 nm can be easily and precisely deposited on a variety of substrate materials. For the nitride superlattice systems of TiN/NbN and TiN/VN, the hardness of these materials exceeds 50 GPa when the superlattice period is in the 4-8 nm range [3,4]. These films are polycrystalline, and the X-ray diffraction peaks are for the superlattice film and not for the individual layers. Satellite X-ray peaks are observed in many of the films.

Not only are these materials hard, but they also appear to resist fracture better than most hard coatings. When they are indented with a Vickers indentor in a Palmquist type test, they do not show any cracks forming at the corners of the indent at loads up to 1000 g. Typical hard TiN coatings, show cracks when the load reaches 150 to 200 g for a 5 μm thick film.

The high hardness in the superlattice films is achieved by controlling the superlattice period, the substrate bias voltage and current, and the composition of each layer. The partial pressure of the reactive gas at each target is determines the composition of each layer material, which in turn affects its hardness. For the TiN/NbN system if equal partial pressures of the reactive gas are supplied to each gas manifold in front of the Ti and Nb targets, the hardness of the coating only reaches 38 GPa; whereas if the partial pressures are split to supply the amount needed to produce stoichiometric layers, the hardness can reach over 50 GPa [3].

The deposition rate is very sensitive to the degree of compound formation or poisoning on the target, which is also a function of the reactive gas partial pressure. The partial pressure should be set at the minimum value that will still give the desired composition of the compound yet still achieve the high hardness. For the nitrides, the deposition rate approaches that of the pure metal, which can only be achieved with partial pressure control of the reactive gas.

Substrate bias is also very crucial in achieving high hardness in the superlattice films, and both the flux and energy of ions are important. When the substrate ion current density was less than 2 mA cm^{-2}, the hardness of the TiN/NbN films did not exceed 35 GPa; but when it was greater than 4 mA cm^{-2}, the hardness could reach its maximum value [3], which occurred when the substrate bias voltage was in the -125 to -150 V range. Values higher or lower than this led to softer films [3-5].

For the superlattice films to be hard, they must maintain their distinct layer structure. If a superlattice film is used on a cutting tool, it will become hot during the cutting operation. In high speed machining, it is not uncommon for the tip of the tool to reach temperatures in excess of 800°C. For the TiN/NbN system, this temperature is high enough for the layers to intermix and form an alloyed hard coating, which will be hard in its own right but not as hard as the superlattice film.

Other superlattice systems such as titanium nitride and aluminum nitride (TiN/AlN) are being evaluated to determine if they can operate at higher temperatures. AlN and TiN are immiscible in one another to temperatures exceeding 1000°C, and this combination may work well for high speed cutting operations. A Japanese company is now marketing this type superlattice coating for high speed ferrous machining operations [14].

Oxide Superlattice Coatings

The success with the deposition of polycrystalline nitride superlattice films and the development of the superlattice strength enhancement model has opened the way for other superlattice systems to be explored. Although multilayered oxide films have been used for years in the optics industry, they have not been widely explored as possible superlattice films with enhanced strength properties. It was not until recent advancements in power supply technology with the introduction of pulsed direct current (dc) power that it was practical to consider the deposition of oxide films at rates that would be economical. Radio frequency (rf) power had always been used in the past for the sputter deposition of nonconducting oxide films, but the deposition rates for oxide films with rf power are very slow. For example, the deposition rate for reactive rf sputter deposited Al_2O_3 is only 2-3% of the metal deposition rate for the same target power.

In the past few years, it has been shown [8-12, 15-17] that pulsed dc power can be used for the reactive sputter deposition of oxides. With pulsed dc power supplied to the sputtering target, the polarity of the voltage is switched back and forth between negative and positive. During the positive pulse, any charging of the oxide layer on the target surface is discharged when electrons are attracted to the positive surface. During the negative pulse, sputtering takes place from all target surfaces, both metallic and oxide. Such sputtering of the oxide surfaces minimizes their build-up and slows down the rate of poisoning of the target. Pulsed dc power supplies come with either fixed frequency or variable frequency. The typical frequency range for variable frequency pulsed dc power supplies is 0 to 100 kHz, but one pulsed dc power supply has a range up to 250 kHz [18].

The interaction of oxygen as a reactive gas with the target surface is much more rapid than is nitrogen in the deposition of nitrides, and this rapid interaction of oxygen makes partial pressure control of the reactive gas more difficult, but certainly not impossible. The interaction of oxygen can be see in Figure 1 for the reactive sputtering of titanium in an argon/oxygen atmosphere with flow control of the reactive gas. When conventional dc power is used along with flow control of the reactive gas as is shown in Figure 1, the partial pressure of the reactive gas increases rapidly when the flow reaches a certain level such as at point A where an oxide layer forms on the target surface. The deposition rate for the oxide is much lower than it is for the titanium metal; and as the deposition rate slows down, and the partial pressure of oxygen rises rapidly since less oxygen is being consumed by the process. The target is fully poisoned at point B, and there is a range of compositions from points A to B that are forbidden with flow control of the reactive gas.

When pulsed dc power is used, the rate of poisoning of the sputtering target is less than it is when conventional dc power is applied. Since the target surfaces are discharged during the positive pulse, all surfaces sputter when the power is switched to the negative pulse. The oxided regions as well as the metallic ones are sputtered, and the oxidation of the metallic surfaces is slowed down. Eventually as the flow of the reactive gas is increased sufficiently, the whole target surface will become covered with an oxide at point B, but the oxidation of the target takes place in a much more controlled fashion.

Pulsed dc power has another positive effect on the reactive sputtering process. If a target surface charges up, it can reach a point where it will break down leading to an arc. Arcing is bad for the process in two ways. It can be damage the power supply, and in the worst case it can cause failure of the power supply. Secondly, arcing leads to droplet ejection from the target surface, and if these droplets are incorporated in the growing film, they degrade the quality of the film. They can be a metallic particle trapped in the oxide film, or they can be a seed for a growth defect

Partial pressure control of the reactive gas along with pulsed dc power is a very powerful tool for the reactive sputter deposition of oxide films. Pulsing helps to control the rate of poisoning of the target surface, and the partial pressure control maintains the partial pressure at the desired level. The hysteresis plot for partial-pressure controlled sputtering of titanium oxide is shown in Figure 2, and its shape is quite different that the shape of the hysteresis plot shown in Figure 1 with flow control of the reactive gas. Note that in Figure 2, the independent variable is show on the ordinate instead of the abscissa of the plot to provide continuity between the plots. There is a large negative slop region in Figure 2 from points A to B as the target goes from an open metallic surface to a fully poisoned one, but with the partial pressure control it is possible to operate at any

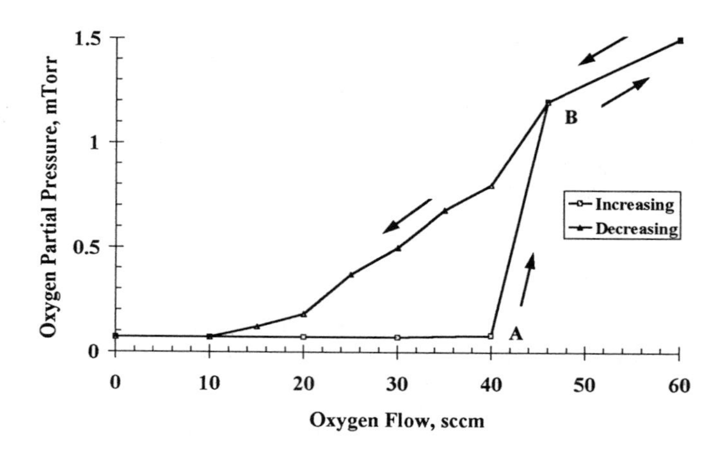

Figure 1. Hysteresis plot for the reactive sputtering of titanium in an argon/oxygen atmosphere with flow control of the reactive gas. The target power was 8 kW, and the total pressure during deposition was 1.1 Pa.

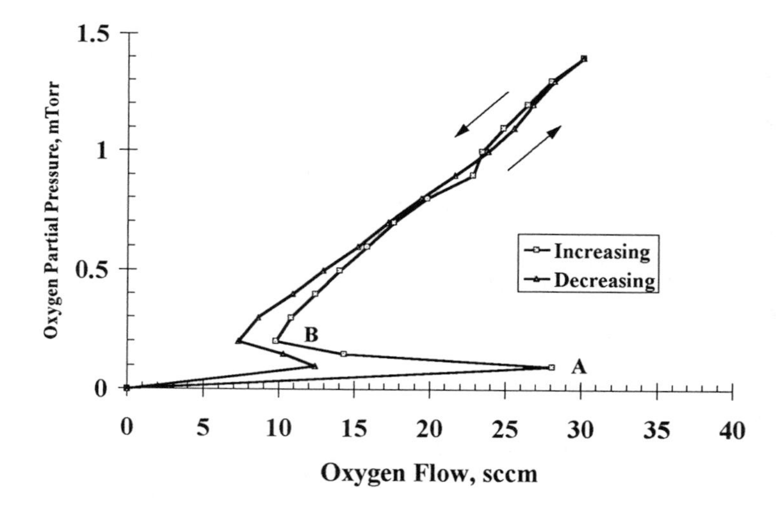

Figure 2. Hysteresis plot for the reactive sputtering of titanium in an argon/oxygen atmosphere with partial pressure control of the reactive gas. The target power was 5 kW, and the total pressure during deposition was 1.1 Pa.

point between points A and B. There are no forbidden compositions as there are in Figure 1 when the partial pressure jumps between points A and B.

The deposition rate drops from the full metal deposition rate at point A in Figure 2 to a much lower rate when it reaches point B. The film produced at point A, even though it has much oxygen in it, is not a clear film as is desired for an oxide film. Clear films of titanium oxide are produced in between points A and B, and the goal is to produce a clear optical film with the desired composition at as high a rate as possible. With pulsed dc power and partial pressure control, TiO_2 films have been deposited at rates equalling 80% of the metal deposition rate.

Similar results have been found for the reactive sputter deposition with partial pressure control and pulsed dc power for aluminum oxide and zirconium oxide films. The shape of the hysteresis curves for these two materials is the same as is shown in Figure 2 for the reactive deposition of titanium in an argon/oxygen atmosphere. Partial pressure control of the reactive gas along with pulsed dc power is the key to the controlled, high-rate deposition of these films, and Al_2O_3 films can be deposited at rates approaching that of the metal . Single layer as-deposited Al_2O_3 films deposited at substrate temperatures less than 300°C have an amorphous structure whereas the ZrO_2 films have a crystalline monoclinic structure.

Multilayer Al_2O_3/ZrO_2 films have been deposited in the opposed-cathode, unbalanced magnetron system. Pulsed dc power was supplied to both targets and the substrate holder, and partial pressure control of the oxygen was used. No external heating of the substrate was used during deposition. The as-deposited multilayer Al_2O_3/ZrO_2 films are amorphous, but low angle X-ray diffraction shows a bilayer thickness of about 9 nm. Optically the films are very clear. The deposition rate for these films was 4 $\mu m\ hr^{-1}$ when there was 3 kW on the Al target and 2 kW on the Zr target. This work is just in its infancy, but the initial results are very encouraging.

CONCLUSIONS

Polycrystalline nanometer-scale multilayer nitride superlattice coatings such as TiN/NbN or TiN/VN with hardnesses exceeding 50 GPa can now be routinely deposited by high-rate-reactive-unbalanced-magnetron sputtering. Control of the reactive sputtering process is crucial to achieving the high deposition rate and the high hardnesses. Partial pressure control results in high deposition rates for the nitride layers, and it is also important for maintaining the desired composition of each layer, which has a direct effect on the hardness of the films. Other key factors affecting the hardness of the superlattice films are the superlattice period and the degree of ion-assisted deposition. Both the substrate ion flux and the ion energy directly affect the hardness of the films. There is good correlation between the Chu and Barnett model for the strengthening effect in superlattice films and what has been observed experimentally. With the use of this model and with the advancement in dc power supply technology, oxide superlattice systems are being explored. It is now possible to deposit nonconducting oxide films at high deposition rates when pulsed dc power and partial pressure control of the reactive gas are used. Multilayered Al_2O_3/ZrO_2 oxide films have been prepared at high deposition rates, which would not have been achievable without the pulsed dc power and partial pressure control. A new high-rate deposition tool in now available for the study of oxide films.

ACKNOWLEDGMENTS

The polycrystalline nitride superlattice work was sponsored by the US Department of Energy (DOE), Division of Materials Sciences, Office of Basic Energy Science, grant number DE-FG02-926ER45434. The DOE project officer was Dr. Allan Dragoo. The oxide superlattice work is being sponsored by the US Air Force Office of Scientific Research (AFOSR), grant number F49620-95-0177, and the AFOSR program manger is Dr. Alexander Pechenik.

The work reported in this paper on both the oxide and nitride superlattice coatings represents the combined efforts of the whole Vapor Deposition Coatings Group at BIRL. Many people have made significant contributions to these two efforts over a number of years, and everyone in the group, both past and present, should be proud of their contributions.

REFERENCES

1. U. Helmersson, S. Todorova, S. A. Barnett, J.-E. Sundgren, L. C. Markert, and J. E. Greene, J. Appl. Phys. **62**, 481 (1987).

2. M. Shinn, L. Hultman, and S. A. Barnett, J. Mater. Res. **7**, 901 (1992).

3. X. Chu, Ph.D. thesis, Northwestern University, June 1995.

4. X. Chu, M. S. Wong, W. D. Sproul, S. L. Rohde, and S. A. Barnett, J. Vac. Sci. Technol. A, **10**, 1604 (1992).

5. X. Chu, M. S. Wong, W. D. Sproul, and S. A. Barnett, Mat. Res. Soc. Proc. **226**, 379 (1993).

6. X. Chu, S. A. Barnett, M. S. Wong, and W. D. Sproul, Surf. Coat. Technol. **57**, 13 (1993).

7. X. Chu and S. A. Barnett, J. Appl. Phys. **77** (8), 4403 (1995).

8. W. D. Sproul, "Reactive Sputter Deposition of Polycrystalline Nitride and Oxide Superlattice Coatings," submitted for publication in Surface and Coatings Technology.

9. M. S. Wong, W. J. Chia, J. M. Schneider, P. Yashar, W. D. Sproul and S. A. Barnett, "High-Rate Reactive DC Magnetron Sputtering of ZrO_x," submitted for publication in Surface and Coatings Technology.

10. J. M. Schneider, M. S. Wong, W. D. Sproul, A. A. Voevodin, A. Matthews, and J. Paul, "Deposition and Characterization of Alumina Hard Coatings by DC Pulsed Magnetron Sputtering," submitted for publication in Surface and Coatings Technology.

11. W. D. Sproul, M. E. Graham, M. S. Wong, S. Lopez, D. Li, and R. A. Scholl, J. Vac. Sci. Technol. A., **13**(3) 1188 (1995).

12. William D. Sproul, Michael E. Graham, Ming-Show Wong, and Paul J. Rudnik, "Reactive DC Magnetron Sputtering of the Oxides of Ti, Zr, and Hf," accepted for publication in Surface and Coatings Technology.

13. William D. Sproul, Paul J. Rudnik, Michael E. Graham, and Suzanne L. Rohde, Surface and Coatings Technology, **43/44** 270 (1990).

14. M. Setoyama, A. Nakayama, T. Yoshioka, T. Nomura, A. Shibata, M. Chudou, and H. Arimoto, Sumitomo Electric Industries, **146** 91 (1995).

15. S. Schiller, K. Goedicke, J. Reschke, V. Kirchhoff, S. Schneider, and F. Milde, Surf. Coat. Technol. **61**, 331 (1993).

16. P. Frach, U. Heisig, Chr. Gottfried, and H. Walde, Surf. Coat. Technol. **59**, 177 (1993).

17. M. E. Graham and W. D. Sproul, 37th Annual Technical Conference Proceedings, Society of Vacuum Coaters, Albuquerque, New Mexico, p. 275 (1994).

18. J. Sellers, "Asymmetric Bipolar Pulsed DC," ENI Tech Note, ENI, Division of Astec America, Inc., 100 Highpower Road, Rochester, NY 14623.

TRIBOLOGICAL PROPERTIES OF Ti/TiN NANOMULTILAYERS

Ph.HOUDY *, P.PSYLLAKI **+, S.LABDI *, K.SUENAGA **, M.JEANDIN **
* LMN, Université d'Evry Val d'Essonne, boulevard des Coquibus, 91025 Evry, France.
** CMPMF, Ecole des Mines de Paris, B.P. 87, 91003 Evry, France.
+ Present address, DCES III, National Technical University of Athens, 15780 Zografou, Greece.

ABSTRACT

The tribological behaviour of Ti/TiN amorphous nanometric multilayers is reported in comparison with that of single Ti and TiN layers, in order to study the wear mechanism of nanostructures submitted to that one may call "macroscopic loading". Ti/TiN nanolayers were deposited onto Si substrate by high vacuum diode r.f. sputtering assisted by in-situ kinetic ellipsometry. Transmission Electron Microscopy (T.E.M.) characterization exhibited the multilayered structure of the films, ascertained by grazing angle X-ray reflectometry and ellipsometry. Sliding wear tests against alumina in dry air showed the ceramic-typed behaviour of the multilayers, the wear of which was partly governed through a microfracture mechanism. Their wear lifetime was found to be higher than that of Ti and TiN single layers and increased with the number of layers.

INTRODUCTION

The use of thin titanium nitride films obtained using P.V.D and CVD[1] is currently increasing for a wide range of hard coating applications such as diffusion barriers and electrical contacts in microelectronics, wear-resistant layers onto cutting tools[2] and erosion-resistant coatings in aircraft industry[3]. However, high residual stresses, which characterise these coatings, can lead to the formation of intragranular microcracks and influence the soundness of the sputtered TiN films to be involved in the fracture mechanism[4]. Ti/TiN multilayers, where the metallic layer (Ti) shows a high dislocation-line energy and could exhibit some resistance to plastic deformation and brittle fracture, have been suggested by several workers to be suitable for tribological applications[5, 6] and for corrosion[7, 8] and erosion[9] protection.

The study of mechanical properties of thin films made of successive nanolayers of metallic and/or ceramic phases (oxides, nitrides) are becoming of great interest because of the high strength levels which can be attained. Several years ago, J.S. Koehler[10] proposed a model based on the dislocation formation and motion theory involving the layer thickness of microlaminate thin films. A few years after, S.L. Lehoczky[11] applied this model to Al/Cu laminates (metal/ metal type) and confirmed the dependence of the tensile properties on the layer thickness. More recently, L.S. Wen et al.[12] showed that the microhardness of Ti/TiN films (metal/ ceramic type) obeys a Hall-Petch's Law involving the Ti layer thickness, or more precisely, the distance between the reinforcing layers, i.e. those of TiN. This work deals with the tribological behaviour of amorphous thin Ti/TiN multilayers deposited by r.f. sputtering technique on Si wafers.

EXPERIMENT

Coating elaboration and structure characterization

Titanium, titanium nitride and Ti/TiN nanolayered films were deposited onto (100) silicon wafers by high-vacuum diode r.f. sputtering technique. A sputter target input of 280W was used at a nitrogen partial pressure of 4 10^{-1} Pa (without nitrogen for the deposition of pure titanium

Mat. Res. Soc. Symp. Proc. Vol. 434

films) and a total pressure of 10^{-1} Pa. The total thickness of the films ranged from 64 to 250 nm, depending on sputtering time. For multilayers, the thickness of every layer was varied to obtain films with 5, 10, 20 and 40 pairs (periods) of Ti/TiN layers. In Situ Kinetic Ellipsometry operating at a wavelength of 632.83 nm (ISKE)[13] permitted the observation of the deposition evolution in situ. Grazing angle X-ray reflectometry (GXR) showed the amorphous nature of these films[14].and Atomic Force Microscopy (AFM) their low surface roughness. The discontinuity of the ellipsometric trajectories corresponding to multilayer indicates that a stratified structure has been achieved[14], as ascertained from T.E.M. post-deposition observation of thin foils of cross-sections (Fig. 1). A.F.M. characterization of surface topography gave an average roughness of 1 nm, which was in good agreement with GXR (0.8 nm) and TEM results .

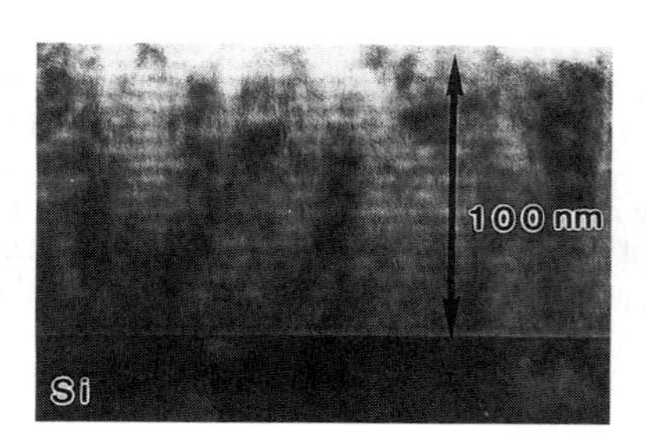

Fig 1 : T.E.M. view of Ti/TiN multilayers a) 20 periods (2.5 nm Ti + 2.5 nm TiN) are visible
b) 14 periods (5 nm Ti + 5 nm TiN) are visible

However, both methods showed that no pure Ti metallic layer was formed. The stratified structure of the film consisted of successive Ti layers wich were more or less rich in nitrogen. The first layer exhibited the stoichiometry of the TiN at equilibrium (e.g. 50 at. % Ti and 50 at. % N). The subsequent layer corresponded to a metallic layer with a gradient of nitrogen concentration (always inferior to 50 at. %) due to N diffusion through the interface. Nitrogen diffused from the saturated layer to the metallic one due to plasma temperature deposition. This decreased the thickness of the metallic layer to a critical value, below which no stratified structure could be obtained without using cooling when depositing.

Tribological characterization

Dry wear testing was performed at room temperature using a "CSEM (Centre Suisse d'Electrotechnique et de Microtechnique)" pin-on-disc apparatus. The discs were made of the materials to be tested, while a 6 mm in diameter alumina ball with a hardness of 1900 HV_{10} and a roughness R_a=0.5 µm was used as counterbody. Preliminary friction tests were carried out in order to determine the tribological behaviour of sputtered titanium and titanium nitride single layers. For Ti films, tests were conducted for sliding speeds of 0.003, 0.005, 0.01, 0.025 and 0.05 m/s, applying normal loads of 1 and 5 N. For TiN films, the normal load was limited to 1 N because of the fracture of the ceramic part of the coating when using more severe experimental conditions. Tribological testing of multilayers were carried out for a sliding speed of 0.005 m/s, using a normal load of 1 N. As a basis for comparison, further tests were applied to the 20-period multilayers at 0.003 and 0.05 m/s. Time until the complete worn out of the films, i.e. wear lifetime, was determined as that from wich the friction coefficient reached the typical value of the Si substrate (i.e. 0.68 $\pm$ 0.05).

Results on Ti and TiN single-layers

The friction coefficient of an homogeneous Ti sputtered film was found to be 0.77 $\pm$ 0.05, when applying a load of 1 N, for all the sliding speeds used. Under 5 N, a 14% decrease of the friction coefficient value (0.66 $\pm$ 0.05) and a 40% decrease of the wear lifetime of the film could be observed. The friction evolution of TiN film showed two stages: the first one with a low friction coefficient (0.27 $\pm$ 0.05) and the second stage with a higher coefficient (of 0.40 $\pm$ 0.05) (Fig. 2). Even though the ceramic film (i.e. TiN) had half of the thickness of the metallic film, it showed a higher wear resistance , which almost doubled the wear lifetime.

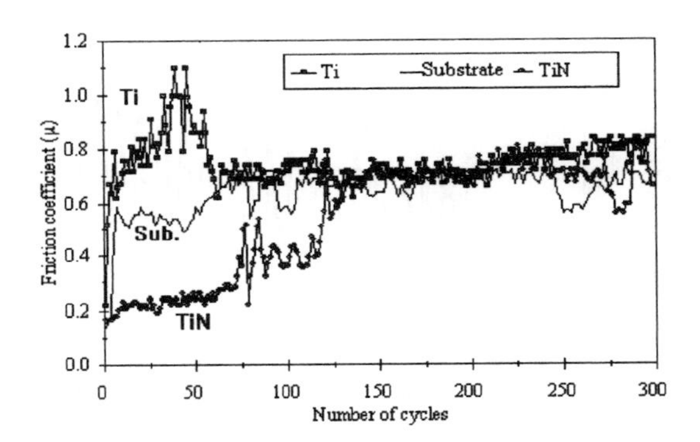

Fig 2 : Friction coefficient evolution for Ti (134nm) and TiN (64nm) films and Si substrate. (Sliding wear conditions: sliding speed of 0.005 $m.s^{-1}$ and applied load of 1N).

Assuming that the wear evolution remained uniform, the wear rate was 2.23 nm/cycle for the metallic and 0.56 nm/cycle for the ceramic single-layered film, when applying 1 N.

Results on Ti/TiN nanometric multilayers

The friction coefficient of the multilayers exhibited the same evolution as that of TiN single-layer, i.e. with two stages (fig. 3).

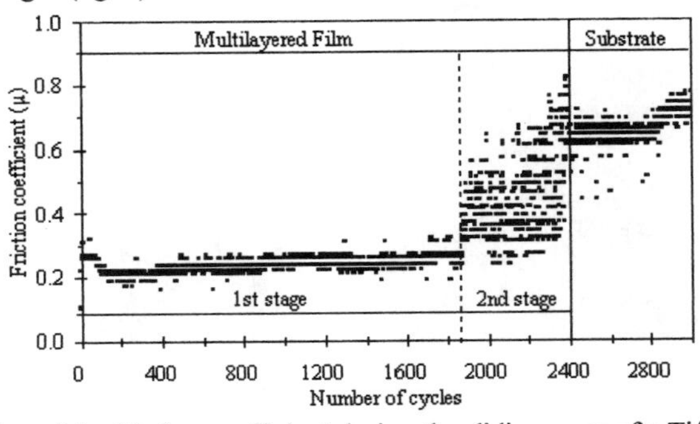

Figure 3 : Evolution of the friction coefficient during the sliding wear of a Ti/TiN multilayer. (Same sliding wear conditions than for fig. 2, for a 40-period multilayer).

(a) The first stage shows a quasi constant friction coefficient (0.20 ± 0.05) whatever the number of multilayers. In the sliding wear of ceramics, debris form by cracking, first, on the scale of the surface roughness, due to interaction with surface asperities. Since Ti/TiN sputtered films showed a very low surface roughness, the influence of the debris on the wear mechanism was not predominant and the values of the friction coefficient recorded corresponded to direct contact interaction between the counterbody (Al_2O_3) and the multilayer.

(b) During the second stage, the friction coefficient shows an average value of 0.35 ± 0.05 with many random oscillations between the minimum and the maximum value (fig 4).

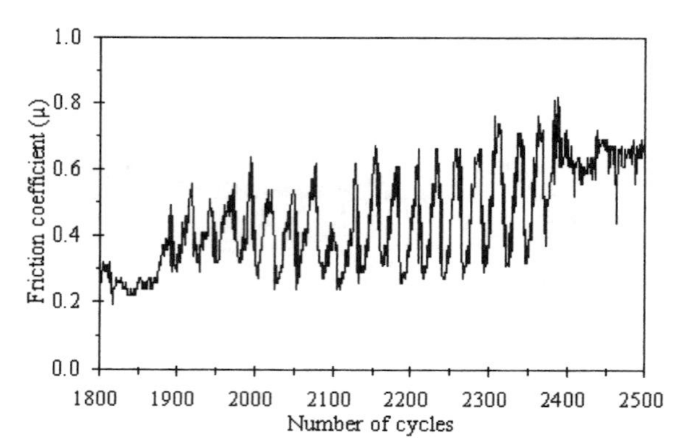

Figure 4 : Evolution of the friction coefficient during the sliding wear of a 40-period Ti/TiN multilayer.(fig. 3 magnified for the second stage).

In this stage, surface and sub-surface cracks initiated, due to fatigue phenomena in the film, which led to a rather big amount of wear debris (Fig. 5 a). These debris could be rather coarse initially due to a sort of local delamination phenomenon to become finer and finer due to pressure effect when testing. Repeated formation-removal of debris, because of the sliding procedure, was the main mechanism governing the wear evolution. Brittle fracture during the sliding wear of Ti/TiN multilayers gave steep wear tracks (Fig. 5 b), which confirm the predominance of the ceramic typed behaviour of the film (Fig. 6).

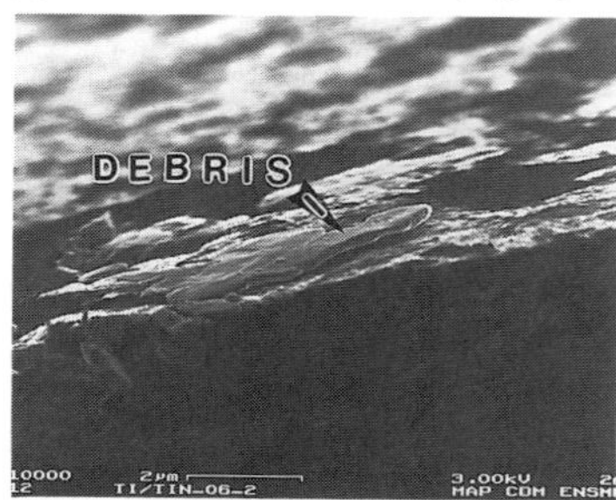

Figure 5 : a) SEM view of a poorly-adhered debris onto the wear track after complete worn out of the multilayer.
b) SEM view of a brittle fracture formed during the sliding wear (which led to a wear track with step edges).

Moreover, in some cases, isolated cracks of ~ 1mm in length were formed unexpectedly. This occurred always during the second stage perpendicularly to the wear track, due to wear debris of the film and/or to fatigue. The widely varying friction coefficient coupled with involving of the "third body" makes this stage undesirable. The duration of the above stages, corresponding to the time till complete worn out of the film, depended on the number of periods and on sliding speed. The first stage, during which a rather mild wear was observed, increased rapidly with the increase of the number of periods. For applications, promoting this stage is of a high interest because of the stability of the friction forces and preventing of the "third-body" influence on wear mechanisms. (One can claim some analogy with the results already obtained from excimer laser-processed cast iron [15]).

Comparing with metallic (Ti) and ceramic (TiN) single-layers, the first stage of Ti/TiN multilayers showed a higher life time. For a 5-period multilayer only, the wear resistance was found to be of the same order as TiN for a given thickness (Fig. 6).

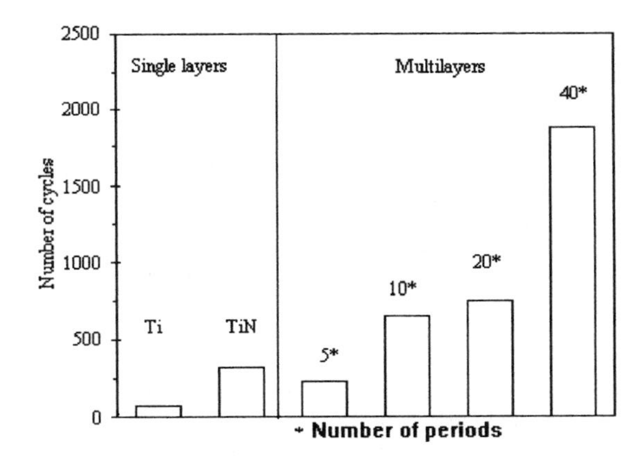

Figure 6 : Wear lifetime of Ti and TiN films and of Ti/TiN multilayers.

In order to determine the influence of sliding speed on the wear lifetime further tests showed that increasing the sliding speed resulted in increasing the total wear lifetime of the film. Moreover, the duration of the first stage increased, while that of the second decreased wich was beneficial. For example, for a 20-period multilayer, using a sliding speed 16 times higher led to a reduction of the second stage from 64 % to 30 % of the total wear lifetime of the film. The use of these films can be therefore recommended for applications involving high sliding speeds.

DISCUSSION

Sputtering deposition of Ti/TiN multilayers, without any cooling conditions, is characterized by slight diffusion of nitrogen to Ti metallic layer. The resulting structure can be termed as a stratified structure where the layer sequence is made of "slightly nitrited Ti" and TiN, repeated for a certain number of periods. Because of the major role of cracking on the wear mechanism, the film can be considered as a ceramic and it could be supposed that its behaviour under static loading will be also that of a ceramic. The delay in the final failure when increasing the number of periods features the beneficial role of the stratified structure under dynamic loading, where the "less ceramic" interlayers seem to play a role similar to that of pure metallic ones in the Koehler's model[10]. The promoting of the wear lifetime with the number of periods or with the decrease of the thickness of the elementary layers, suggests the existence of a critical thickness, for which the stratified structure can be obtained and the wear lifetime optimized.

The increase of the wear lifetime for multilayered coatings can be explained by changes in the mechanical behaviour due to the presence of interfaces. First, the lifetime increase follows a law of mixture for a low number of periods, corresponding to an average value between that for titanium and that for titanium nitride. Second, for a Ti-based/TiN higher number of periods, therefore a higher number of interfaces, the wear lifetime is better than that of titanium nitride and is proportional to the number of interfaces. The two stages observed during the sliding process correspond to respectively a plastic deformation of the coating and then craking in the coating with the formation of debris coupled with great variations of the friction coefficient. The first phenomenon is due to interaction with the stratified nano-structure of the multilayered coating. For this, the ellipsometry technique could be helpful for the determination of the critical thickness to obtain a higher lifetime and optimize the elaboration conditions for multilayers.

CONCLUSION

The wear resistance of the Ti/TiN multilayers has been shown to be higher than that of Ti and TiN single layers of the same total thickness. By increasing the number of pairs of the elementary layers, the wear resistance of the coating increases also. Under controlled atmosphere and temperature, the elimination of diffusion rates of nitrogen in the metallic layer can permit the deposition of a "more metallic" Ti layer in Ti/TiN multilayers. The study of these films will allow to optimise the thickness of the elementary layers to achieve the best wear behaviour for Ti/TiN. Nanotribological testing should permit to investigate more thoroughly wear mechanisms, while nanoindentation examination would allow to go into the study of mechanical properties and on the role of each elementary metallic and ceramic layer on the behaviour of the film. Next experiments (deposition at various temperature) would give information on the interdiffusion process and the quality of the interfaces in order to relate interface mechanical properties and "third-body" formation to interface structure on a nanometric scale.

REFERENCES

1. M.Staia and al, 9th Surf. Modif. Tech. Int. Conf., Cleveland, October 29 (1995).
2. D. Hofmann, B. Hensel, M. Yasuoka and N. Kato, Surf. Coat. Technol., **61** 326 (1993).
3. S. Hogmark and P. Hedenqvist, Wear, **179** 147 (1994).
4. A.R. Pelton and al, Ultramicroscopy, **29** 50 (1989).
5. Y. Ding, Z. Farhart, D.O. Northwood, A.T. Alpas, Surf. Coat. Technol., **68/69** 459 (1994).
6. E. Vancoille, J.P. Celis and J.P. Ross, Tribol. Int., **26** (2) 115 (1993).
7. B. Enders, H. Martin and G.K. Wolf, Surf. Coat. Technol., **60** 556 (1993).
8. R. Hübler and al, Surf. Coat. Technol.,**60** 551 (1993).
9. A. Leyland and A. Matthews, Surf. Coat. Technol., **70** 19 (1994).
10. J.S. Koehler, Phys. Rev. B, **2** (2) 547 (1970).
11. S.L. Lehoczky, J. Appl. Phys., **49** (11) 5479 (1978).
12. L.S. Wen and al, J. Magn. Magn. Mater., **126** 200 (1993).
13. Ph. Houdy and P. Boher, Le Vide, les Couches Minces, **259** 15 (1991).
14. S. Labdi, Ph.Houdy, P.Psyllaki and M.Jeandin, E-MRS, Strasbourg, Thin Solid Film, (1995).
15. C. Papaphilippou and al, 6th Int Conf on Tribology, Budapest, **3** 26 (1993).

STRUCTURAL AND PHASE TRANSFORMATIONS IN THIN FILM Ti-ALUMINIDES AND Ti/Al MULTILAYERS

R. BANERJEE, S. SWAMINATHAN, R. WHEELER AND H. L. FRASER
Department of Materials Science and Engineering, The Ohio State University, Columbus, OH 43210

Abstract

Multilayered Ti/Al thin films (with nominally equal layer thickness of Ti and Al) have been sputter deposited on oxidized silicon substrates at room temperature. Transmission electron microscopy (TEM) and high resolution electron microscopy have been used to characterize the structure of these multilayers as a function of the layer thickness. Ti changed from an *hcp* to an *fcc* and back to an *hcp* structure on reduction of the layer thickness. Al too changed from an *fcc* to an *hcp* structure at a layer thickness of 2.5 nm. The observed structural transitions have been explained on the basis of the Redfield-Zangwill model. Subsequently Ti-aluminide thin films were deposited using a γ-TiAl target. These films were found to be amorphous in the as-deposited condition with crystallites of α-Ti(Al) embedded in the amorphous matrix. On annealing under a protective Ar atmosphere at a temperature of 550 ^{0}C, the Ti-aluminide film crystallized into a nanocrystalline two phase microstructure consisting of γ-TiAl and α_2-Ti_3Al. The crystallization of the aluminide film has been investigated in detail by *in-situ* annealing experiments on a hot stage in the TEM. The results of this investigation have been discussed in this paper.

Introduction

Laminated intermetallic composites based on Ti-aluminides are potentially important for use as high temperature structural thin film coatings in aerospace applications [1]. These laminated composites are expected to have attractive mechanical properties. The ability to tailor the microstructure of such thin films on a nanoscale makes magnetron sputtering a suitable technique for processing these multilayers. Towards this end a preliminary study has been conducted on two different systems. The first one consists of multilayered thin films of pure elemental Ti/Al and the second one consists of alloy Ti-aluminide films. Both types of films have been deposited on oxidised Si substrates using a DC Magnetron sputtering unit. A series of structural transitions were observed on reducing the layer thickness in the Ti/Al multilayers [2,3]. Both Ti and Al have bulk close packed structures with Ti being *hcp* and Al being *fcc*. Ti transforms from *hcp* to *fcc* on reducing the layer thickness to about 5 nm and it transforms back to *hcp* on further reduction of the layer thickness to about 2.5 nm. Al also undergoes a transition from *fcc* to *hcp* on reducing the layer thickness to about 2.5 nm. Though the formation of *fcc* Ti in thin films of Ti has been reported previously [4,5,6], the reversal of *fcc* Ti to *hcp* Ti on reducing the layer thickness and the formation of *hcp* Al has not been reported earlier. An attempt has been made to rationalize these observations on the basis of the Redfield-Zangwill model [7]. The Ti-aluminide thin films which have been deposited using a γ-TiAl based target, were found to be primarily amorphous in the as-deposited condition with a small volume fraction of a crystalline phase. The phase and

Mat. Res. Soc. Symp. Proc. Vol. 434 © 1996 Materials Research Society

microstructural evolution during the crystallization of the amorphous phase has been investigated in detail by *in situ* hot stage TEM. Microstructures of the films annealed externally in a furnace have been compared to those developed as a result of annealing in the hot stage inside the TEM and in this context the effect of film thickness on the crystallization process has been discussed.

Experimental Procedure

The thin films have been deposited using a custom designed UHV magnetron sputtering system. The base pressure was 5 x 10^{-8} torr prior to sputtering and the argon pressure was 2 x 10^{-3} torr during the sputtering process. The Ti/Al multilayered thin films have been deposited using pure Ti and Al targets, which were 3" in diameter and 0.25" thick. Films with bilayer thicknesses (or compositionally modulated wavelengths (CMW)) of 108, 21, 9.8 and 5.2 nm have been deposited. For the Ti-aluminide films a γ-TiAl target of the same dimensions has been used. The nominal composition of the γ target was Ti-48Al (all compositions are in atomic percent) and the film thickness was approximately 1.5 µm. All films have been deposited on oxidised Si (100) wafer substrates with a 200 nm oxide layer on the surface. The Ti layers were deposited using 200 W and the Al layers using 160 W DC power. The Ti-aluminide films were deposited using 180 W DC power. All the depositions have been carried out at 313 K. The Ti/Al multilayers were studied by cross-section TEM. Details of the cross-section sample preparation procedure are described elsewhere [8]. The Ti-aluminide films were studied primarily by plan view TEM. Sections of the Si-wafer with the film were mechanically ground from the substrate side to a thickness < 130 µm. Subsequently, disks of 3 mm diameter were ultrasonically drilled from these ground sections, dimpled from the substrate side until the Ti-aluminide film was exposed and finally ion-milled from both sides to achieve electron transparency. Such a sample preparation procedure results in the formation of regions of free standing film free of any Si and SiO_2,, near the edge of the perforation. Furthermore, it gives a gradient of thickness within the electron transparent region. The samples were characterized in a Phillips CM200 TEM operating at 200 kV. For in situ annealing experiments the thin foil samples were heated in a GATAN 652 hot stage. The external annealing was carried out by encapsulating pieces of the film (prior to TEM sample preparation) in quartz tubes under a protective argon atmosphere followed by a furnace heat treatment at 823 K for 10 minutes.

Results and Discussion

Ti/Al multilayered thin films

The Ti/Al multilayers exhibited a series of interesting structural transitions on reduction of the layer thickness of Ti and Al layers [2,3]. Nominally the layer thickness of the Ti layer was equal to that of the Al layer in each of these multilayers. In the 108 nm multilayer, the structure of the both Ti and Al was found to be the same as their bulk stable structures; Ti being *hcp* and Al being *fcc*. The first evidence of formation of *fcc* Ti was found in the 21 nm multilayer. In this multilayer, the structure of the Ti layer was found to be a mixture of both *hcp* and *fcc* stacking sequences. The *fcc* regions formed small pockets within the predominantly *hcp* stacking sequence. Fig. 1. shows a high resolution electron micrograph of a cross section of the 21 nm multilayer. The presence of *fcc* Ti within the *hcp* stacking sequence giving rise to an *hcp* + *fcc* structure within the Ti layers is clearly observed. Further reduction of the layer thickness to a CMW value of 9.8 nm results in a complete transformation of the Ti layers to an *fcc* structure as shown in Fig. 2. No transformation occurs in the Al layers at this layer thickness. When the CMW is reduced to 5.2 nm, the Al layers transform to an *hcp* structure and the Ti layers revert back to an *hcp* structure. Fig. 3. shows a high resolution electron micrograph from the 5.2 nm multilayer. The *hcp* stacking sequence in both the Ti as well as the Al layers is clearly seen.

It is possible to transform an *fcc* ABCABC... stacking sequence into an *hcp* ABABAB... stacking sequence or vice versa by introducing successive intrinsic stacking faults in the atomic layers of the parent structure. The Redfield-Zangwill model [7] predicts that a bicrystal interface introduces a Friedel like oscillatory modulation of wavelength λ_F to the bulk stacking fault chemical potential. A metallic superlattice can be considered to be a system which consists of a sequence of bimetallic interfaces and the stacking sequence is determined by the resultant of the superposition of the primary bounding oscillatory potentials in each layer. For simplicity, assume a linear superposition of these bounding modulations. Consequently, the net potential of stacking faults in a layer is determined by the phase difference between the two bounding modulations which in turn is a function of the layer thickness. Since the bounding modulations are basically damped sinusoidal waves, as the layer thickness reduces, the superposition of the modulations oscillates between constructive and destructive interference regimes. At large layer thickness values, much larger than the value of λ_F, the effect of superposition is not felt and the bulk crystal structures dominate; for example, this applies to the 108 nm multilayer. As the layer thickness is reduced to the extent that superposition effects contribute to the stacking sequence, constructive interference can result in certain regions of a layer having a net negative stacking fault potential while other regions in the same layer have a net positive value. This leads to a mixed *fcc* + *hcp* structure within the same layer, such as is observed in the Ti layers of the 21 nm multilayer. It should be noted that the layer thickness is still greater than the wavength λ_F, of the oscillatory modulations. Futher reduction of the layer thickness below λ_F will intially result in a constructive interference such that the stacking

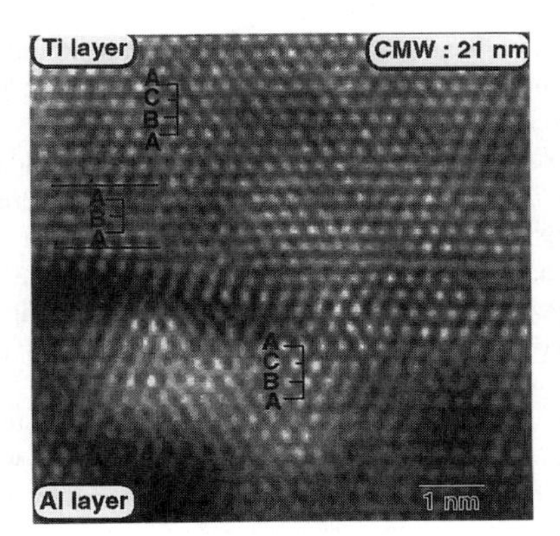

Fig.1. High resolution TEM micrograph showing the stacking sequence in the 21 nm multilayer. The mixed *hcp* + *fcc* structure of the Ti layer is clearly visible.

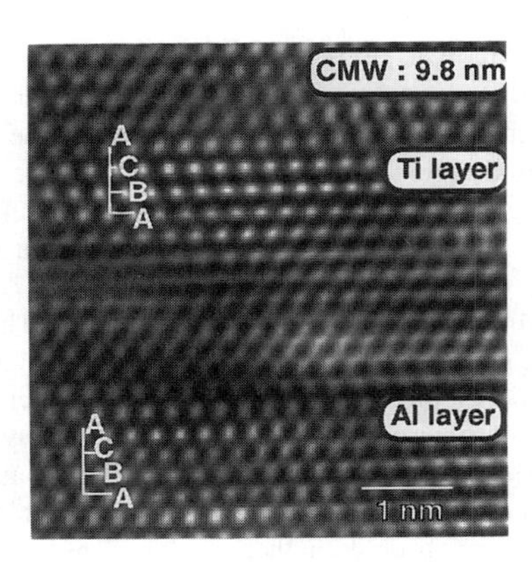

Fig.2. High resolution TEM micrograph showing the *fcc* Ti/ *fcc* Al stacking sequence in the 9.8 nm multilayer.

fault potential is negative throughout the layer, converting an entire layer of *hcp* to *fcc* or vice versa [7]. This is observed in the Ti layer of the 9.8 nm multilayer and in the Al layer of the 5.2 nm multilayer suggesting that the λ_F for Ti is larger than that for Al. However, on further reduction of the layer thickness, the interference again becomes destructive, stabilizing the bulk crystal structure in the layer; this is observed in the Ti layers of the 5.2 nm multilayer. It appears from the above discussion that the Redfield-Zangwill model is able to predict, in principle, a sequence of transformations in a multilayer with reducing layer thickness as follows :

$$hcp \rightarrow fcc + hcp \rightarrow fcc \rightarrow hcp, \text{ or}$$
$$fcc \rightarrow hcp + fcc \rightarrow hcp \rightarrow fcc.$$

Ti-aluminide thin films

In the as-deposited condition, the Ti-aluminide thin films consisted of an amorphous matrix with embedded particles of a crystalline phase. Fig.4. shows the general microstructural features of the as-deposited Ti-aluminide film. Selected area diffraction (SAD) patterns from the embedded crystalline particles indicate that the second phase is a hexagonal phase isostructural to α-Ti with lattice parameters of a=0.290 nm and c=0.469 nm [9,10]. The composition of this phase is Ti-48Al which indicates that it forms polymorphically in the amorphous matrix. Annealing the Ti-aluminide film in a furnace when it was sealed in a quartz tube under a protective argon atmosphere at 823 K resulted in a two phase microstructure consisting of γ-TiAl and α_2-Ti_3Al. This observation is consistent with the equilibrium binary phase diagram.

In order to observe the sequence of phase evolution during crystallization of the amorphous matrix in the Ti-aluminide film, *in situ* annealing experiments were conducted on a hot stage in the TEM. It is important to note at this point that the *in situ* experiments were conducted on those regions of previously thinned TEM specimens which were free of Si/SiO_2. Furthermore, the gradient in thickness of the foil from the edge of the perforation inwards into thicker regions presents an ideal opportunity to study the effect of thickness on the crystallization behaviour. However, it should be noted that since the film under study is amorphous, it is extremely difficult to determine the thickness of the film in the TEM due to the absence of extinction contours. Therefore, only an approximation of the thickness values has been presented in this paper. On heating the sample no changes were observed till a stage temperature of 753 K. Above this temperature two different phenomena were observed. In the very thin regions of the specimen (less than 20 nm) near the edge of the perforation, the crystallites of α-Ti started to grow. The growth occurred with the intially formed crystallites acting as seeds. This gave rise to a faceted microstructure consisting of large facets of α-Ti (Fig.5(a)). EDS results indicated that the composition of the α-Ti is the same as that of the amorphous matrix in which it grows suggesting that the transformation is a polymorphic one. However, in the relatively thicker, greater than 50 nm regions (still electron transparent) of the specimen, independent nucleation and growth in the amorphous matrix was observed to occur above 753 K (Fig.5(b)). A high rate of nucleation and a large number of nucleation sites were observed in these regions. The resultant microstructure was a nanocrystalline one with very fine grain sizes (50-200 nm). Moreover the microstructure is very similar in appearance to that observed in the Ti-aluminide film crystallized externally in a furnace. Ring diffraction patterns from these regions indicate it to be a intimately mixed two phase microstructure consisting of γ-TiAl and α_2-Ti_3Al. The different microstructures in the thick and the thin regions are shown in Fig.5(c).

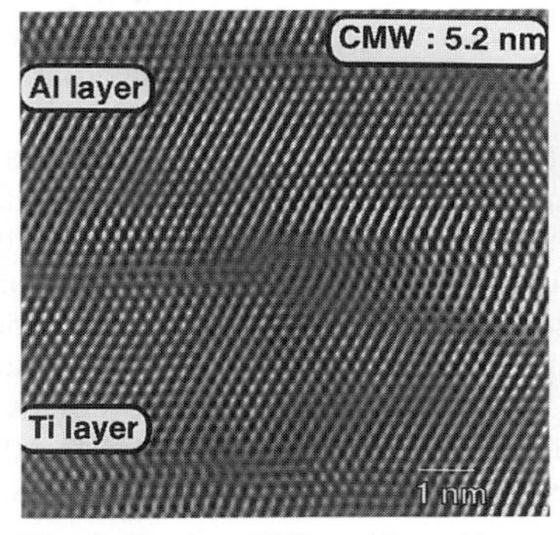

Fig. 3. The hcp Ti/ hcp Al stacking sequence in the 5.2 nm multilayer.

Fig.4. Crystallites of a-Ti embedded in an amorphous matrix in the as-deposited Ti-aluminide film.

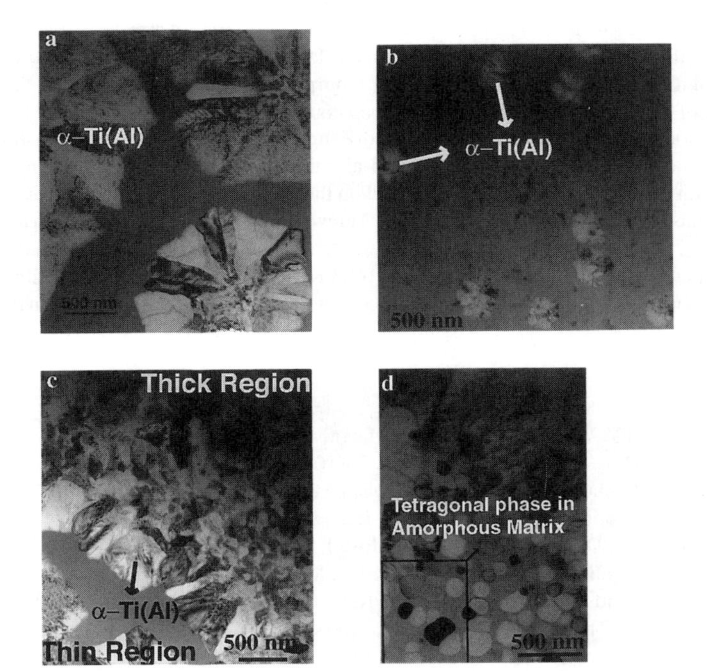

Fig.5. (a) Faceted growth of α-Ti at 753 K in the very thin regions of the film from the seeds present in the as-deposited condition. (b) Nucleation and growth in the amorphous matrix in relatively thicker regions of the Ti-aluminide specimen at 753 K. (c) Significantly grown crystallites of α-Ti(Al) in the relatively thinnner regions of the film at 853 K together with the two phase microstructure in the thicker regions. (d) Formation of the surface crystallized tetragonal phase in the film at 883 K.

The above observations clearly indicate an effect of film thickness on the crystallization process in these films. It is interesting to note that the α-Ti phase of composition Ti-48Al can be stabilized and grown in very thin regions of the sample, without any visible signs of decomposition. A similarity is found between the crystallization behaviour of the thicker regions in the *in situ* annealed specimen and the externally furnace annealed sample. This is expected considering the fact that when the film was heated in a furnace, it was of a uniform nominal thickness of 1.5 μm, reflecting the characteristics of a thick film throughout. When the temperature was raised to 883 K, the formation of a new crystalline phase was observed primarily in those thin regions of the film which were still amorphous. The grains of this crystalline phase were found to be defect free and had a near circular projected view during the initial stages of growth before grain impingement occurred (Fig. 5(d)). Furthermore the nucleation sites for this phase seem to be at the surface of the film making it a surface crystallized phase [9]. Electron diffraction evidence suggests that this phase is an ordered tetragonal phase with lattice parameters of a = b = 0.688 nm, c = 0.710 nm.

Summary and Conclusions

The present study focuses on two different aspects of structural and phase transitions in sputter deposited thin films in the Ti-Al system; the first one being a series of structural transitions observed in Ti/Al multilayers as a function of decreasing layer thickness and the second one being the phase evolution during crystallization of an amorphous phase in Ti-aluminide thin films. The transition of Ti from *hcp* to *fcc* and back to *hcp* coupled with the transition of Al from *fcc* to *hcp* has been modeled on the basis of the Redfield-Zangwill theory for structural transitions in close-packed metallic superlattices. Regarding Ti-aluminide films, an interesting effect of the film thickness on the crystallization behaviour and on the stability of competing phases was observed. An α-Ti phase of nominal composition Ti-48Al grew in very thin regions of the film at temperatures as low as 753 K which is significantly lower than the thermodynamically predicted stability regime for an α phase of this composition (>1623 K). The thicker regions of the film transformed into a two phase $\gamma+\alpha_2$ microstructure by a process of direct nucleation and growth in the amorphous matrix.

References

1. K. S. Chan and Y. W. Kim, Acta Metall. Mater., 43(2), 439 (1995).
2. R. Ahuja and H. L. Fraser, J. Elec. Mater., 23(10), 1027 (1994).
3. R. Banerjee, R. Ahuja and H. L. Fraser, to appear in Phys. Rev. Lett.
4. A. F. Jankowski and M. A. Wall, J. Mater. Res., 9(1), 31 (1994).
5. D. Schechtmann, D. van Heerden and D. Josell, Mater. Lett., 20, 329 (1994).
6. D. Josell, D. Shechtmann and D. van Heerden, Mater. Lett., 22, 275 (1995).
7. A. C. Redfield and A. M. Zangwill, Phys. Rev., B34(2), 1378 (1986) and references therein.
8. R. Ahuja, Ph. D. Thesis, The Ohio State University, 1994.
9. R. Banerjee, S. Swaminathan, R. Wheeler and H. L. Fraser to appear in Metastable Phases and Microstructures, Mater. Res. Soc. Symp. Proc. (1996).
10. R. Banerjee, S. Swaminathan, J. M. K. Wiezorek, R. Wheeler and H. L. Fraser, to appear in Metall. Trans. A

THE STRUCTURAL INVESTIGATION OF ALUMINA AND ALUMINUM NITRIDE MIXED THIN FILMS PREPARED BY D.C. PLASMA PROCESSES UNDER DIFFERENT CONDITIONS

PAUL W. WANG*, SHIXIAN SUI
Department of Physics and Materials Research Institute, The University of Texas at El Paso, El Paso, Texas 79968, * pwang@utep.edu

ABSTRACT

Composite films of aluminum nitride and alumina were fabricated on 6061 aluminum alloys in a d.c. plasma chamber. Samples were treated by three main processes. They were 1) Ar plasma etching, 2) NH_3/Ar plasma with low pressure and low current density, and 3) NH_3 plasma with high pressure and high current density. The oxygen-free Al surface was obtained after 10 min. 2.8 keV Ar^+ sputtering in a UHV analysis chamber after the sample was treated by processes 1 and 2. Composite films of aluminum nitride and alumina were obtained on samples treated by processes 1, 2, and 3. The surface compositions and bonding environments of the composite films were characterized by AES and XPS. Composite films containing Al-N, Al-O and Al-Al bonds were formed but no nitrogen-oxygen bonds were observed. The thicknesses of the films were estimated by argon sputtering in the UHV chamber. The surface morphologies of samples after fabrication processes in d.c. plasma were investigated by SEM. A possible formation mechanism of the composite film in the ammonia plasma is proposed.

INTRODUCTION

Improvements in Al and Al alloy surface properties such as hardness and corrosion resistance are important for current industrial applications [1]. Both Al_2O_3 and AlN are hard ceramic materials [2,3] with high melting temperatures [2,4]. The Al_2O_3 film can resist attack by dilute acid, by solutions containing ions of metals which are less electropositive than aluminum, and by atomic oxygen [5]. The chemical stability of AlN may contribute to the 50~100 Å thin Al_2O_3 layer formed on it [4].

Multilayer thin film structures often exhibit properties not present in the bulk form and attract great attention in scientific and technological communities. Therefore, a AlN/Al_2O_3 multilayer composite system was fabricated and tested in this study in order to adopt the unique characteristics of both materials. However, embedding the AlN into an Al_2O_3 matrix is not an easy task due to the fact that aluminum and its alloys are very chemically reactive to oxygen and form an aluminum oxide layer on their surfaces. This surface oxide layer prevents the nitrogen from reacting with the metallic aluminum. Hence, a clean Al surface without oxide is the critical requirement in order to grow AlN onto the surface of the aluminum alloys. A d.c. plasma was generated to fabricate the AlN/Al_2O_3 composite system onto a 6061 Al alloy. The plasma was used to not only remove the native oxide layer on the Al surface in the initial growth of AlN but also to provide energetic nitrogen ions penetrating into the Al substrate. The surface compositions and bonding environments were analyzed by the surface sensitive techniques of Auger Electron Spectroscopy (AES) and X-ray Photoelectron Spectroscopy (XPS) in a UHV chamber. The morphology of the surface after each process was investigated by Scanning Electron Microscopy (SEM).

EXPERIMENT

The schematic drawing of the experimental set-up of the d.c. plasma system is shown in Fig.1. The specimen worked as a cathode and the stainless steel plasma chamber as an anode. Ammonia and argon gases were used for the plasma discharge. A 40 l/s turbomolecular pump was used and a thermocouple gauge was used to control the pressure inside the chamber. The 6061 Al alloys were mechanically polished, cleaned by acetone, methanol, and de-ionized water in an

Mat. Res. Soc. Symp. Proc. Vol. 434

ultrasonic cleaner and blown dry by nitrogen gas. Then sample was mounted on top of the electrode inside the plasma chamber. The plasma processing procedure is listed below:

a. The chamber was evacuated by a turbomolecular pump and heated up to 350K for one hour to outgas the adsorbed gas molecules on the inner walls of the plasma chamber. We then

b. Cooled the chamber down to room temperature while pumping.

c. Filled the chamber with pure argon to 0.1 Torr and maintained this pressure under pumping. Then we connected 2000 V d.c. between specimen and chamber for half an hour.

d. Pumped out the argon in the chamber and refilled the chamber with argon and ammonia. The partial pressure ratio of argon to ammonia was about 1:9 and we kept the pressure at about 0.5 to 0.8 Torr.

e. Connected a 1000 V d.c. voltage between specimen and chamber, and controlled the discharge current density on the sample at about 3 mA/cm^2. We maintained this condition for one hour.

f. Pumped the used NH_3/Ar mixed gas out and refilled the plasma chamber with pure ammonia. We maintained the pressure at ~30 Torr, current density at 6 to 7 mA/cm^2, and d.c. voltage at 800 V for ten hours.

After plasma processing, the specimen was taken out of the plasma chamber, exposed in air, and either sent into a UHV chamber for surface analyses or sent into a high vacuum chamber for SEM measurements.

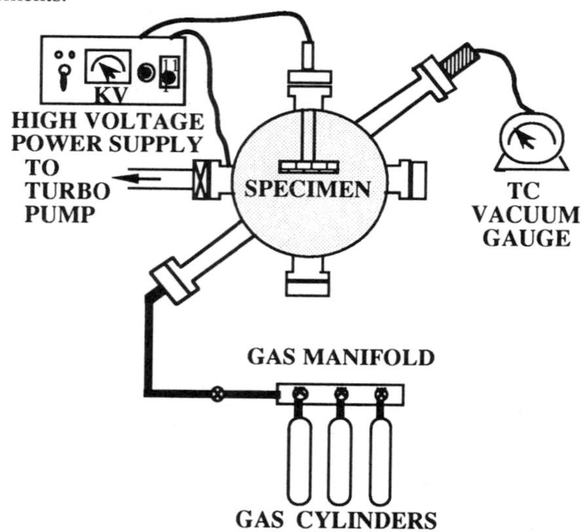

Figure 1. The schematic of the d.c. plasma set-up

EXPERIMENTAL RESULTS

AES Results

Since to have an oxygen-free Al surface is a critical step in growing the AlN on the Al surface, a sample treated after step e, i.e., low current, low pressure treatment, was analyzed by AES. Reports from the literature [6,7] and our previous sputtering studies indicated that Ar^+ sputtering alone without sample heating cannot produce an oxygen-free Al surface. Because the surface oxide was formed during sample transport from the plasma chamber to the surface analysis UHV chamber, 2.8 keV Ar^+ ions were used to sputter off the oxide layer. The oxygen-free Al surface was indeed seen after 10 min. sputtering as shown in Fig. 2b. It is noticed that the kinetic energies of Al LVV Auger signals in Fig. 2a and 2b are different. The

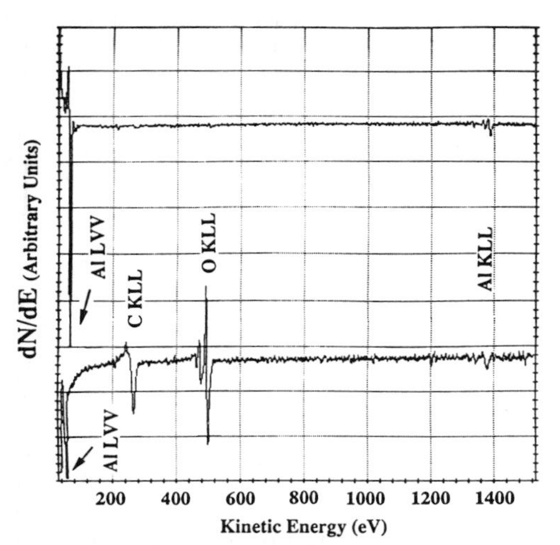

Figure 2 The Auger spectra of Al after process steps a, b, c, d and e. Spectrum obtained after plasma treatment (a), and after 10 min. 2.8 keV Ar^+ sputtering (b).

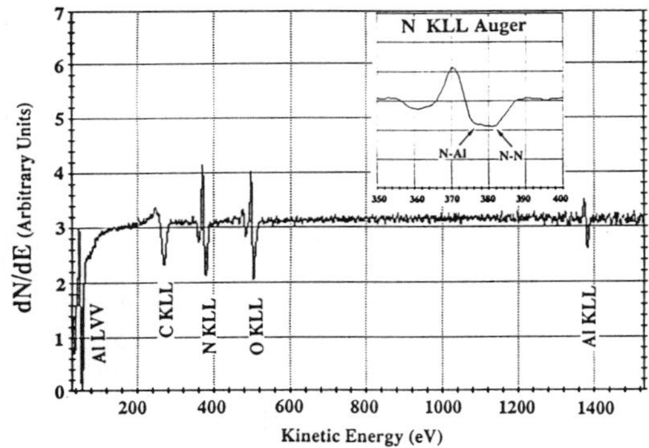

Figure 3 The Auger spectrum of Al after treatment by all the process steps. The insert is the N KLL signal; one peak is due to the N-N bonds and the other results from the N-Al bonds.

high kinetic energy Al LVV signal corresponds to the metallic aluminum and the low one is due to the oxidized aluminum [8]. The AES spectrum of the composite film obtained by the entire series of process steps is shown in Fig. 3 where a strong nitrogen KLL Auger signal located approximately at 380 eV is observed. The relative surface compositions of the AlN/Al_2O_3 composite system are 46%Al, 23%N, 15%O, and 16% contaminant C. The insert in Fig.3 is the expanded nitrogen signal where two clear components of the nitrogen signal approximately located at 376 eV and 382 eV, respectively, are seen. They are assigned to nitrogen-aluminum and nitrogen-nitrogen bonds, respectively, which are consistent with the XPS results as described below.

XPS Results

Four elements, Al, N, O and C were found in the XPS spectrum of the composite film obtained after all the process steps which is consistent with the AES result. In order to identify the bonding environment of each elemental component, a binding energy scan containing the major photoelectrons from the individual components was performed by XPS. Fig. 4 shows the N 1s photoelectron signal where two components were resolved after the signal was deconvoluted by a curve fitting program. The high bonding energy (BE) component is due to nitrogen atoms bonded to nitrogen atoms. The lower BE component is caused by nitrogen atoms bonded to aluminum atoms. No nitrogen-oxygen bonds were found. Two components were also observed in the Al 2p signal after curve fitting as shown in Fig. 5. The high BE peak corresponds to the Al-O bonds and the low BE one is due to the Al-N bonds. The assignment of a component to a specific bond is based upon previous studies [9,10] and our sputtering results. For example, three components were obtained in the Al 2p XPS signal after a 1500 Å surface layer was removed by the 2.8 keV Ar ions. These three components, assigned to Al-O, Al-N and Al-Al bonds in Fig. 6, are consistent with the BEs of Al-O and Al-N bonds in Fig. 5. The high BE peak corresponds to the Al-O bonds and the low BE one is due to the Al-N bonds. The assignment of the component to a specific bond is based upon the previous studies [9,10] and our sputtering results. For example, three components were obtained in the Al 2p XPS signal after 1500 Å surface layer was removed by the 2.8 keV Ar ions. These three components assigned to Al-O, Al-N and Al-Al bonds in Fig. 6, are consistent with the BEs of Al-O and Al-N bonds in Fig. 5.

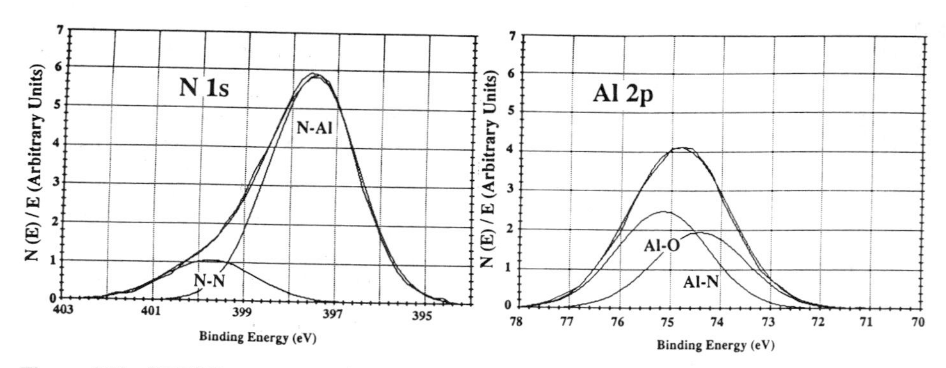

Figure 4 The XPS N1s spectrum of the composite film before sputtering which consists of two components. High BE component is due to the N-N bonds and low BE one indicates N-Al bonds.

Figure 5 The XPS Al 2p spectrum of the composite film before sputtering. The Al-O and Al-N bonds were resolved.

The 2.8 keV Ar ions were used to sputter off the surface layers of the composite film gradually. After each sputtering, the XPS spectrum was taken and the relative concentrations of Al-N, Al-O and Al-Al bonds were calculated by integrations of the areas under the components with the atomic sensitivity factors of the elements being taken into account. The relative concentrations of Al-N, Al-O, and Al-Al bonds versus the depth of the composite film are plotted in Fig. 7. It is clearly seen that the Al-N bond reaches a maximum concentration at 600 Å below the surface whereas the Al-O bonds decrease and Al-Al bonds increase with depth.

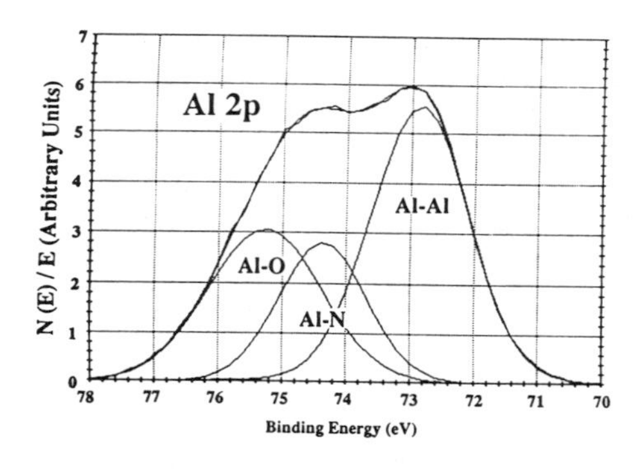

Figure 6 The Al 2p XPS spectrum after the composite film was sputtered for 30 min. Three components of the Al signal indicate Al-O, Al-N and Al-Al bonds.

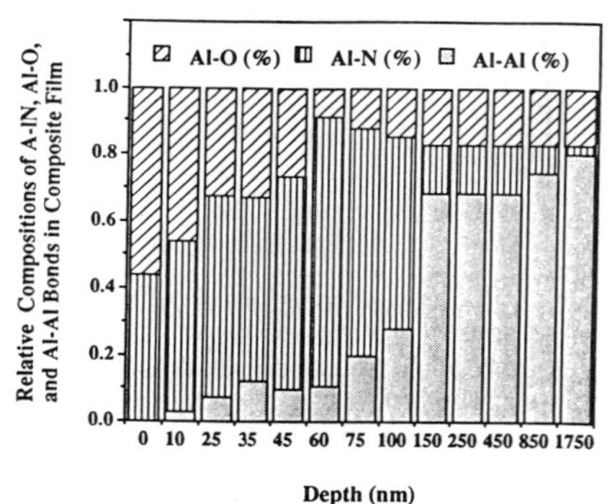

Figure 7 The relative concentrations of Al-N, Al-O and Al-Al bonds as functions of depth.

SEM Results

The morphologies of the sample surfaces after different process steps were investigated by the SEM where a 15 keV primary electron beam was used. Fig. 8a shows a rough sample surface obtained after a sample was etched by the Ar plasma, i.e., process steps a, b, and c. Small islands, 10 to 30 μm in diameter, were formed on the surface after a sample was treated by the entire series of process steps with only 4 hours in step f. Those islands grew, enlarged and therefore smoothed their edges as shown in Fig 8c after a sample was treated for 10 hours in step f compared to 4 hours in Fig 8b. The magnification of these three pictures are almost the same, *ca.* 400, as indicated at the bottom of the each picture.

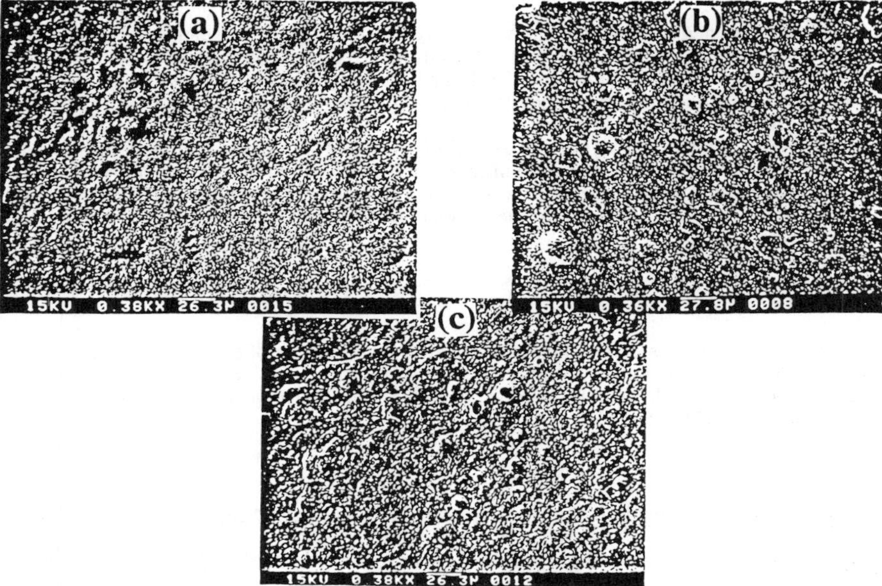

Figure 8 SEM pictures of the Al surfaces (a) before plasma treatment, (b) after steps a, b, and c, and(c) after all the process steps.

DISCUSSION

The sample without oxygen reported here was prepared by steps a, b, c, d, and e in the plasma chamber and sputtered clean by 2.8 keV Ar^+ ions inside the characterization UHV chamber. The XPS depth profile of the AlN on Al substrate indicated that a peak concentration of N was observed at 600 Å below the surface after the sample was treated by the whole series of process steps. Even though there is a difference between step e and step f, both of the steps are mainly NH_3 plasma treatments. Therefore, nitrogen passivation may be the reason why oxygen-free surfaces can be obtained on Al samples treated with the Ar/NH_3 plasma. Once 2.8 keV Ar^+ ions sputter off the nitrogen passivated surface layer, a fresh metallic Al surface without oxygen is seen.

Because of the removal or thinning of the oxide layer on the sample surface by the energetic Ar ions in the plasma chamber, nitrogen ions can react with metallic Al and form AlN. A 400 Å thick surface layer containing carbon contaminant was observed in the sputtering/XPS experiments.

No evidence shows that nitrogen-oxygen bonds are formed on surface. This implies that two separate phases coexist on the film surface. Since two components, Al-N and Al-O, were resolved in the Al 2p XPS spectrum as shown in Fig. 5, there are two co-existing phases on the surface. They are aluminum nitride and aluminum oxide. Similar AES results were reported in $(AlN)_x(Al_2O_3)_{1-x}$ films prepared by reactive magnetron sputtering deposition [11]. Eventually, three phases, aluminum nitride, aluminum oxide and aluminum metal, were formed inside the film which was observed in the sputtering/XPS experiments as shown in Fig. 7. This multi-phase feature in the composite film needs further investigation.

SEM studies also demonstrated that two phases coexist on the surface. The aluminum nitride islands and aluminum oxide flat are consistent with previous work [12].

CONCLUSION

In this study, we demonstrate that the $AlN/Al_2O_3/Al$ composite system can be fabricated onto a 6061 Al alloy by d.c. plasma treatments. No nitrogen-oxygen bonds were observed which implies two separate phases, aluminum nitride and aluminum oxide, coexist on the surface. Sputtering/XPS results indicate three kinds of bonds , Al-N, Al-O and Al-Al bonds, with different concentrations formed below the surface.

The formation of this multi-layer composite film involves 1) removal or thinning of the oxide layer by energetic Ar ions in the plasma, 2) nitrogen ions generated in the ammonia plasma either penetrating through the thin oxide layer to form AlN or directly reacting with metallic Al to form AlN, and 3) once sample is taken out of the plasma chamber and exposed to air, an aluminum oxide layer formed.

ACKNOWLEDGMENT

This work was mainly supported by the Marshall Space Flight Center of the National Aeronautics and Space Administration under Contract No. NAG8-1021. This work was also partially supported by the Materials Research Center of Excellence at the University of Texas at El Paso under NSF contract HRD-9353547.

REFERENCES

[1] R.F. Farrell, Metal Coating in McGraw-Hill Encyclopedia of Science and Technology, 7th ed. McGraw-Hill, New York, 1992,Vol. 11, p.32.
[2] CRC Handbook of Chemistry and Physics, 72th edition, ed. by D.R. Lide, CRC Press, Boca Raton, Florida, 1991, p.12-133.
[3] C.F. Cline and J.S. Kahn, 'Microhardness of Single Crystals of BeO and Other Wurtzite Compounds", J. Electrochem. Soc. **110**, 733, (1963).
[4] G.A. Slack and T.F. McNelly, "Growth of High Purity AlN Crystals", J. Crys. Growth, 34, **263** (1976) .
[5] S. Wernick, R. Pinner, "The Surface Treatment and Finishing of Aluminum and its Alloys", 5th ed., ASM, Metals Park, Ohio 1982.
[6] A.M. Bradshaw, P. Hofmann and W,.Wyrobisch, "The Interaction of Oxygen with Aluminum (111)", Surf. Sci. **68**, 269 (1977).
[7] J.M. Fontaine, O. Lee-Deacon, J.P. Duraud, S. Ichimura, and C. Le Gressus, "Electron Beam Effects on Oxygen Exposed Aluminum Surfaces", Surf. Sci. **122,** 40 (1982).
[8] P.M. Raole, P.D. Prabhawalkar, D.C. Kothari, P.S. Pawaw and S.V. Gogawale, "XPS Studies of N^+ Implanted Aluminum", Nucl. Instru. and Meth. in Phys. Res. **B23**, 329 (1987).
[9] A. Pushutski and M. Folman, " Low Temperature XPS Studies of NO and N_2O Adsorption on Al (100)", Surf. Sci. **216**, 395 (1989).
[10] C.D. Wagner, W.M. Riggs, L.E. Davis, J.F. Moulder and G.E. Muilenberg, Handbook of X-ray Photoelectron Spectroscopy, Perkin-Elmer Corp., Eden Prairie, MN, 1979, p. 160.
[11] R. Shinar, "Auger and Electron Energy Loss Spectroscopy Studies of AlN and $(AlN)_x(Al2O3)_{1-x}$ Thin Films", J. Vac. Sci. Technol. **A10**, 137 (1992).
[12] E.I. Meletis and S. Yan, "Formation of Aluminum Nitride by Intensified Plasma Ion Nitriding", J. Vac. Sci. Technol. **A9**, 2279 (1991).

LOW TEMPERATURE SYNTHESIS OF MO_2C/W_2C SUPERLATTICES VIA ULTRA-THIN MODULATED REACTANTS

CHRISTOPHER D. JOHNSON, DAVID C. JOHNSON
Chemistry Department, University of Oregon, Eugene, OR 97403

ABSTRACT

We report here a synthesis method of preparing carbide superlattices using ultra-thin modulated reactants. Initial investigations into the synthesis of the binary systems, Mo_2C and W_2C using ultra-thin modulated reactants revealed that both can be formed at relatively low temperature(500 and 600°C respectively). DSC and XRD data suggested a two step reaction pathway involving interdiffusion of the initial modulated reactant followed by crystallization of the final product, if the modulation length is on the order of 10 Å. This information was used to form Mo_2C/W_2C superlattices using the structure of the ultra-thin modulated reactant to control the final superlattice period. Relatively large superlattice modulations were kinetically trapped by having several repeat units of each binary within the total repeat of the initial reactant. DSC and XRD data again are consistent with a two step reaction pathway leading to the formation of carbide superlattices.

INTRODUCTION

Interest in transition metal carbides comes from the fact that they have many important technological applications. Most of these applications take advantage of their high hardness and chemical stability at elevated temperatures. The largest area which makes use of these properties is as abrasives and cutting tools. They also find use in thin film form as protective coatings against erosion and tribiological wear, again because they are very hard, but also because they have good adhesion to substrates and a low coefficient of friction.[1] Thin films of tungsten carbide also show promise as diffusion barriers in microelectronics because of its chemical stability and low conductivity.[2] Other areas of interest include carbides as high temperature catalytic materials,[3] and superconductivity in the molybdenum carbides. Our interest in forming superlattice carbide materials stems from the potential ability to tailor the physical properties of these systems by changing the layer thicknesses of the constituents.

Traditional methods of carbide synthesis, however, are not suitable for the preparation of refractory carbide superlattices. Typical synthesis routes involve direct reaction of the elements either by arc melting or by heating mixed powders of the elements to temperatures exceeding 1000°C for long periods of time. In these methods the high reaction temperatures are necessary to break the strong bonds in the reactants, particularly the carbon-carbon bonds in the carbon containing material, and to overcome the low diffusion rate of reactants through the resulting product layers. More recently developed methods of synthesis which overcome these difficulties include self-propagating high temperature reactants[4] and decomposition of organometallic precursors.[5] The first of these methods, self-propagating high temperature reactants, takes advantage of the large heats of formation of the carbides from elemental reactants. In this technique compacts of intimately mixed powders of the reactant elements are locally heated to initiate the reaction process. Once initiated, the heat produced by the reaction provides enough heat to sustain the reaction in the form of a reaction wave. Typical temperatures obtained in this process are on the order of 2000-3000°C. In the other method, decomposition of organometallics

Mat. Res. Soc. Symp. Proc. Vol. 434 © 1996 Materials Research Society

precursors on heated substrates eliminates diffusion as a concern permitting relatively low synthesis temperatures (350-600°C).

Our approach to carbide synthesis involves the use of ultra-thin modulated reactants to overcome the typical difficulties with stable reactants and intermediates and slow diffusion rates. We have recently found that the binary carbides, TiC, Mo_2C and W_2C can be formed at low temperatures (350, 500 and 600°C, respectively) by use of multilayer reactants, if the elemental layer thicknesses are below a critical value. For example, Mo and C layers will interdiffuse and nucleate Mo_2C at 500°C when 3 Å C and 7 Å Mo layer thicknesses are used. If much thicker layers are used, no reaction occurs below 600°C. Our goal in this study is to investigate the feasibility of using ultra-thin modulated reactants as a method of synthesizing carbide superlattices by controlling the initial reactant structure.

EXPERIMENT

Sample Preparation

The carbon/metal multilayer samples used in this study were prepared using a custom built ultra-high vacuum deposition chamber.[6] The elements were evaporated at a rate of 0.5 Å/s using Thermionics e-beam gun sources which were independently controlled by Leybold-Inficon XTC quartz crystal monitors. The layer thicknesses of the individual elements were held constant at 3 Å C, 6 Å W and 7 Å Mo. The appropriate number of C-Mo layers were then alternated with the appropriate number of C-W layers to obtain the desired multilayer structure. A series of samples were made in which the intended superlattice repeat layer thicknesses ranged from 76 to 608 Å. Samples were simultaneously deposited on a PMMA (polymethylmethancrylate) coated silicon wafer and uncoated silicon wafer. To obtain sample free of the substrate for thermal analysis, the PMMA coated wafers were soaked in acetone to dissolve the PMMA and the "floated" sample collected on a Teflon filter. The sample deposited on silicon was used for low-angle x-ray and high-angle x-ray analysis.

Thermal Analysis

DSC data was collected on portions of samples which were separated from the substrate. This was accomplished using a TA Instrument's TA 9000 calorimeter fitted with a 910 DSC cell. An initial scan ramped from 50 to 600°C at 10°C/min was done and high-angle x-ray diffraction used to determine phase formation, additional portions of samples were heated to corresponding points of interest as shown by the DSC data and x-ray data collected. In this way correlation between features present in the DSC scans and product structure can be made.

X-ray Diffraction

High-angle diffraction data were used to determine whether the as-deposited, floated and annealed samples contained crystalline elements or compounds. Low-angle diffraction data were collected to determine if the samples contained a periodic layered structure as deposited and after annealing. These data were collected on a Scintag XDS 2000 θ-θ diffractometer using Cu K_α radiation with a sample stage modified for rapid and precise alignment.

RESULTS AND DISCUSSION

In order to evaluate the ability to use ultra-thin modulated reactants as a method of forming carbide superlattices, four different intended superlattice thicknesses were prepared as part of this study. They were made by alternately depositing molybdenum and carbon followed by tungsten and carbon as shown in Figure 1.

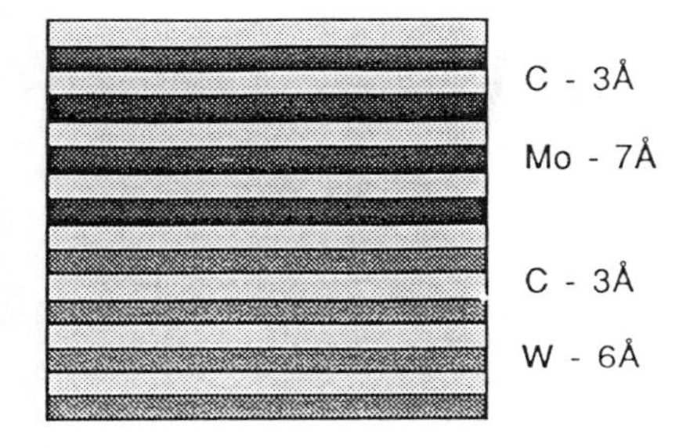

Figure 1. Illustration of layering scheme used to create the initial modulated reactants.

This $[(Mo/C)_n(W/C)_m]$ unit was repeated to give films of approximately 600 Å total thickness. Low-angle x-ray diffraction data for all of the samples contained diffraction maxima resulting from the elemental modulation and front surface-back surface interference out to 3 degrees. This data was used to determine the thicknesses of the deposited layers. No diffraction maxima were observed in high angle x-ray diffraction scans, suggesting that the samples were amorphous with respect to x-ray diffraction as deposited. Table I summarizes the intended and the experimentally determined layer thicknesses.

Table I. Summary of intended and measured superlattice thicknesses.

	SAMPLE			
	WMC1	WMC2	WMC3	WMC4
Intended Mo/C Thicknesses	40 Å	80 Å	160 Å	320 Å
Intended W/C Thicknesses	36 Å	72 Å	144 Å	289 Å
Total Intended Layer Thicknesses	76 Å	152 Å	304 Å	609 Å
Measured Total Layer Thicknesses	75 Å	155 Å	308 Å	620 Å

DSC and x-ray diffraction were used to investigate the reaction between the elemental layers. Figure 2 shows the DSC data collected on sample WMC1, which is representative of all of the samples studied. This data contains a distinct exotherm at approximately 550°C, which is midway between DSC exotherms found for the binary compounds (Mo_2C and W_2C have exotherms at 500°C and 600°C, respectively). Diffraction data was collected as a function of temperature to determine the structural changes occurring during this exotherm. Figure 3 shows high-angle x-ray data collected on sample WMC1 before and after the exotherm. The high-angle diffraction maxima which appear after the exotherm suggest that the exotherm corresponds to the crystallization of the sample. The positions of the high-angle diffraction maxima are consistent with those expected from both Mo_2C and W_2C, which have the same structure and nearly identical lattice parameters.

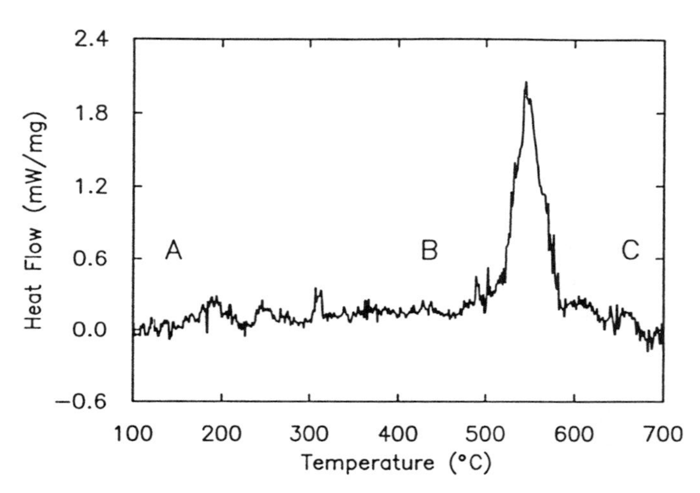

Figure 2. Differential scanning calorimetry data of Mo/C and W/C sample WMC1 with a 76 Å intended repeat layer thickness.

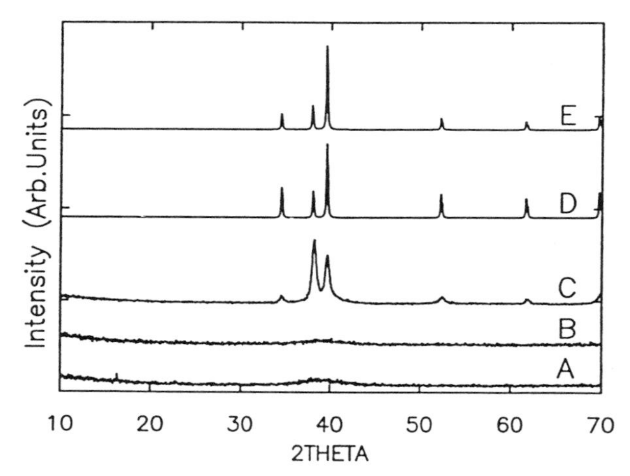

Figure 3. High-angle x-ray diffraction data showing the evolution of phases corresponding to the DSC scan in Figure 2: A = as-deposited, B = 400°C, C = 600°C, D = JCPDS simulation of W_2C, and E = JCPDS simulation of Mo_2C.

In order to more clearly determine the structures of the materials obtained in this experiment, samples deposited directly on silicon substrates were annealed at various temperatures to determine their behavior as thin films. Figure 4 shows the low-angle x-ray diffraction data collected on WMC1 sample, again as deposited and annealed to 400, 600, and 800°C. This data clearly shows the presence of Bragg maxima corresponding to the layered structure of this sample even when heated to 800°C. The relative intensity of these peaks remain constant in the samples heated to 600°C, but begin to decrease when the sample is heated to 800°C. This suggests that mixing of the tungsten and molybdenum containing layers only occurs at temperatures considerably higher than the 550°C exotherm seen in the DSC.

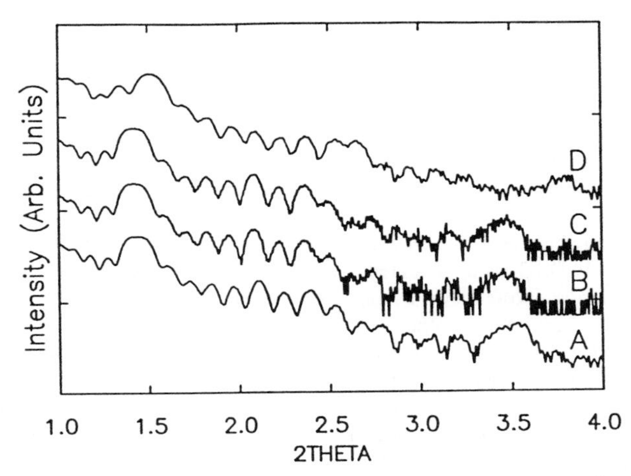

Figure 4. Low-angle x-ray diffraction data showing the layered structure of sample WMC1 annealed to various temperatures: A = as-deposited, B = 400°C, C = 600°C, and D = 800°C.

Figure 5 shows the high-angle x-ray data collected on sample WMC1 on the substrate as a function of annealing temperature.

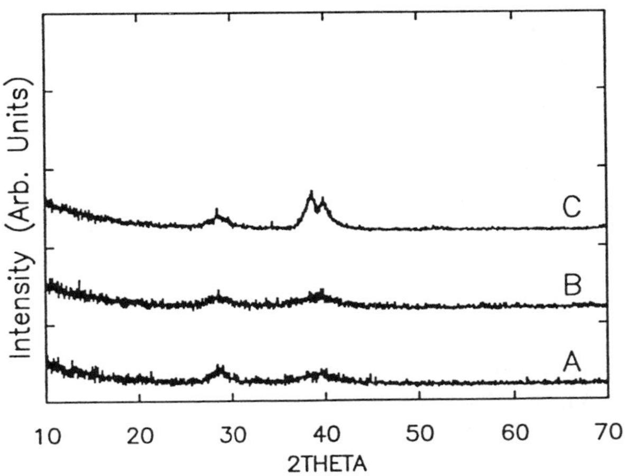

Figure 5. High-angle x-ray diffraction data of sample WMC1 annealed on silicon substrate: A = as-deposited, B = 400°C, C = 600°C.

The data shows peaks that correspond to the (002) and (101) lines of both W_2C and Mo_2C. The relative peak heights and the lack of additional diffraction maxima suggests that the crystallites are preferentially oriented with respect to the substrate. This was confirmed by rocking-angle diffraction scans of the (002) maxima which showed a FWHM of approximately 9° on the sample annealed at 600°C. Crystallite size of approximately 150 Å was determined using the Scherrer equation. This size is twice the thickness of the superlattice period determined from the low-angle diffraction data, suggesting that the crystallites have grown through several layers. The fact that crystallites of larger size than the original layer thickness are obtained, along with the lack of attenuation of the Bragg maxima in the low-angle diffraction data suggests that a layered Mo_2C/W_2C superlattice has been made. The lack of splitting of the high-angle diffraction data, however, suggests that there is no long range coherence between the crystalline layers.

Upon examining the thicker superlattice samples in the same manner, several interesting trends are observed. DSC data collected on the thicker samples shows that the exotherm temperature decrease monotonically as layer thickness increases. The two thickest samples, WMC3 and WMC4, have DSC exotherms very close in temperature to that found in the binary molybdenum carbide system. This suggests that nucleation may be taking place primarily within the Mo-C layers. The samples with thicker layering also showed a decrease in preferential orientation as layer thicknesses increased. This may be the result of a higher percentage of nucleation events occurring further from the Mo-C/W-C interface. In addition, there was an increase in crystallite size as the thickness of the original superlattice increased. In all samples, the crystallites were larger than the original superlattice period. We are currently undertaking a long term annealing study to improve the crystallinity of the samples.

SUMMARY

DSC and XRD data provide evidence that ultra-thin modulated reactants are a viable method of creating nanoscaled layered composites of refractory carbides. This is accomplished by controlling the relative layer thicknesses of elemental reactants vs. the superlattice period in the initial modulated reactant structure.

ACKNOWLEDGMENTS

This work was supported by the National Science Foundation (DMR-9213352 and DMR-9308854), the Office of Navel Research (N0014-93-1-0205), and by the U.S. Department of Energy, Pacific Northwest Laboratories through the Energy Research Fellowship Program (C.D.J.).

REFERENCES

1. A. S. Salnikov, Met. Term. Obr. Met. **35** (4) 15-19.
2. S. Ghaisas, J. Appl. Phys. **70** (12) 7626-7628.
3. I. Nikolov, T. Vitanov, V. Nikolova, J. Power Sources **5** (2) 197-206.
4. Z. A. Munir, U. Anselmi-Tamburini, Mat. Sci. Rep. **3** (7, 8) (1989).
5. J. E. Parmeter, D. C. Smith, M. D. Healy, J. Vac. Sci. Tech. A, Vac. Surf. Films **12** (4) pt. 2 2107-2113.
6. L. Fister, X. M. Li, T. Novet, J. McConnell, D. C. Johnson, J. Vac. Sci. Tech. A **11** 3014-3019 (1993).

FGMs by Sedimentation

Y. HE, V. SUBRAMANIAN, J. LANNUTTI
Ohio State University, Columbus, OH 43210, lannuttj@KCGL1.eng.ohio-state.edu

ABSTRACT

Simple sedimentation in organic solvents followed by hot-pressing is used to produce alumina-NiAl functionally graded materials (FGMs). Varying degrees of agglomeration influenced the phase arrangement in the mixed layer(s) producing microstructural variation. We discovered a pronounced structural dependence on NiAl stoichiometry. Slight variations in Al content are known to influence T_{db} and these apparently lead to large increases in residual stress. Zirconia - NiAl FGMs were better able to accommodate these levels of residual stress possibly due to accommodation by enhanced tetragonal phase retention. However, these FGMs undergo transformation of the tetragonal phase to the monoclinic form, starting from the surface. Finally, variable microstructures result in detectable changes in the stress-strain behavior of alumina-NiAl FGMs.

INTRODUCTION

Functionally graded materials (FGMs) are made using a variety of techniques[1-38]. Each new pathway for the production of ordered layers carries with it characteristic compositional and architectural capabilities. We use simple sedimentation in organic solvents followed by hot-pressing to produce ceramic-intermetallic FGMs. Compositional flexibility is combined with variable microstructural parameters. This process generated intriguing property data[9-11] that assisted in the identification of key fabrication and processing concerns[12,39].

At the heart of this approach is the agglomeration of powders of different size and density. This occurs by the generation of water-based mensici that 'glue' the particles together. This decreases gravity-induced phase segregation. Post-formation solvent replacement can achieve acceptable (>50% of theoretical) presintered densities after the desired architecture has been 'fixed' in place.

During sedimentation of equisized two phase powder mixtures net downward segregation of denser particles is favored. Modification of particle interactions can control both segregation and phase distribution in the resulting sediments. Many mechanisms (van der Waals interactions, capillary condensation and polymer/surfactant bridging) contribute to these forces in organic solvents and influence segregation.

Capillary condensation can produce particularly strong attractive forces between two surfaces[40]. In mixtures of immiscible liquids, small menisci composed of the minor phase can condense out between particles. For ceramic surfaces in hydrophobic organic solvents[41], water menisci have been shown to obey the Kelvin equation[42]. Capillary forces offer an effective, clean mechanism to minimize powder segregation.

EXPERIMENT

FGM Processing

The NiAl powder we used had an average particle size of approximately 17 μm. We employed AKP-30 α-alumina** having an as-received ultimate particle size of 0.41 μm. These materials have a net density difference of 34% and vary in particle size by factors of 20 to 50. For mixed alumina - NiAl powders in a hydrophillic solvent (e.g., methanol) the NiAl particles settle quickly, leaving much of the alumina in suspension. This produces a settled compact

** Sumitomo Chemical/America Inc., New York, New York 10154

Mat. Res. Soc. Symp. Proc. Vol. 434 © 1996 Materials Research Society

having two distinct layers. In hexane, larger van der Waals forces and the agglomerative effects of adsorbed water result in more limited segregation[9]. Drying the powders beforehand made no difference in the behavior of the methanol system but greatly decreased the degree of agglomeration in hexane.

To avoid cracks during composite drying and decrease subsequent densification increases in sediment density were needed. This was achieved by replacing hexane with alternate solvents after sediment formation. These relax the sediments primarily due to the dissolution of 'bridging' water. Higher densities result and reduce drying shrinkage.

Through a series of simple mixing experiments methanol and water were found to be the only solvents to show incomplete solubility. Both were found to be at least partially insoluble in hexane. With the addition of minor fractions of acetone, methanol became completely miscible with hexane. We use an apparatus involving a semi-permeable membrane that allows gradual introduction of solvents or solvent mixtures into hexane without disturbing the presettled sediments.

In an experiment illustrating the 'relaxing' effect of these substitutions on sediment density, twenty µl of water per gram AKP were added to each of several hexane-sedimented AKP-30 specimens. The particle-particle contacts in this fine ceramic powder are far more numerous and dominate downward motion in the alumina-NiAl mixtures. The prepared samples were then settled through a 1 inch diameter perforated plate into a settling column. Hexane was removed and exchanged with the solvent mixtures indicated in Figure 1. After 24 hours, the new height of the sediment was noted Additional solvent replacement followed. This process was repeated until the final solvent combination was reached. The solvent was then drawn off and the membrane removed to allow the specimen to dry.

Figure 1 compares the densities obtained by solvent replacement approximately 24 hours after each exchange. In specimens 1-6, hexane was replaced by methanol or a 50/50 volume percent mixture of methanol and water. The maximum density achieved was approximately 15 percent solids, a fairly typical density.

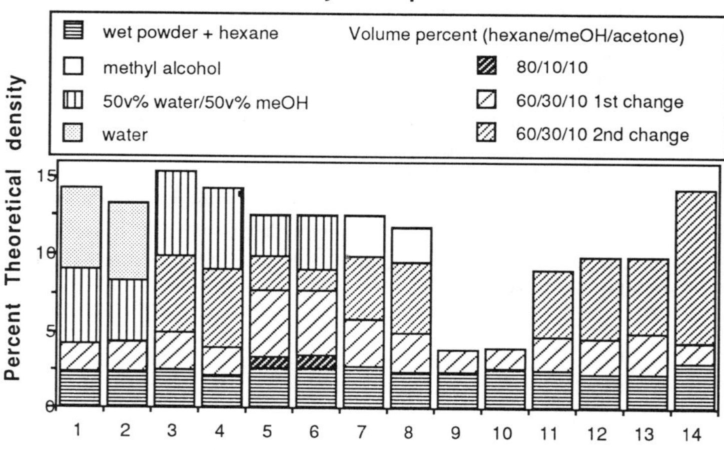

Figure 1. Sediment densities 24 hours after replacement by the indicated solvent mixtures. The highest sediment densities are seen in the methanol/water mixtures. Volume percents of hexane, methanol and acetone are listed as hexane/methanol/acetone.

Less cracking was generally observed for specimens in which the density increased least during drying (an initially higher wet density). In addition, specimens which cracked often did so due to adherence to the glass tube. The change in density and tendency to adhere during drying was observed to be smallest with either the hexane/methanol/acetone solvent mixtures or pure methanol.

Our current process is based on these simple observations. We first choose the number of layers, their thickness and composition. The required quantities of dried powder are then weighed out and added to dry hexane in 30 ml glass vials inside a glove box. The powder mixtures for each layer are then removed from the glove box and ultrasonicated. After sonication, water is added to the composite layer powder mixtures by micropipet. These mixtures are then ultrasonicated once more and are settled sequentially using a perforated plate device mounted on a hot-press die. The die is lined with a combination of graphoil®, boron nitride and molybdenum foil. The graphoil® and boron nitride act as sealants and reduce die wear. Molybdenum is chemically compatible with both NiAl and alumina and reduces powder contact with the graphoil®. After settling is complete, excess solvent is drawn off and the perforated plate apparatus removed.

The fully dried specimens were hot-pressed[11] at 1500°C and 8000 psi for 2 hour. A hold period of 1 hr at 800°C was incorporated into the cooling ramp to relieve residual stresses in the NiAl. Additionally, cooling from 800°C to room temperature was limited to 3°C per minute. Pressure was applied at 800-1000°C and released at 800-500°C.

Ceramic-intermetallic FGMs produced by this process undergo graceful failure at moderate to high temperatures (527 and 727°C) [11]. Multiple cracks propagate and are blunted in the intermetallic NiAl layer. At no point was loss of the alumina layer observed, giving rise to a ceramic surface displaying graceful failure at T~500°C.

RESULTS

Effects on Microstructure

Varying degrees of agglomeration influence the phase arrangement in the mixed layer(s). This produces considerable microstructural variation. Figure 2 shows examples of the degree to which a lack of agglomeration can affect phase distribution. Figure 2a shows the unfired powder resulting from sedimentation without any added moisture. The clean, uncoated NiAl particle surfaces indicate that the two phases display little mutual adherence. A typical final microstructure (Figure 2b) shows fine NiAl segregation to the top of the mixed layer.

Figure 3 describes a system to which 20 μl of moisture was added. The unheated powder (Figure 3a) shows scattered adherence of the ceramic powder to the NiAl. A final microstructure (Figure 3b) is reasonably well-mixed and, within the boundaries of the layer, can be considered compositionally uniform. However, continuous areas of NiAl represent networks of presintered NiAl particle contacts where little or no alumina is present.

Different levels of uniformity can be achieved if we alter the initial moisture distribution. The basic goal of these additions is to 'glue' the two different types of particles together during downward motion. If we localize moisture on NiAl surfaces before co-sonication this maintains contact with the neighboring ceramic particles. Figure 4a shows that the ceramic particles are much better adhered to the NiAl particles. The NiAl particles are barely visible in the presintered sediment. The resulting fired microstructure (Figure 4b) clearly suggests better final separation of the NiAl particles in the mixed layer.

Figure 2a. Presintered arrangement of intermetallic (dark phase) and ceramic powders (light phase) resulting from sedimentation without introduced moisture. Clean, uncoated NiAl particle surfaces demonstrate the lack of adherence of the ceramic particles to the NiAl.

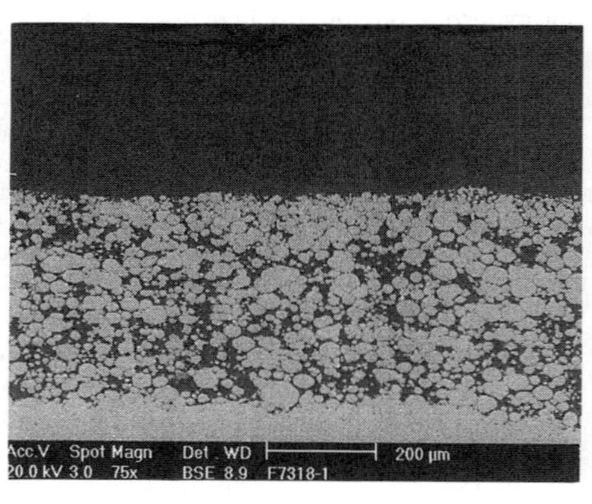

Figure 2b. Hot-pressed microstructure resulting from an FGM prepared without added moisture. Segregation of fine NiAl to the top of the mixed layer is clearly evident.

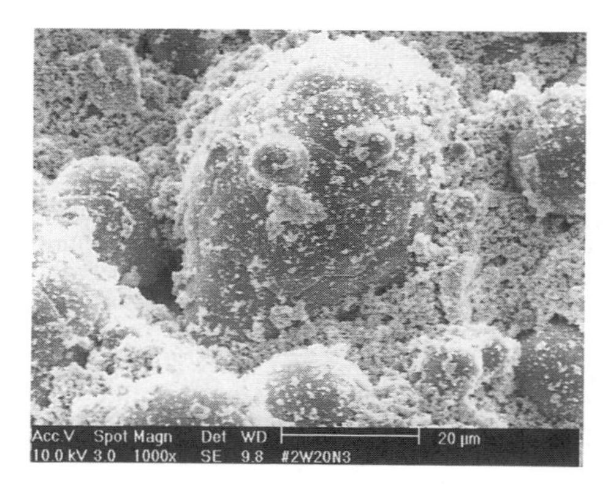

Figure 3a. Presintered arrangement of intermetallic and ceramic powders resulting from sedimentation with 20 µl of introduced moisture. Scattered adherence of the ceramic powder to the NiAl is evident.

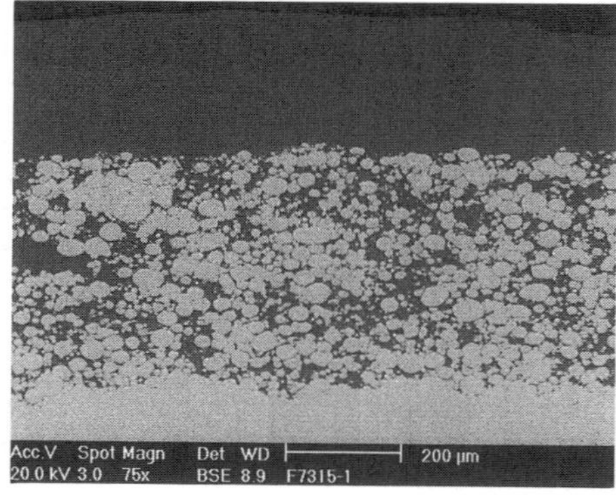

Figure 3b. Hot-pressed microstructure resulting from an FGM prepared using 20 µl of added moisture. The microstructure is reasonably well-mixed; however, continuous areas of NiAl represent networks of presintered NiAl particle contacts.

Figure 4a. As-sedimented arrangement of intermetallic and ceramic powders using 20 µl of NiAl-localized moisture. The ceramic particles are much better adhered to the NiAl.

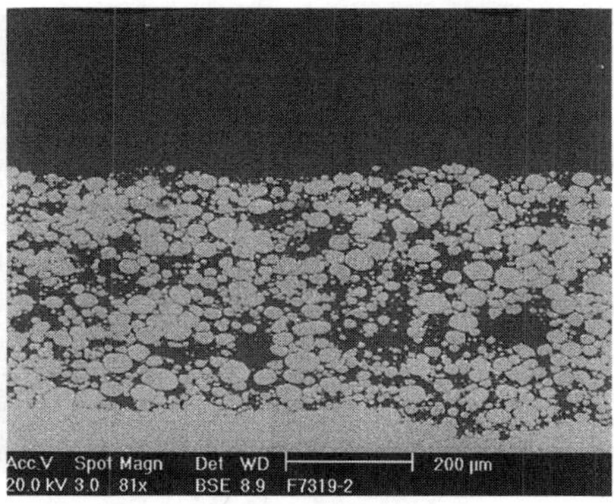

Figure 4b. Hot-pressed microstructure resulting from an FGM prepared using NiAl-localized moisture. Improved separation of the NiAl particles is evident.

The technique can also be used to induce other microstructural variations. Figure 5 shows a dense mixed layer formed by sedimentation of NiAl and presintered AKP 30 agglomerates. During densification at 1500°C the NiAl has intruded into the surface of these agglomerates prior to overall densification. The resultant structure is intimately interlocked only at the alumina boundary. Simple interfacial control can also be achieved by third-phase coatings (e.g., from a simple salt or alkoxide solution) on the surface of the NiAl particles.

Finally, the technique requires careful monitoring. Figure 6a shows contamination of the darker alumina layer by NiAl. This is produced when NiAl particles adhere to the walls of the sedimentation apparatus before final layer deposition. They are then engulfed by the alumina sediment and co-deposit in the ceramic layer. The relative lack of chemical interaction between alumina and NiAl during densification ensures that these particles will degrade the tensile abilities of this layer.

Figure 6b shows the microstructure of a system in which the sediment was allowed to stand for several days before the solvent was removed. A fine network of alumina particles outlining original NiAl particle boundaries near the mixed layer - NiAl interface is now evident. Brownian motion was apparently sufficient to cause fine particle migration down into the NiAl layer. Under some circumstances this may be desirable to promote NiAl grain size minimization.

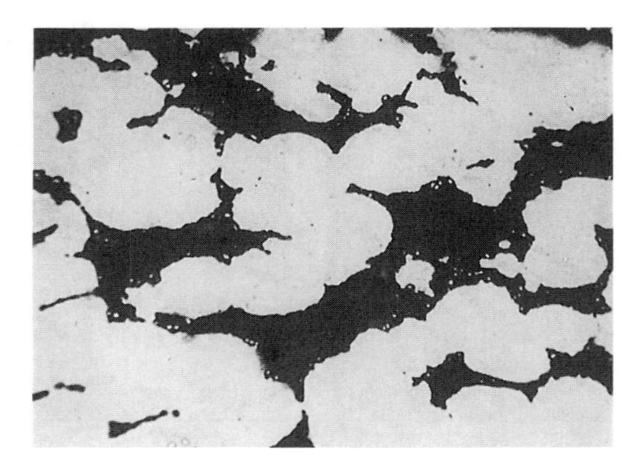

Figure 5. Mixed layer formed by sedimentation of NiAl and presintered AKP 30 agglomerates (500X). The NiAl (light phase) has intruded into the surface of the agglomerates prior to overall densification. The resultant structure is intimately interlocked only at the alumina boundary.

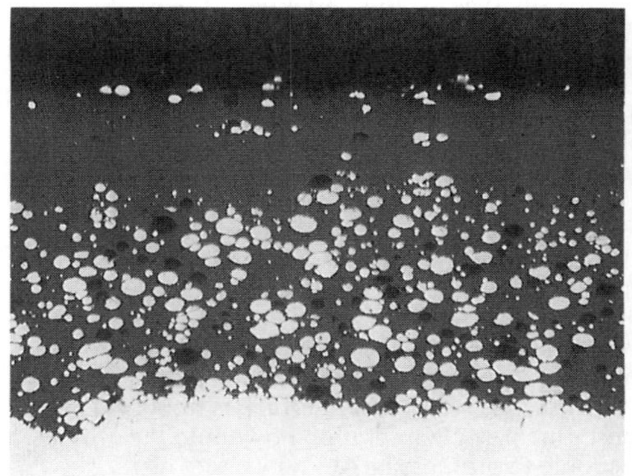

Figure 6a. Contamination of the alumina layer by NiAl. NiAl particles adhered to the walls of the sedimentation apparatus were co-deposited in the ceramic layer. 100 X

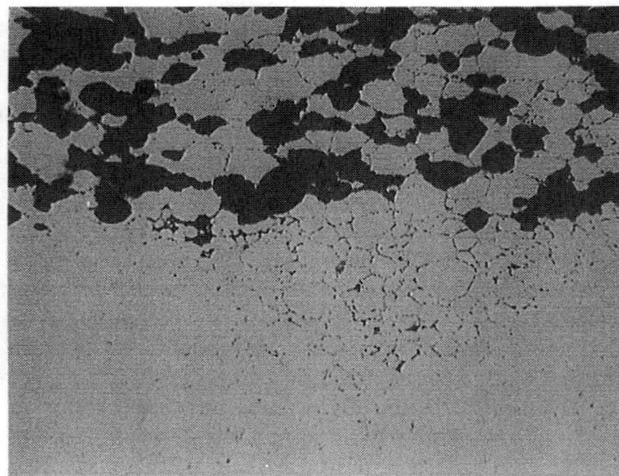

Figure 6b. The microstructure of a composite in which the sediment was allowed to stand for several days. A fine network of alumina particles outlining original NiAl particle boundaries near the mixed layer - NiAl interface is now evident. 200 X

Al content

During our research architectures that had previously been fabricated without failure began to exhibit cracking during cooling in the hot press. This had previously been associated with the development of too-large levels of residual stress[11]. We launched an investigation to determine the cause of these failures. Composition/architecture had not been deliberately varied so we assumed that a fault in the processing had developed. We focused on the following: temperature/pressure variations, carbon contamination, annealing time/pressure release, AKP-30 diffusion into the NiAl layer, and oxide 'stringers' caused by incidental NiAl oxidation.

These tests did not identify a hidden flaw in the processing. We re-examined the NiAl itself and determined that a different lot number was involved. In examining the composition of this new lot (50.50 at% Al) we found that the stoichiometry had varied slightly from the original alloy (50.05 at% Al). Small compositional variations in NiAl are known to result in large variations in T_{db}[43-45]. Increases in T_{db} increase the resultant room temperature residual stress levels[12] which can lead to fracture. While T_{db} variation is generally neglected in FRCs, in this composite form it clearly cannot be due to residual stress constraints.

Zirconia-NiAl FGMs

During this period of difficulty we investigated the fabrication of zirconia-NiAl FGMs. We believed that zirconia would be less sensitive to the residual stresses apparently being generated. This rationale was developed in considering phase transformation-induced fracture toughness in PSZ. This depends upon the critical transformation stress σ_c^T, zone size, and the influence of these parameters on the tetragonal to monoclinic transformation.

The prime factor possibly influenced by FGM architecture-induced compressive stresses will be σ_c^T. We can write:

$$\sigma_{eff} = \sigma_a - \sigma_r$$

$$\sigma_{eff} = \sigma_a - (\sigma_{rm} - \sigma_{rfgm}^C)$$

where σ_{eff} is the effective stress needed to initiate fracture, σ_a is the applied stress, σ_{rm} is the normal radial stress surrounding each tetragonal particle in the monoclinic matrix. σ_{rfgm}^C is the radial compressive stress surrounding each tetragonal particle produced by the FGM architecture. The matrix toughness of the zirconia ceramic, K_{IC}^M, increases linearly with increases in σ_{rfgm}^C.

The elastic buildup of σ_{rfgm}^C begins when the NiAl T_{db} is reached during cooling. We anticipate that the change in strain free energy during the transformation, ΔG_ε, will be increased by this and lead to the retention of more tetragonal phase. This will favorably influence the volume fraction of tetragonal phase susceptible to transformation.

We believed that the enhanced retention of tetragonal phase in zirconia-NiAl FGMs would accommodate those stresses that lead to failure of the alumina-NiAl composites. This effect should complement that due to the greater thermal expansion coefficient of zirconia.

Processing and fabrication of 'standard' architecture zirconia-NiAl composites was identical to that of the alumina-NiAl form. As anticipated, destructive fracture during cooling did not occur. With 0 mole % yttria, we did observe limited crack development. With 2 mole% yttria, however, no such fracture occurred. From this we concluded that the proposed mechanism for stress absorption via enhanced tetragonal phase retention seemed to be operative.

Toughness characterization of these composites was carried out. We have, however, become aware of the difficulties involved in applying this form of property data to FGMs[46]. For a notch in the zirconia layer a maximum toughness of 10.62 MPa $m^{0.5}$ was obtained. A fractographic view of the failure process indicates that as fracture moves into the mixed layer NiAl grains just beneath the interface decohere from the zirconia matrix. This is identical to the behavior seen for alumina-NiAl[10]. Further intermetallic fracture takes place via a mixture of mostly transgranular NiAl fracture and some NiAl decohesion out of the zirconia matrix.

An unfortunate development soon became apparent. The 0 mole% yttria composite exhibited spallation of the ceramic layer 4 months after its fabrication. The 2 mole% yttria composite retained its ceramic layer for many months before a few small isolated areas began to spall off. However, optical microscopy (Figure 7) and x-ray diffraction (Figure 8) both show substantial room temperature conversion of the retained tetragonal phase to the monoclinic form. While the literature does document the environmental sensitivity of zirconia[47], ZrO_2-based FGMs have been fabricated by a variety of investigators[2,8,19,20,36].

Figure 7. Optical micrograph showing the transformation zone of a ZrO_2-NiAl FGM several months after fabrication. This process begins at the upper surface of the composite and moves towards the mixed layer. 100 X

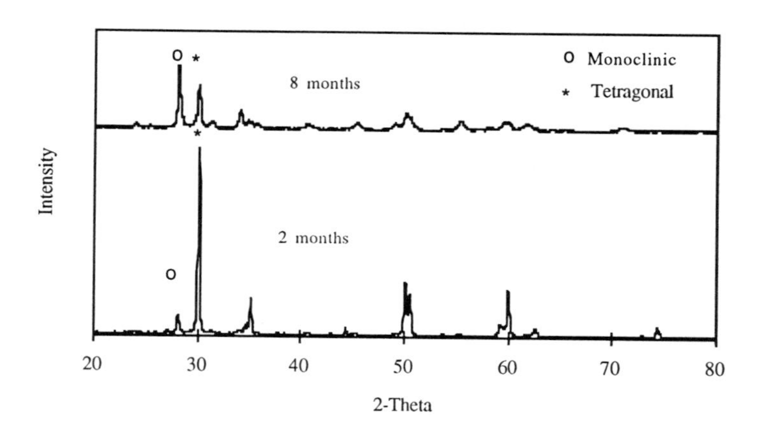

Figure 8. XRD showing the degree of tetragonal to monoclinic transformation in the zirconia layer of a ZrO_2-NiAl FGM after several months of room temperature exposure.

Microstructure and Properties

For the different microstructures found in Figures 3b and 4b, Figure 9 shows the force-displacement curves. Figure 10 displays the lateral thermal expansion behavior. These indicate significant differences in the behavior of these composites due to the initial microstructural differences.

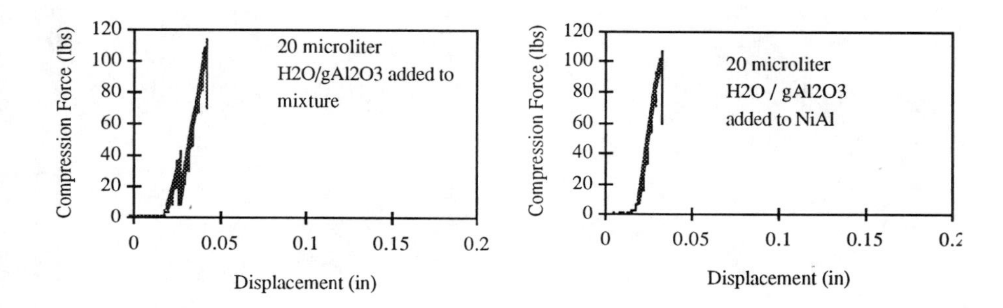

Figure 9. Force displacement curves obtained during four point bend test of samples having different microstructures.

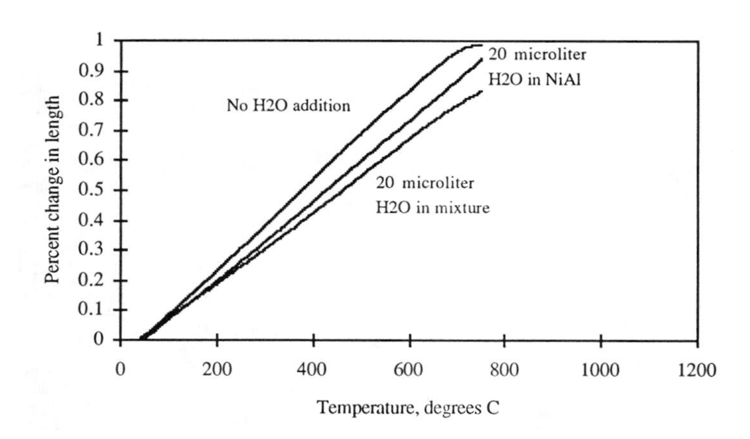

Figure 10. Lateral linear thermal expansion as a function of temperature for samples having different microstructures.

Figure 11 shows cross-sectional SEM of the fracture surfaces generated during the four point bend test. In the alumina layer fracture takes place via a mixture of intergranular, transgranular and step cleavage. In the mixed layer substantial decohesion of the NiAl phase from the alumina matrix has occurred. The degree of decohesion is much greater than that seen during toughness evaluations. In the NiAl layer, fracture takes place via a mixture of intergranular and transgranular modes.

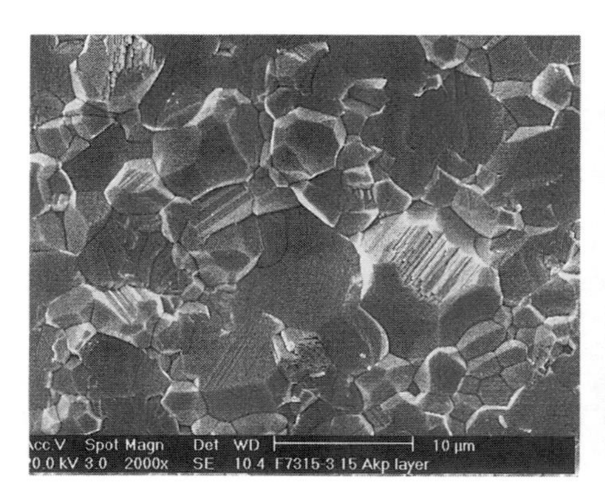

Figure 11a. Fracture surface of the alumina layer showing transgranular, intergranular and step cleavage

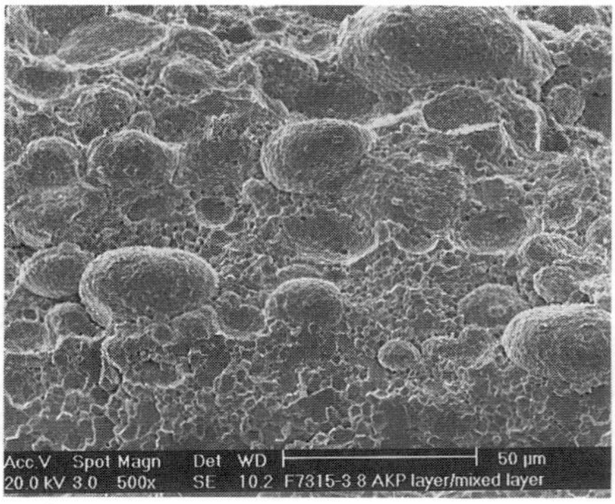

Figure 11b. Fracture surface of the alumina/alumina-NiAl interface showing decohesion of the NiAl phase from the matrix.

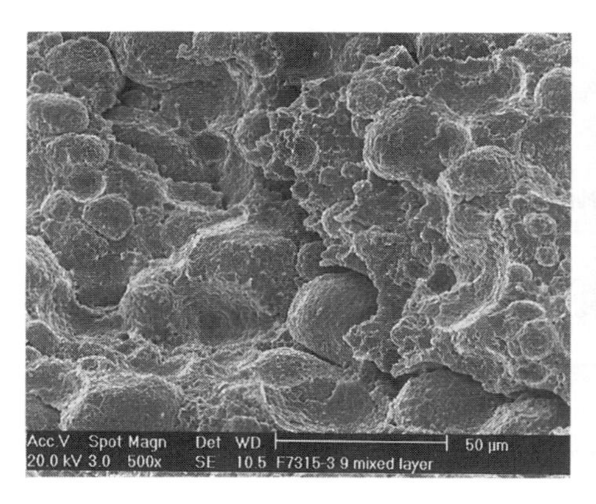

Figure 11c. Fracture surface of the mixed layer showing substantial decohesion of the NiAl phase.

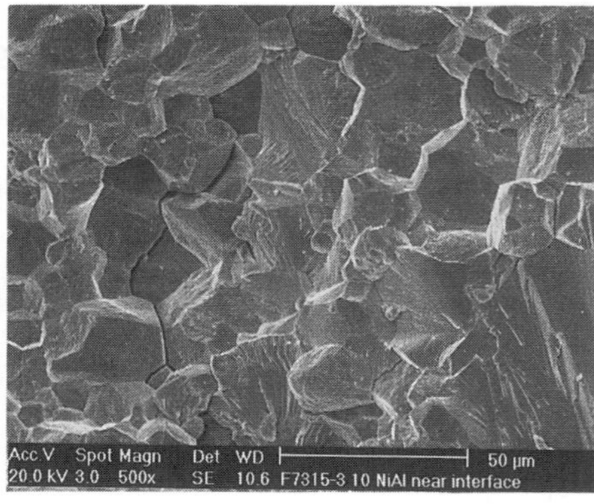

Figure 11d. Fracture surface of the NiAl layer showing a mixture of intergranular and transgranular fracture.

CONCLUSIONS

1) Controlled moisture addition during sedimentation of powders of different density is an effective means of avoiding segregation. Pathways for the production of different microstructures are also generated.
2) Slowly replacing hexane with a mixture of 60 volume percent hexane, 30 volume percent methanol and 10 volume percent acetone promoted relaxation and decreased problems associated with drying shrinkage. During drying this mixture exhibited the smallest density increase and less tube wall adherence.
3) These graded structures are sensitive to slight variations in intermetallic stoichiometry. These can lead to large variations in residual stress.
4) Zirconia - NiAl FGMs are better able to accommodate high levels of residual stress.
5) Tetragonal to monoclinic transformation occurs in the Zirconia-NiAl FGMs, starting from the surface. This results in spallation of the ceramic layer.
6) Variable microstructures result in detectable changes in the behavior of these alumina-NiAl FGMs.

ACKNOWLEDGMENTS

The authors gratefully acknowledge the support of the Office of Naval Research under contract# N00014-95-1-0336.

REFERENCES

(1) Chu, J.; Ishibashi, H.; Hayashi, K.; Takebe, H.; Morinaga, K. *Journal of the Ceramic Society of Japan* **1993**, *10*, 841-844.
(2) Takebe, H.; Morinaga, K. *Materials and Manufacturing Processes* **1994**, *9*, 721-733.
(3) Boch, P.; Chartier, T.; Huttepain, M. *Journal of the American Ceramic Society* **1986**, *69*, 191-192.
(4) Niino, M.; Maeda, S. *ISIJ International* **1990**, *30*, 699-703.
(5) Katsuki, H.; Ichinose, H.; Shiraishi, A.; Takagi, H.; Hirata, Y. *Journal of the Ceramic Society of Japan* **1993**, *101*, 1068-1070.
(6) Capurso, J. S.; Alles, A. B.; Schulze, W. A. *Journal of The American Ceramic Society* **1995**, *78*, 2476-2480.
(7) Hehn, L.; Zheng, C.; jr, J. J. M.; Hubbard, C. R. *J. Mats. Sci.* **1995**, 1277-1282.
(8) Ham-Su, R.; Wilkinson, D. S. *J. Am. Ceram. Soc.* **1995**, *78*, 1580-1584.
(9) Miller, D. P.; Lannutti, J. J.; Yancey, R. N. In *16th Annual Conference on Composites and Advanced Ceramic Materials*; American Ceramic Society, Westerville, OH.: Cocoa Beach, FL., 1992; pp 365-373.
(10) Miller, D. P.; Lannutti, J. J.; Soboyejo, W. O.; Noebe, R. D. In *International Symposium on Structural Intermetallics*; Minerals, Metals and Materials Society, Warrendale, PA.: Seven Springs, PA., 1993; pp 783-790.
(11) Miller, D. P.; Lannutti, J. J.; Noebe, R. D. *J. Mat. Res.* **1993**, *8*, 2004-2013.
(12) Lannutti, J. J. *Comp. Eng.* **1994**, *4*, 81-94.
(13) Koizumi, M.; Urabe, K. *Journal of the Iron and Steel Institute of Japan* **1989**, *75*, 887-893.
(14) Koizumi, M. In *16th Annual Conference on Composites and Advanced Ceramics*; American Ceramic Society, Westerville, OH: Cocoa Beach, Florida, 1992; pp 333-347.
(15) Atarashijya, K.; Kazuya, K.; Nagai, T.; Uda, M. In *16th Annual Conference on Composites and Advanced Ceramics*; American Ceramic Society, Westerville, OH: Cocoa Beach, Florida, 1992; pp 400-407.
(16) Drake, J. T.; Williamson, R. L.; Rabin, B. H. *J. Appl. Phys.* **1993**, *74*, 1321-1326.
(17) Williamson, R. L.; Rabin, B. H.; Drake, J. T. *J. Appl. Phys.* **1993**, *74*, 1310-1320.
(18) Bishop, A.; Lin, C. Y.; Navaratnam, M.; Rawlings, R. D.; McShane, H. B. *Journal Of Materials Science Letters* **1993**, 1516-1518.
(19) Zhu, J.; Yin, Z.; Lai, Z. *Journal of Materials Science Technology* **1994**, *10*, 188-192.

(20) Omori, M.; Kawahara, M.; Sakai, H.; Okubo, A.; Hirai, T. **1994**, 649-652.
(21) Wantanabe, R. *MRS Bulletin* **1995**, 32-34.
(22) Wantanabe, S.; Hayashi, N.; Kinoshita, Y.; Ohashi, A.; Uchida, Y.; Dykes, D.; Touchard, G. **1994**, 185-188.
(23) Jones, S. A.; Burlitch, J. M. *Materials Letters* **1994**, 233-235.
(24) Kudesia, R., Niedzialek S.E., Stangle G.C., McCauley J.W., Spriggs R.M., Kaieda Y. In *16th Annual Conference on Composites and Advanced Ceramics*; American Ceramic Society, Westerville, OH: Cocoa Beach, Florida, 1992; pp 374-383.
(25) Bhaduri, S. B.; Radhakrishnan, R. In *16th Annual Conference on Composites and Advanced Ceramics*; American Ceramic Society, Westerville, OH: Cocoa Beach, Florida, 1992; pp 392-399.
(26) Sata, N. In *16th Annual Conference on Composites and Advanced Ceramics*; American Ceramic Society, Westerville, OH: Cocoa Beach, Florida, 1992; pp 384-391.
(27) Niedzialek, S. E.; Stangle, G. C.; Kaieda, Y. *J. Mat. Res.* **1993**, *8*, 2026-2034.
(28) Feng, H. J.; Moore, J. J. *Journal of Materials Engineering and Performance* **1993**, *2*, 645-650.
(29) Atarashiya, K.; Uda, M. In *International Conference on Advanced Composite Materials*; 1993; pp 1351-1355.
(30) Kang, Y.-S.; Miyamoto, Y.; Muraoka, Y.; Yamaguchi, O. *Journal of Society of Materials Science Japan* **1995**, *44*, 705-709.
(31) Ma, X.; Tanihata, K.; Miyamoto, Y.; Kumakawa, A.; Nagata, S.; Yamada, T.; Hirano, T. In *16th Annual Conference on Composites and Advanced Ceramics*; American Ceramic Society, Westerville, OH: Cocoa Beach, Florida, 1992; pp 356-364.
(32) Hirano, K.; Maruyama, S.; Watanabe, O. **1993**, *59*, 194-199.
(33) Sasaki, M.; Hirai, T. *Journal of the European Ceramic Society* **1994**, *14*, 257-260.
(34) Araki, M.; Saski, M.; Kim, S.; Suzuki, S.; Nakamura, K.; Akiba, M. *Journal of Nuclear Materials* **1994**, 1329-1334.
(35) Kitaguchi, S.; Hamatani, H.; Saito, T.; Shimoda, N.; Ichiyama, Y. *Nippon Steel Technical Report* **1993**, 28-32.
(36) Nakashima, S.; Arikawa, H.; Chigasaki, M.; Kojima, Y. *Surface and Coatings Technology* **1994**, *66*, 330-333.
(37) Matsuzaki, Y.; Kawamura, M.; Fujioka, J.; Okazaki, S. *Journal of Japan Inst. Metals* **1994**, *58*, 697-706.
(38) Sampath, S.; Herman, H.; Shimoda, N.; Saito, T. *MRS Bulletin* **1995**, 27-31.
(39) Lannutti, J. J. *MRS Bulletin* **1995**, 50-51.
(40) Christenson, H. K. *Journal of Colloidal and Interface Science* **1985**, *104*, 234-249.
(41) Bloomquist, C. R.; Shutt, R. S. *Ind. Eng. Chem.* **1940**, *32*, 827-831.
(42) Israelachvili, J. N.; McGuiggan, P. M. *Science* **1988**, *241*, 795-800.
(43) Noebe, R. D.; Bowman, R. R.; Cullers, C. L.; Raj, S. V. *Materials Research Society Symposium* **1991**, *213*, 589-596.
(44) Noebe, R. D.; Bowman, R. R.; Nathel, M. V. "Review of the Physical and Mechanical Properties and Potential Applications of the B2 Compound NiAl, NASA TM 105598," NASA, 1992.
(45) Noebe, R. D.; Bowman, R. R.; Nathal, M. V. "Physical and Mechanical Metallurgy of NiAl," NASA Lewis Research Center, 1993.
(46) Erdogan, F. *Composites Engineering* **1994**, *5*, 753-770.
(47) Sato, T.; Shimada, M. *J. Am. Cer. Soc.* **1985**, *68*, 356-59.

Pressureless Co-Sintering of Al_2O_3/ZrO_2 Multilayers and Bilayers

Peter Z. Cai, David J. Green and Gary L. Messing
Department of Materials Science and Engineering, The Pennsylvania State University, University Park, PA 16802

Abstract

Various types of damage were observed in pressureless-sintered Al_2O_3/ZrO_2 symmetric laminates (multilayers) and asymmetric laminates (bilayers) fabricated by tape casting and lamination. These defects included channel defects in ZrO_2-containing layers, Al_2O_3 surface defects parallel to the layers, decohesion between the layers, and transverse damage within the Al_2O_3 layer in the bilayers. Detailed microscopic observation attributed the defects to a combined effect of mismatch in both sintering rate and thermal expansion coefficient between the layers. Crack-like defects were formed in the early stages of densification, and these defects acted as pre-existing flaws for thermal expansion mismatch cracks. Curling of the bilayers during sintering was monitored and the measured rate of curvature change, along with the layer viscosities obtained by cyclic loading dilatometry, was used to estimate the sintering mismatch stresses. The extent of damage could be reduced or even eliminated by decreasing the difference in layer sintering rate. This was accomplished by reducing the heating rates or by adding Al_2O_3 in the ZrO_2 layers.

Introduction

Cracks and crack-like defects are commonly observed in film/substrate systems and ceramic hybrid laminates during processing [1-3]. One source of such defects is the thermal expansion mismatch between the constituent layers. Residual stresses usually arise during the cooling stage, and for brittle systems, this can cause cracking. Mismatch stress can also be generated during the sintering process when the co-sintering layers have different densification kinetics. Although sintering mismatch and thermal expansion mismatch are two mechanisms operative at different stages of processing, both can contribute to defect formation in laminated structures [4]. In general, the exact origin for damage and the morphology of these defects in a laminate system depend on the laminate geometry, layer sintering characteristics, layer thermal expansion coefficients, and the fabrication process. In this study, hybrid ceramic laminates consisting of Al_2O_3 and ZrO_2 layers of comparable layer thicknesses were investigated. The origin of defect generation, as well as the evolution of these defects, was studied by careful observation of the defect morphology after sintering. Efforts were made to find means of eliminating processing damage by mixing one component into another, and by controlling both heating and cooling rates. The aim of this paper is to review the results of three recent papers [4-6] from which more detailed information can be found.

Experimental

An MgO-doped Al_2O_3 powder (Premalox, Alcoa, particle size≈0.3 μm) and a CeO_2-stabilized ZrO_2 powder (TZ-12Ce, Tosoh, particle size≈0.3 μm) were used in this study. The powders were mixed in different ratios with binders and modifiers, to form the five slip formulations used in this study: 100% Al_2O_3 (A100), 100% ZrO_2 (100Z), 90 wt% ZrO_2-10 wt%

Mat. Res. Soc. Symp. Proc. Vol. 434 © 1996 Materials Research Society

Al_2O_3 (90Z), 80 wt% ZrO_2-20 wt% Al_2O_3 (80Z) and 70 wt% ZrO_2-30 wt% Al_2O_3 (70Z). The slip was cast onto a glass substrate and the dried tapes had an average thickness of about 150 μm. After binder burnout, the A100, Z100, Z90, Z80 and Z70 green tapes had relative density of 57%, 46%, 49%, 50% and 52%, respectively. The stamped tapes were laminated to form two types of laminates: symmetric laminates (multilayers) and asymmetric laminates (bilayers). In the multilayers, the Al_2O_3 and ZrO_2 layers were stacked in an alternating sequence to a total of 23 layers; in the asymmetric laminates, three Al_2O_3 layers were on one side and three ZrO_2 layers were on the other, forming a laminate resembling a bilayer. In addition to the laminates made of A100 and Z100 tapes, multilayers and bilayers were fabricated with the Z100 layers replaced by Z90, Z80 and Z70 layers to form a total of four types of laminates: A100-Z100, A100-Z90, A100-Z80, and A100-Z70 [4]. Following lamination (90°C and 48 MPa for about 12 min) and binder burnout (450°C for 8 hours), the laminates were sintered at 1530°C for 90 minutes. Two heating rates were used: 5°C/min and 1°C/min. The sintered samples were either furnace cooled or cooled at a rate of 3°C/min. The difference in layer densification rates led to curling in the bilayers starting at about 1000°C with further densification. The curvature development was monitored by telephotography.

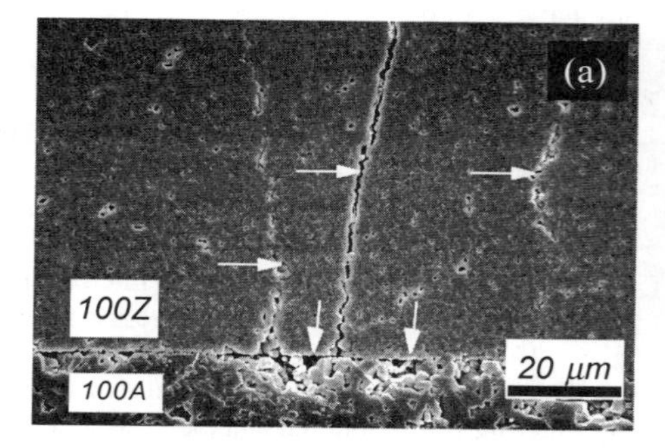

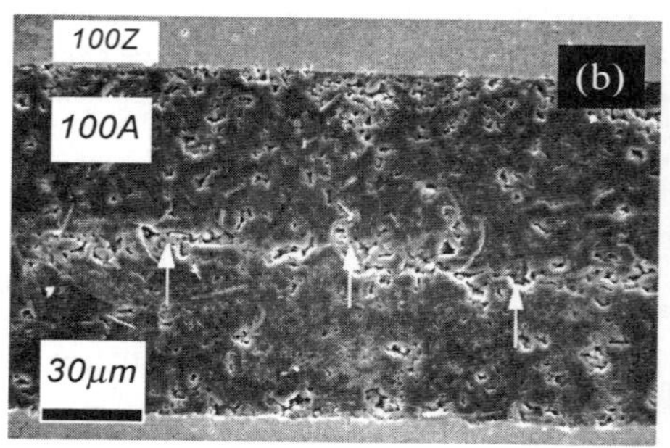

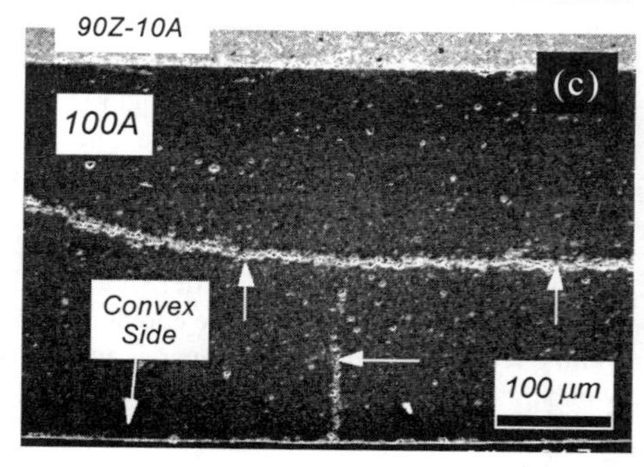

Fig. 1 Typical defects observed in ceramic multilayers and bilayers: (a) Channel and interfacial cracks in the ZrO_2*-containing layer and (b) surface defects in the* Al_2O_3 *layer of a multilayer; (c) intralayer and transverse defects in the* Al_2O_3 *layers of a bilayer*

Results and Discussion

Processing Defects

Three types of processing defects were observed in the multilayers. The first type was the channel crack in the ZrO_2-containing layers. Channel cracks, as depicted in Fig. 1(a), arise as a result of the in-plane tensile stress in the ZrO_2-containing layer due to its more rapid densification rate and a higher thermal expansion coefficient than the Al_2O_3. The second type, also shown in Fig. 1(a), was the debonding crack between the Al_2O_3 and ZrO_2-containing layers. These debonding cracks occasionally deviated from the interface into the Al_2O_3

layer and ran parallel to, but in close vicinity of, the interface. The third type of defect observed in the multilayers was the surface crack within the Al_2O_3 layers, as shown in Fig. 1(b). The intralayer decohesion is a result of the surface tensile stresses in the biaxially compressive Al_2O_3 layers due to edge effects [3]. In addition to the three defect types observed in the multilayers, the bilayers exhibited a unique type of damage, viz., the transverse defects within the Al_2O_3 layers (Fig. 1(c)). These defects can be attributed to the tensile bending stresses developed on the convex side of the bilayer during curling.

Most of the defects observed in these laminates had crack opening displacements much less than 1 μm. The small crack opening displacement seems to suggest that thermal shrinkage misfit during cooling was the origin of damage. However, closer examination of the damage by scanning electron microscopy revealed that defect formation and evolution is a much more complex process. The crack in the center of Fig. 1(a) clearly is brittle in nature, evidenced by the well-defined crack path and the uniformity of the crack opening displacement. The other two crack-like features, on the other hand, appear to consist of either a succession of pores or a series of low density areas that outline a damage path. Because the cavitational defects represent the weakest regions of the ZrO_2-containing layer after densification, they would be the most susceptible to the residual stresses that occur during the cooling stage when the ZrO_2-containing layer is in tension. In fact, there is evidence [4] that it was indeed these low density regions and cavitational crack-like defects that acted as pre-existing flaws for the generation of thermal expansion mismatch cracks.

Curling of the Bilayers During Sintering

For the ceramic sintering compacts used in this study, it has been demonstrated by cyclic loading dilatometry that their mechanical response is linear viscous for the majority of the sintering process, except for the early stages of heating when the compacts are still elastic [5]. Such viscous mismatch stresses in ZrO_2-containing layers can be calculated using the following relationships [4]

$$\sigma_1 = \frac{1}{1+mn}\eta_1 \Delta\dot{\varepsilon} \qquad (1)$$

and

$$\sigma_1 = \frac{m^4 + mn}{n^2 + 2mn(2m^2 + 3m + 2) + m^4}\eta_1 \Delta\dot{\varepsilon} \qquad (2)$$

for the multilayers and bilayers, respectively. In Eqs. (1) and (2), $\Delta\dot{\varepsilon} = d\varepsilon / dt$ is the strain rate mismatch, and m and n are the layer thickness ratio and the layer viscosity ratio, respectively. The above viscous relationships are deduced from the well known viscous-elastic analogy (VE analogy) [4]. In a curling linear viscous bilayer, the corresponding curvature equation can be derived as [4]:

$$\dot{k} = \frac{6(m+1)^2 mn}{\eta_1\left(m^4 n^2 + 2mn(2m^2 + 3m + 2) + 1\right)} \cdot \eta_1 \Delta\dot{\varepsilon} \qquad (3)$$

where $\dot{k} = dk / dt$ is the rate of curvature change. The above equations suggest that the viscous mismatch stresses can be calculated from the rate of curvature change if the viscous properties of both constituent layers are known. In this study, the A100, Z100, Z90, Z80 and Z70 uniaxial viscosities were measured with cyclic loading dilatometry [5]. The layer thickness ratio was assumed to be constant.

The bilayer curvature and the calculated ZrO_2-containing layer stress in the multilayers are shown in Figs. 2(a) and 2(b), respectively, for the A100-Z100, A100-Z90, A100-Z80 and A100-Z70 laminates. It can be seen that, with increasing amounts of Al_2O_3 in the ZrO_2 layer, the observed bilayer curvature and the calculated layer stress in the multilayers both decreased. Integration of Eq. (3) with respect to time shows that the bilayer curvature reflects the magnitude of mismatch in sintering strain. Therefore, the decrease in curvature with addition of Al_2O_3 in ZrO_2 suggests a decrease in sintering mismatch between the layers. This is consistent with the shrinkage behavior of the monolithic monoliths. The surface tensile stresses in the Al_2O_3 layers responsible for the surface defects in Fig. 1 can be shown to have the same order of magnitude as the biaxial stress in the ZrO_2-containing layers. [6]

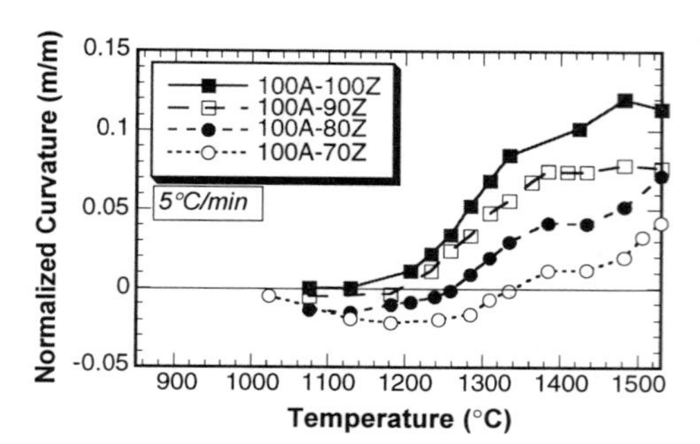

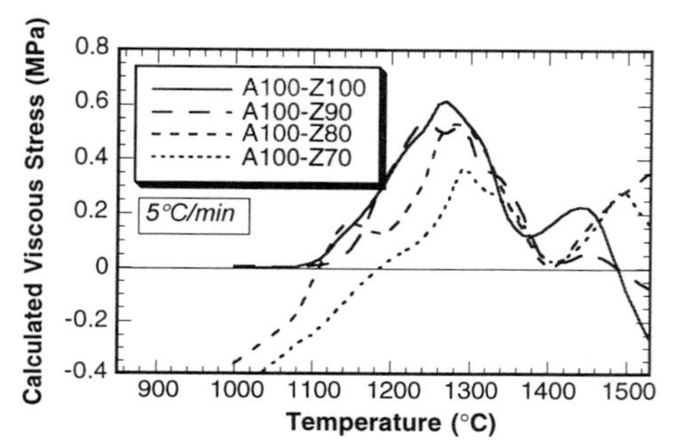

Fig. 2 (a) The curvature development in the bilayers, and (b) the calculated layer mismatch stresses for the multilayers during sintering.

Residual Stresses During Cooling

The biaxial tensile stresses in the ZrO_2-containing layer from the thermal expansion mismatch were calculated from the thermal expansion coefficients measured for all the Al_2O_3 and ZrO_2 materials [4]. Viscoelastic formulations [6] were used to study the effect of cooling rate on possible stress relaxation. Fig. 3 shows the residual stress development during cooling for the A100-Z100 multilayers under both furnace cooling and 3°C/min cooling conditions. As seen in Fig. 3, the residual stress is significantly smaller for the 3°C/min cooling, and the difference results from the initial cooling period (above 1200°C), during which no appreciable residual stress develops for the slower cooling rate. The absence of residual stress development for the slow cooling rate at temperatures higher than 1200°C demonstrates that the elastic mismatch stress can be almost entirely relaxed in this temperature range, provided that the cooling rate is sufficiently slow. Therefore, slower cooling rates would result in smaller residual stresses after cooling, and thereby reduce the amount of cracking in the cooled samples.

Since defect generation in the laminates is attributed to the mismatch in sintering rate and thermal expansion, reduction of the heating and cooling rates can be used to decrease the extent of damage. Another means of reducing the mismatch in sintering rate would be to mix Al_2O_3 in the ZrO_2 layers. In doing so, the more rapid sintering rate of the ZrO_2 layer would be decreased, and the thermal expansion coefficients of the two layers would be more similar. The overall effect of Al_2O_3 addition in ZrO_2 layers and the rate of heating and cooling on cracking of the multilayers can be appreciated in Fig. 4, in which the number of channel cracks in the ZrO_2-containing layers per unit thickness is plotted for A100-Z100, A100-Z90, A100-Z80 and A100-Z70 multilayers under the two cooling conditions and the two heating conditions. It is clear that both the mixing of Al_2O_3 in ZrO_2 layers, as well as slower heating and cooling rates were effective in decreasing the amount of damage in the laminates.

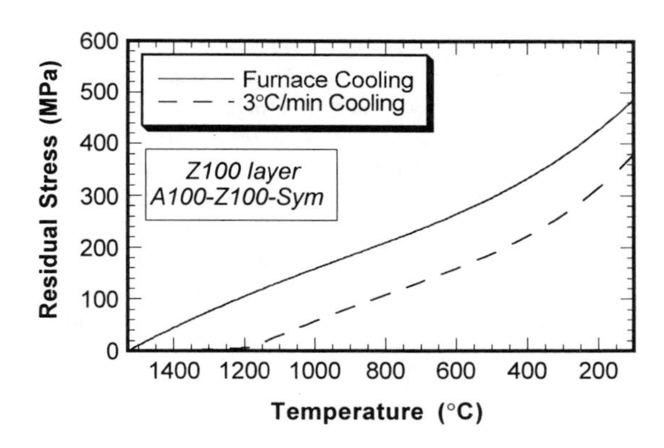

Fig. 3 Calculated viscoelastic residual stresses in Z100 layers for the A100-Z100 multilayers during cooling under furnace cooling, and at a 3°C/min cooling rate.

Summary

Viscous stress analysis showed that the biaxial tensile stress in the ZrO_2-containing layer (and the surface tensile stress in Al_2O_3 layer) were on the order of 1 MPa during densification. The small tensile stresses, however, must be sufficient in retarding densification and creating cavitational damage in the direction perpendicular to the biaxial stresses. These defects can either further develop during sintering or act as pre-existing sites for thermal expansion mismatch cracks. This is supported by detailed microscopic observation of the sintered laminates. Due to the dual origin of the processing defects, both the sintering process and the cooling process need to be carefully controlled in order to achieve defect-free laminates. Indeed, slower heating and cooling rates led to a decrease in the extent of damage. Blending Al_2O_3 in ZrO_2 layers was proven to be another effective method in reducing the mismatch in layer sintering rates. It was also found that the magnitude of the residual stress was dependent on the cooling rate, and further, that smaller cooling rates were capable of completely relaxing the residual stress at temperatures above 1200°C.

Acknowledgments

Financial support from the Center for Advanced Materials at the Pennsylvania State University is gratefully acknowledged.

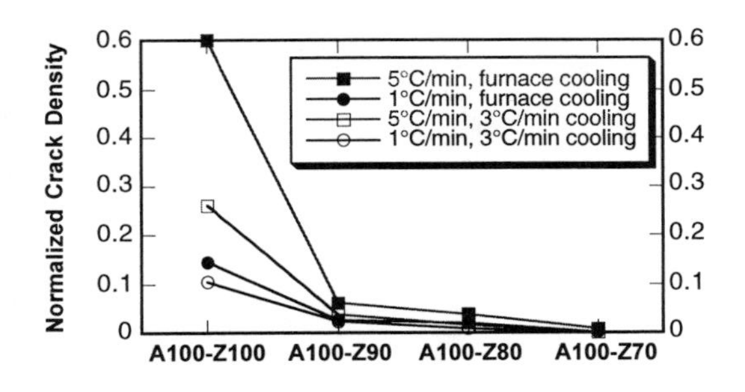

Fig. 4 Crack density, or the number of cracks per unit length, of the channel cracks on a cross-section of A100-Z100, A100-Z90, A100-Z80 and A100-Z70 multilayers, sintered under different combinations of heating and cooling rates. The unit length is normalized by the layer thickness.

References

[1] R. K. Bordia and R. Raj, *J. Am. Ceram. Soc.*, **68** [6] 287-92 (1985)

[2] R. K. Bordia and A. Jagota, *J. Am. Ceram. Soc.*, **76** [10] 2475-85 (1993)

[3] C. Hillman, Z. Suo, and F. F. Lange, "Cracking of Laminates Subjected to Biaxial Tensile Stresses", to be published in *J. Am. Ceram. Soc.*

[4] P. Z. Cai, D. J. Green, and G. L. Messing, "Constrained Densification of Al_2O_3/ZrO_2 Hybrid Laminates: I. Experimental Observations", manuscript in preparation.

[5] P. Z. Cai, G. L. Messing, and D. J. Green, "Determination of the Mechanical Response of Sintering Compacts by Cyclic Loading Dilatometry", submitted to *J. Am. Ceram. Soc.*

[6] P. Z. Cai, D. J. Green, and G. L. Messing, "Constrained Densification of Al_2O_3/ZrO_2 Hybrid Laminates: II. Viscoelastic Stress Calculations", manuscript in preparation.

Part III

Stability Issues

MECHANICAL PROPERTIES OF METAL-INTERMETALLIC MICROLAMINATE COMPOSITES

J. HEATHCOTE,* G. R. ODETTE,** G. E. LUCAS,** R. G. ROWE***
*Materials Department, University of California, Santa Barbara, CA 93106, johnh@engineering.ucsb.edu
**Department of Mechanical Engineering, University of California, Santa Barbara, CA 93106
***GE CRD, 1 River Road, K-1, MB265, Schenectady, NY 12301

ABSTRACT

Tensile strengths, static and dynamic fracture toughness, and fatigue crack propagation were measured for different combinations of Nb metal-intermetallic microlaminate composites. Metal layer bridging produced toughening by factors of 2 to 5 under static conditions. Dynamic testing reduced the toughness significantly. Fatigue crack propagation rates were comparable to data for pure Nb. A key composite property, the stress -displacement function $\sigma(u)$ of the constrained metal layers, was evaluated by several techniques and used in a bridging-crack stability analysis to predict tensile strengths in agreement with experimental values. The results provide guidelines for improving microlaminate performance.

INTRODUCTION

A microlaminate comprised of alternating microns-thick layers of intermetallic and ductile metal is a composite architecture that may be attractive for multi-layer coating applications in advanced airfoils. Intermetallic-based systems have the potential to exhibit high temperature properties superior to superalloys for these applications.[1] Compositing intermetallics with ductile reinforcements can produce toughening by several mechanisms, thus overcoming the intrinsic low temperature brittleness of the intermetallic [2-6]. A major advantage of the microlaminate architecture is the availability of attractive manufacturing routes based on physical vapor deposition techniques. In addition, the small size scales can lead to increases in the ductile layer strength, and smaller length scales may also increase the matrix cracking stresses by limiting the size of potential processing flaw sizes.

Hence, work has been in progress to assess the mechanical properties of several microlaminate composites based on the Nb_3Al/Nb and Cr_2Nb/Nb systems. Properties of interest in these systems include room temperature strength and fracture resistance which turn out to be strongly dependent on the stress-displacement behavior, $\sigma(u)$, of the constrained metal layers. In this work, the tensile strengths, static and dynamic fracture toughness, and fatigue crack propagation were measured for a set of microlaminate composites, and $\sigma(u)$ functions were evaluated by a combination of experimental and analytical techniques. The $\sigma(u)$ functions can then be used to predict resistance curve behavior in other geometries or the fracture strength in uncracked tensile specimens. This provides a fundamental framework for improving the mechanical behavior of the microlaminate system.

Mat. Res. Soc. Symp. Proc. Vol. 434

EXPERIMENT

Materials

Microlaminate composites about 132 μm thick were fabricated by Magnetron® sputter deposition of alternating layers of metal and intermetallic precursor onto steel substrates. Four microlaminate composites have been examined to date and are described in Table 1. Following deposition the microlaminates were heat treated for 2 hours at 1000°C for Nb_3Al and $(Nb,Ti)_3Al$ and 1200°C for the Cr_2Nb systems, which transformed the precursor layers into well ordered intermetallic structures [7,8]. In the Nb_3Al and $(Nb,Ti)_3Al$ composites, a bcc metal precursor phase transformed to the Nb_3Al (A15) structure; in the Cr_2Nb composites, a nearly amorphous, nanocrystalline metal phase transformed to the Cr_2Nb (Laves) structure. Further details of the processing and microstructure of the resulting microlaminates are described elsewhere [7].

Table 1
Description of Microlaminate Composites

ID	Intermetallic	Metal Composition (atom %)	Layer Thickness (μm)
L9	Nb_3Al	Nb	2
L20	$(Nb,Ti)_3Al$	Nb - 27.7 Ti - 6.7 Al	2
L17	Cr_2Nb	Nb - 4.7 Cr	2
L60	Cr_2Nb	Nb - 3.3 Cr	6

Mechanical Tests and Fractography

All mechanical properties were evaluated at room temperature. Fracture strength was measured on flat tensile specimens with gauge section 6.4 mm wide by 10 mm long at displacement rate of 0.05 μm/s. In all cases, elastic loading was terminated by fast fracture.

Static resistance curves (toughness, K_r, versus crack extension, Δa) were measured in three-point loading tests on single edge notched bend (SENB) specimens and in tension tests on center-cracked tension (CCT) specimens. In all cases, crack propagation was parallel to the metal-intermetallic interface. SENB specimens were electro-discharge machined (EDM) to dimensions 8 mm x 25.4 mm, with a small starter notch on one side of the long dimension. The samples were glued into a small aluminum bend frame that allowed loading and prevented buckling of the thin sheet. The composite specimens were fatigue pre-cracked to a crack length (a) to specimen width (W) ratio of about a/W ~ 0.3. CCT specimens were fabricated by EDM to dimensions 51mm x 12.7mm with center notches of length 1.6mm on either side of a 1.6mm diameter hole at the center of the gauge section. Sharp cracks were grown beyond the ends of the notches by fatigue pre-cracking to a/W ratios of about 0.4. In both cases the resistance curve tests were carried out at a displacement rate of 0.005μm/s, and crack growth was monitored by optical microscopes mounted on each side of the specimen. At each sign of crack growth, the load was decreased by 10%, and the crack length was remeasured. The tests self terminated when crack propagation became unstable. ABAQUS finite element calculations were used to evaluate $K_I(a/w)$ for the specific test geometry. Details are described elsewhere [9].

Dynamic fracture toughness was also measured on SENB specimens at a displacement rate of 25.4μm/s. Only the load-displacement data were recorded for this test. To date dynamic toughness has only been assessed for $Cr_2Nb/Nb(Cr)/6\mu m$ microlaminate L60.

Fatigue crack propagation was also examined in the L60 microlaminate on an SENB specimen with an initial a/W of 0.36 at a frequency of 10 Hz and a stress ratio, R, of 0.2. Crack length was monitored by optical microscopes on either side of the specimen, and the load-crack length data were used to determine crack growth rate da/dN as a function of stress intensity range ΔK.

The in-situ strengths of the metal layers were estimated by microhardness measurements. Details of these measurements are described elsewhere [9].

Fracture surfaces of selected specimens were examined by both optical and high resolution scanning electron microscopy (SEM). In addition, confocal microscopy was used to obtain quantitative three-dimensional images of conjugate fracture surfaces. These were combined with a fracture reconstruction technique originally proposed by Kobayashi et. al. [10,11] and recently extended by Edsinger and co-workers[12] to determine the deformation patterns in the metal layers and to establish the distribution of crack opening displacements at which the ductile metal layers fractured (u*). Details of this analysis are described elsewhere [9,12].

RESULTS

Fracture Strengths

Measured values of fracture strength determined from tensile tests (from two testing lots) are shown in Table 2. As will be shown below, the lot-to-lot variation for the L17 microlaminate can be attributed to differences in the size of the critical growth defect.

Table 2
Tensile Test Results

ID	System	Fracture Stress (MPa)	
		Lot 1	Lot 2
L9	Nb_3Al/Nb	473-474	451-501
L20	$(Nb,Ti)_3Al/Nb(Ti,Al)$	580-664	
L17	$Cr_2Nb/Nb(Cr)/2\mu m$	576-616	725-750
L60	$Cr_2Nb/Nb(Cr)/6\mu m$		441-500

Resistance Curves

The static resistance curves for the three SENB and CCT tests are plotted in Figure 1. The curves are offset from $\Delta a=0$ by the length of a pre-existing bridge zone present after fatigue pre-cracking. Estimates of initiation toughness are about 4 $MPa\sqrt{m}$. The region of stable crack growth is generally smaller in the CCT tests. All of the 2 μm thick layer microlaminates showed increasing toughness over the first 50-100μm of crack growth, leveling off to steady state levels: about 10-11 $MPa\sqrt{m}$ for Nb_3Al/Nb; about 12-14 $MPa\sqrt{m}$ for $(Nb,Ti)_3Al/Nb(Ti,Al)$ and $Cr_2Nb/Nb(Cr)$. The resistance curve for the thicker 6μm layer $Cr_2Nb/Nb(Cr)$ microlaminate rose to the highest toughness, reaching 20 $MPa\sqrt{m}$ after about 200μm of crack growth; however, the

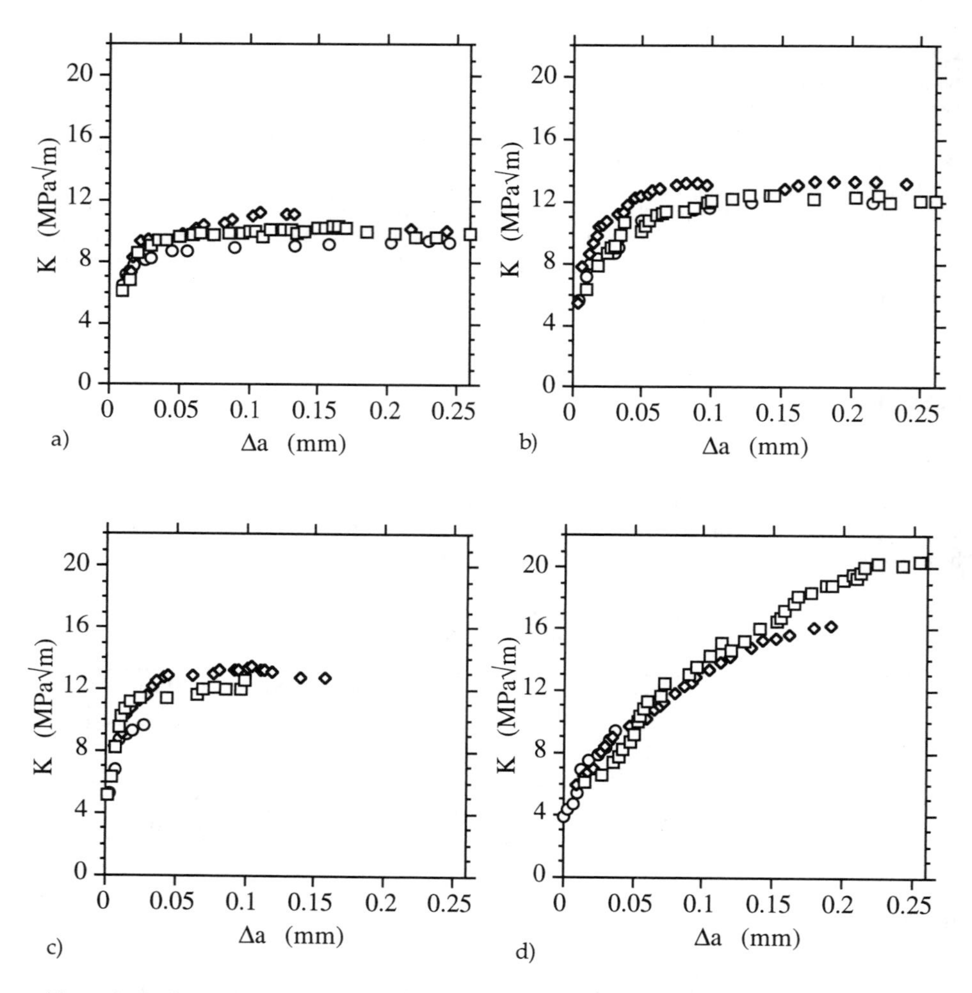

Figure 1. Resistance curves (K_r vs. Δa) for: a) L9 -- Nb_3Al/Nb; b) L20 -- $(Nb,Ti)_3Al/Nb(Ti,Al)$; c) L17 -- $Cr_2Nb/Nb(Cr)/2\mu m$; and d) L60 -- $Cr_2Nb/Nb(Cr)/6\mu m$. The squares and diamonds are for SENB and the circles for CCT tests.

transient slopes (dK/da) were significantly lower compared to the 2μm layer microlaminates, and it exhibited a relatively low fracture strength (Table 2).

Contrary to this resistance curve behavior, fracture of the Cr_2Nb/Nb(Cr)/6μm (L60) microlaminate under dynamic conditions appeared to be linear elastic, with unstable crack propagation occurring at a toughness of about 7-8 MPa$\sqrt{m}$.

Fatigue Crack Propagation

The fatigue crack propagation data for L60, Cr_2Nb/Nb(Cr), are shown as da/dN versus ΔK in Figure 2, along with similar data from Murugesh et al [13] for Nb_3Al, Nb, and a Nb_3Al/Nb *in situ* composite containing about 40% Nb equiaxed particulate. All of these data approximately exhibit a Paris law relationship of the form

$$da/dN = A (\Delta K)^m \quad (1)$$

with m in the approximate range 10 to 30. The microlaminate shows greater resistance to crack propagation than the intermetallic and *in situ* composite data and only slightly less resistance than pure Nb.

Fractography

The fracture reconstructions based on confocal microscopy showed that crack growth in the static resistance curve tests occurred by crack tunneling in the intermetallic layers, followed by stretching to failure of the intervening metal layers. The cumulative distributions of the displacement at failure of the metal layers, u*, normalized by layer thickness, t, for each of the microlaminates are shown in Figure 3. The Nb and Nb(Ti,Al) layers had significantly larger u*/t, compared to the Nb(Cr) layers. Preliminary fracture reconstructions for the dynamic fracture toughness test on the Cr_2Nb/Nb(Cr) microlaminate (L60) showed that, unlike the static resistace curve tests, the onset of cleavage and unstable crack propagation was preceded by little or no plastic deformation of the metal layers.

Representative SEM micrographs of fracture surfaces are shown in Figures 4 and 5. A fractograph for Nb_3Al/Nb (L9) static SENB specimen is shown in Figure 4a. The Nb_3Al layers fail by brittle fracture, and the Nb layers fail primarily by chisel point failure near the middle of the specimen. Fractographs representative of the Cr_2Nb/Nb(Cr) microlaminates (L17, L60) are shown in Figure 4b and c. Ductile failure occurred by the growth and coalescence of microvoids in both the 2 and 6μm metal layers as shown in Figures 4b. The reduced ductility (i.e., u*) in the Nb(Cr) layers is associated with this microvoid damage. However, there was a distinct transition from ductile microvoid coalescence to brittle cleavage fracture associated with a corresponding transition between stable (slow) and unstable (fast) fracture as shown in Figure 4c. The fracture surface of the Cr_2Nb/Nb(Cr) microlaminate (L60) tested dynamically showed only cleavage fracture. The fracture surfaces of the Nb(Al,Ti) layers in microlaminate L20 showed a combination of chisel point failure (predominant) along with regions of lower ductility associated with large voids. In all cases, the metal layers appeared to be strongly bonded to the matrix; however, intermetallic matrix cracking parallel to the layers was occasionally observed in Nb_3Al/Nb (L9), more frequently in Cr_2Nb/Nb(Cr) (L17,L60), and most frequently in $(Nb,Ti)_3Al$/Nb(Ti,Al) (L20). Effective debonding by intermetallic cracking appears to be the reason for the large u*/t in the Nb(Ti,Al) layers.

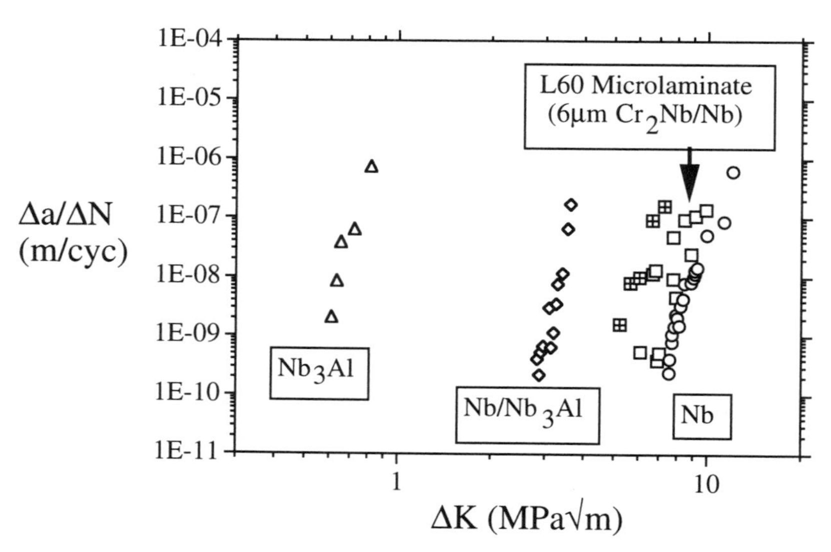

Figure 2. Fatigue crack propagation rates as a function of stress intensity range for L60 -- $Cr_2Nb/Nb(Cr)/6\mu m$. Data from Murugesh et al [13] are shown for comparison.

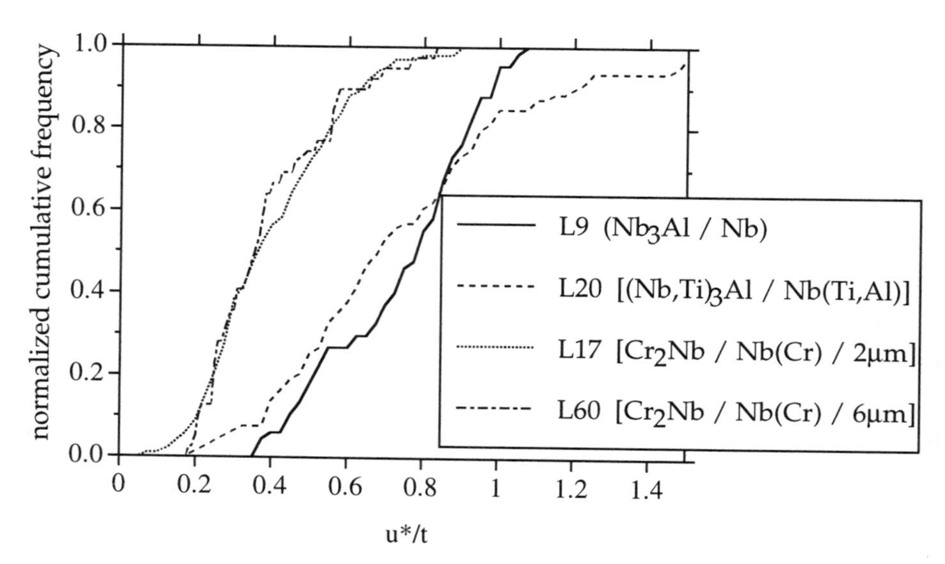

Figure 3. Cumulative frequency distributions of u* measurements from fracture reconstructions for the four microlaminates.

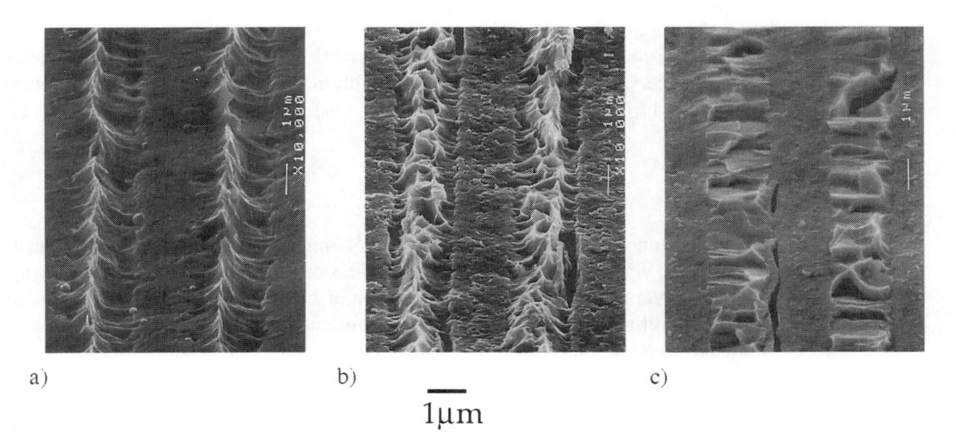

Figure 4. SEM fractographs of a) L9 -- Nb_3Al/Nb, b) L17 -- Cr_2Nb/Nb(Cr) for stable and c) unstable crack propagation.

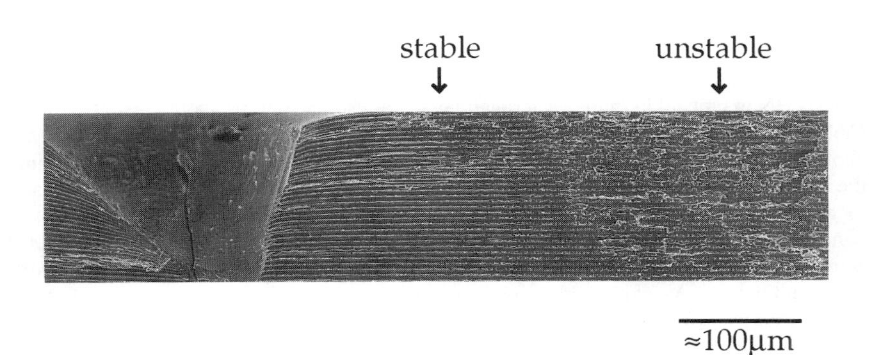

Figure 5. SEM micrograph of tensile specimen fracture surface in L17 -- Cr_2Nb/Nb(Cr) showing the conical growth defect as the crack initiation site and the transition from stable, sub-critical crack growth by ductile failure of the Nb(Cr) layers to unstable crack propagation with attendant cleavage fracture in Nb(Cr) layers.

Fractography examinations of the tensile samples revealed that conical-shaped growth defects acted as the sites for crack initiation. The defects penetrated approximately the entire sample thickness of the 2 μm layer cases, but were part-through in the 6 μm layer composites. The growth defect half widths ranged from about 50 to 105 μm. The sequence of events leading to fracture was most clearly revealed in the Nb(Cr) (L17 and L60) microlaminates as shown in Figure 5. In this case, the initial cracks grew stably out of the growth defects marked by metal layer ductile microvoid coalescence. At a critical size associated with macroscopic fracture at the fracture stress, the crack transitioned to unstable growth. Cleavage fracture facets are evident in the fast fracture region.

DISCUSSION

It appears that the resistance curve behavior in the microlaminate composites is largely due to bridging of the metal layers, which in turn is a function of the stress-displacement behavior, $\sigma(u)$, of the constrained metal layers. While $\sigma(u)$ functions cannot be directly measured, they were estimated for static test conditions by two methods: 1) by construction of $\sigma(u)$ based on estimates of both the maximum stress (σ^*) and crack opening at ligament rupture (u^*) from microhardness and fracture reconstruction measurements, respectively; and 2) by self-consistent fits to the resistance curves using a large scale bridging code .

To construct $\sigma(u)$ functions from microhardness and fracture reconstruction measurements, it was assumed that $\sigma(u)$ exhibited a modified sawtooth shape, with a peak strength followed by a decreasing tail to a final value of u^*. Assuming a maximum constraint factor on the metal layer of 3, which is consistent with theory and experiment for strongly bonded ductile reinforcements [3,14,15], peak strengths σ_p were estimated from microhardness VHN by

$$\sigma_p = 3 * UTS \approx 3 * (3 * VHN) \qquad (2)$$

where the ultimate tensile strength, UTS, in units of MPa was taken to be approximately three times VHN in units of kg/mm^2. Microhardness was estimated to be about $260 \pm 15 kg/mm^2$ for the $Cr_2Nb/Nb(Cr)$ microlaminates (L17, and L60) and the $Nb_3Al/Nb(Ti,Al)$ microlaminate (L20) and about $170 \pm 10 kg/mm^2$ for the Nb3Al/Nb (L9). The extension at peak stress, u_p, which influences the initial slope of the resistance curve, was taken as 5% of the layer thickness, a value that resulted in good fits to the data. Thermal expansion mismatch between the metal and intermetallic components results in residual tensile stresses σ_r in the metal layers, and a value of 500 MPa at zero displacement was estimated for σ_r from curvature measurements on asymmetric, several layer microlaminates. As explained in more detail elsewhere, the tail of the $\sigma(u)$ function was constructed by averaging nine linear sawtooth functions with u^* equal to every 10th percentile intercept in the cumulative u^* distribution (i.e., Figure 3). The resulting $\sigma(u)$ functions are shown in Figure 6 as solid lines.

To fit $\sigma(u)$ functions from the resistance curves, a large scale bridging model developed by Odette and Chao [2] was used. This model computes a resistance curve corresponding to a given test geometry, the $\sigma(u)$ function, the plane strain elastic modulus, E' ($= E/(1 - \nu^2)$ where ν is Poisson ratio and E is Young's modulus) and the effective intermetallic toughness, K_m, by finding a self consistent solution for the crack opening profile and crack face stress distribution for every increment of crack advance. The $\sigma(u)$ function was again modeled as a modified sawtooth function; i. e.,

$$\sigma(u) = \sigma_r + (\sigma_p - \sigma_r)(u/u_p) \qquad u \leq u_p$$

$$\sigma(u) = \frac{\sigma^*}{2}\left[1 - \frac{u - u_p}{u^* - u_p}\right]^n + \frac{\sigma^*}{2}\left[1 - \left(\frac{u - u_p}{u^* - u_p}\right)^{1/n}\right] \qquad u_p < u \leq u^* \tag{3}$$

$$\sigma(u) = 0 \qquad u > u^*$$

A best fit was sought that closely reproduced the initial slope (sensitive to σ^*, σ_r, and u_p), the length of crack extension (sensitive to u*), and the resistance curve shape (sensitive to the post maximum $\sigma(u)$ shape) in the transition regime between initiation and steady state. Figure 6a-d shows the results of this fitting procedure; the ranges of $\sigma(u)$ functions are shown as shaded regions, and the averages as dashed lines. In general, the ranges are fairly narrow, yielding similar values of u_p, u*, and, to a lesser degree, σ^*. Overall, the fit curves show qualitative agreement with curves constructed from microhardness and confocal microscopy measurements (solid lines).

These $\sigma(u)$ functions can be combined with the bridging model to predict resistance curve behavior in other specimen or component geometries. They can also be combined with knowledge of the intrinsic flaw size of the material to predict the fracture stress in geometries of interest. Fracture occurs when

$$K_a(a) > K_r(a) \qquad \text{and} \qquad dK_a/da > dK_r/da, \tag{2}$$

where K_a is the applied stress intensity factor, K_r is the material fracture resistance, and a is the crack length [2,15]. To apply this, tensile specimens were modeled as center cracked or single edge notched panels with the initial crack equal in size to the fracture-initiating conical growth defect. The fit $\sigma(u)$ functions for a given microlaminate were combined with the bridging model and the initial defect size for a given tensile test to predict the resistance curve behavior for that geometry, and the applied K_a versus a was determined from handbook elasticity expressions for the same geometry [16]. Hence, the fracture criteria expressed in equation (4) is equivalent to the point at which the resistance curve $K_r(\Delta a)$ is tangent to the applied K curve $K_a(a)$. The experimental and predicted failure stresses for the samples that have been analyzed to date are shown Table 3. The experimental values fall very close to the predictions for all of the microlaminates that have been analyzed. The lot-to-lot difference in fracture stresses for L17 samples can be explained by the difference in intrinsic flaw sizes in the two sets of specimens. In all cases, fracture occurs before the steady state toughness is achieved. For this reason, the rising portion of the resistance curve mediates the fracture strength of the material. This is particularly interesting for the 6μm microlaminate, L60, which reached the highest toughness of all of the composites but had a lower slope in the transient region and a lower fracture stress. The higher ultimate fracture resistance is offset by the lower slope in the transient region.

Although it has not yet been fully analyzed, the linear-elastic response and loss of fracture toughness under dynamic conditions is due to the loss of ductility of the constrained metal layers at high strain rates. The slightly higher value of initiation toughness may be due to the strain rate sensitivity of the strength of the metal layer. Clearly this type of behavior under dynamic loading conditions has significant implications to design and microlaminate development.

As noted earlier, the resistance to fatigue crack propagation was higher for the microlaminate than a Nb_3Al/Nb *in situ* composite with similar volume fraction (40%) of metal phase. In the case of the *in situ* composite, however, the metal phase was particulate and the fatigue crack tended to avoid the metal particles [13]; whereas, in the microlaminate tested here the fatigue crack was forced to cut through all the metal layers. Moreover, the values of ΔK were well above the matrix toughness K_m, so that the fatigue crack should propagate by crack tunneling in the matrix followed

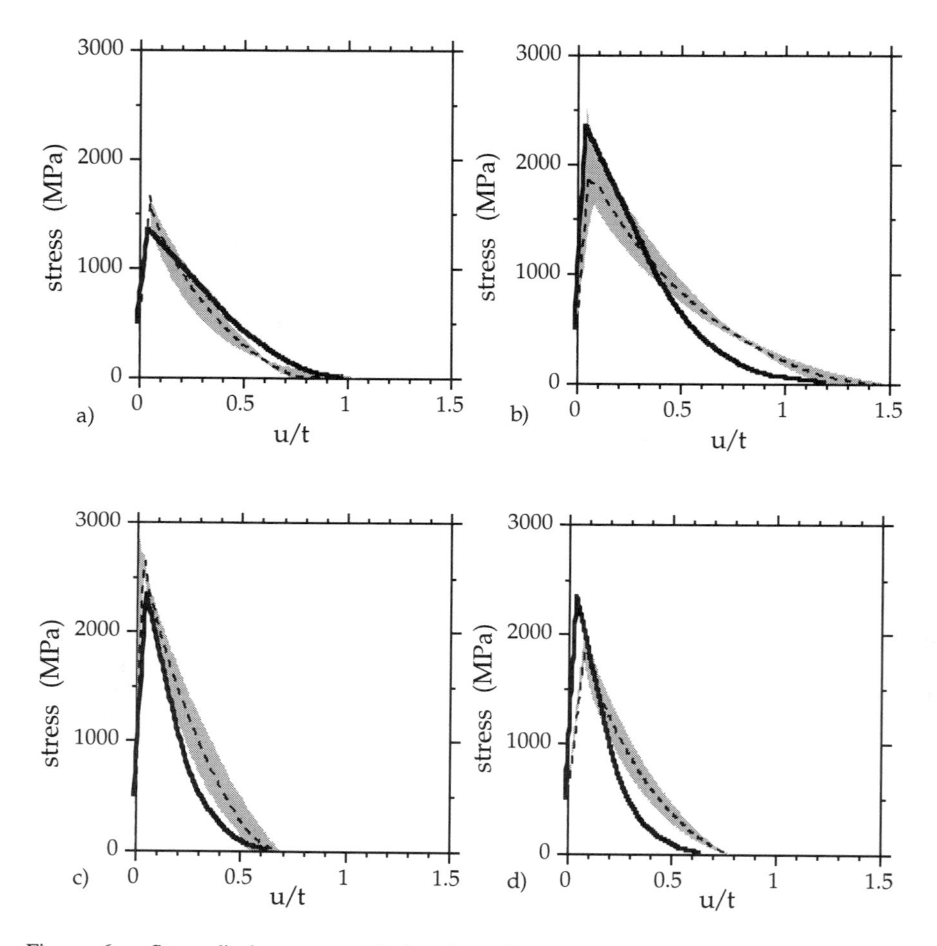

Figure 6. Stress-displacement σ(u) functions for: a) L9 -- Nb_3Al/Nb; b) L20 -- $(Nb,Ti)_3Al/Nb(Ti,Al)$; c) L17 -- $Cr_2Nb/Nb(Cr)/2\mu m$; and d) L60 -- $Cr_2Nb/Nb(Cr)/6\mu m$. The ranges obtained with the fit procedure are shown as shaded regions and the averages as dashed lines. Values obtained from the constructions from microhardness and confocal microscopy are shown as solid lines.

by fatigue of the intervening ligaments. Hence, the fatigue crack propagation is largely controlled by the metal layers, and hence environmental effects (e.g., cyclic cleavage, interstitial impurity pickup, oxidation) may be particularly important considerations for fatigue resistance.

Table 3.
Comparison of predicted and actual fracture stresses

ID -- Lot #	System	Flaw Size, a (μm)	Predicted Fracture Stress (MPa)	Measured Fracture Stress (MPa)
L9 -- Lot 1	Nb_3Al/Nb	58	445-505	473-474
L20 -- Lot 1	$(Nb,Ti)_3Al$/Nb-Ti-Al	70	550-650	580-664
L17 -- Lot 1	Cr_2Nb/Nb(Cr)/2μm	90	520-600	576-616
L17 -- Lot 2	Cr_2Nb/Nb(Cr)/2μm	40	705-820	725-750

CONCLUSIONS

The fracture properties of microlaminate composites based on alternating layers of pure Nb or Nb-alloy and either Nb_3Al or Cr_2Nb have been evaluated. All of the microlaminates with 2μm-thick layers showed a resistance curve with an initiation toughness estimated at 4MPa$\sqrt{m}$ rising to a saturation value of about 8 to 14 MPa$\sqrt{m}$ by 50-100 μm of crack growth. The one microlaminate with 6μm thick layers exhibited a lower slope in the transient regime of the resistance curve but rose to a K_I of about 20MPa$\sqrt{m}$ by about 200 μm of crack growth. For the four microlaminates tested, fracture strengths ranged from a low of about 450 MPa to a high of about 750MPa. While resistance curve behavior was exhibited under static conditions, the tests to date under dynamic conditions show a linear elastic fracture response. Fatigue crack propagation behavior, on the other hand was more similar to Nb metal than to either intermetallic or a comparable volume fraction metal-particle reinforced intermetallic.

A combination of scanning electron microscopy and confocal microscopy/fracture reconstruction showed that fracture in the resistance curve tests proceeded by crack tunneling in the intermetallic layers leaving intact ductile metal ligaments in the crack wake. These ductile ligaments provided tractions on the crack surfaces that led to the observed resistance curve behavior. While Nb and Nb(Ti,Al) failed largely by chisel point fracture, the Nb(Cr) layers failed by void growth and coalescence in the stable crack growth regime and cleavage in the unstable crack growth regime. Under dynamic test conditions, the onset of cleavage fracture occurred much sooner. Fractography also indicated that uncracked tensile specimens failed by the subcritical growth of cracks out of conical growth defects followed by unstable crack propagation.

Constrained metal layer stress-displacement, $\sigma(u)$, functions were constructed from both microhardness tests and on u* distributions and from a bridging model analysis of resistance curves. Both sets of functions were in reasonable agreement. The bridging model was then used with the $\sigma(u)$ functions to evaluate the fracture stress. The resulting predictions were in good agreement with experimental measurements of fracture stress.

The self consistent analysis has shown the relation between the mechanical behavior of these microlaminate composites and their fundamental properties. The results show that fracture strength is mediated by both the resistance curve behavior of the material -- as dictated by the $\sigma(u)$ function of the metal layers -- and the intrinsic, conical growth defect size. The fracture strength has been shown to be strongly dependent on the transient portion of the resistance curve, and hence, strongly dependent on K_m and σ^*. Moreover, as the defect size is reduced, fracture strength becomes increasingly influenced by the magnitude of dK_r/da in this regime. These results suggest that improved processing conditions to reduce the size and density of growth defects and appropriate tailoring of the metal layer composition to achieve the desired $\sigma(u)$ function can improve both the resistance curve behavior and the fracture stress of the in-situ synthesized microlaminate composites. In addition, improvements in the metal layer response to fatigue and dynamic loading should also improve the corresponding properties of the microlaminates.

ACKNOWLEDGMENTS

This work was funded by Wright-Patterson Air Force Base under subcontract from General Electric. (Contract number FY1457-91-01001)

REFERENCES

1. R. L. Fleisher, Mat. Res. Soc. Symp. Proc., **133**, pp. 305-310 (1989).
2. G. R. Odette, B. L. Chao, J. W. Sheckherd, G. E. Lucas., Acta Metall. Mater., **40**, p. 2381 (1992).
3. H. Deve, A. G. Evans, G. R. Odette, R. Mehrabian, M. L. Emiliani, R. J. Hecht, Acta Metall. Mater., **38**, pp. 1491-1502 (1990).
4. H. C. Cao, B. J. Dalgleish, H. Deve, C. K. Elliot, A. G. Evans, R. Mehrabian, G. R. Odette, Acta Matall., **37**, 11, pp. 2969-2977 (1989).
5. D. L. Anton and D. M. Shah, Mat. Res. Soc. Symp. Proc., **194**, pp. 45-52 (1990).
6. F. E. Heredia, M. Y. He., G. E. Lucas, A. G. Evans, D. Konitser, Acta Metall. Mater., **41**, p. 505 (1993).
7. R. G. Rowe, D. W. Skelly, M. Larsen, J. Heathcote, G. E. Lucas, G. R. Odette, 93-CRD-229, General Electric, December 1993.
8. R. G. Rowe, D. W. Skelly, M. Larsen, J. Heathcote, G. E. Lucas, G. R. Odette, Mat.Res.Symp. Proc., **322**, pp. 461-472 (1994).
9. J. Heathcote, G. R. Odette, G. E. Lucas, R. G. Rowe, D. Skelly, Acta Met Metall (in press).
10. K. Ogawa, X. J. Zhang, T. Kobayashi, R. W. Armstrong, G. R. Irwin, ASTM-STP-833, American Society for Testing and Materials, Philadelphia, PA, p. 393 (1984).
11. T. Kobayashi, and D. A Shockey, Met. Trans., **18A**, p. 1941 (1987).
12. K. Edsinger, G. R. Odette, G. E. Lucas, B. Wirth, B., Effects of Radiation in Materials: 17th International Symposium, D. Gelles, R. Nanstad, A. Kumar, E. Little, Editors, ASTM-STP-1270, (in press).
13. L. Murugesh, K. T. Venkateswara Rao, R. O. Ritchie, Scripta Met. et Mater., **29**, pp. 1107-1112 (1993).
14. M. Johnson, Master's Thesis, University of California, Santa Barbara (1992).
15. M. Bannister and M. Ashby, Acta Metall. Mater., **39**, 2575-2582 (1991).
16. H. Tada, The Stress Analysis of Cracks Handbook, 2nd Edition, St. Louis, MO (1985).

STRUCTURAL STABILITY OF Si-O-a-C:H/Si-a-C:H LAYERED SYSTEMS

U. MÜLLER, R. HAUERT
Swiss Federal Laboratories for Materials Testing and Research (EMPA), Überlandstrasse 129, CH-8600 Dübendorf (Switzerland)

ABSTRACT

Amorphous hydrogenated carbon films are of technological interest as protection coatings due to their special properties such as high hardness, chemical inertness, electrical insulation and infrared transparency. However, some applications still suffer from the poor thermal stability and adhesion problems of these coatings. To ensure good adhesion, especially on hardened steels and non-carbide forming substrates, an extra interlayer has to be deposited first. Often a silicon containing interlayer, Si-a-C:H for example, is used for this purpose. This Si-a-C:H interface layer was deposited by rf plasma deposition from tetramethylsilane. Then a-C:H films containing Si-O with a varying silicon content were produced from a mixture of acetylene and hexamethyldisiloxane. The structural changes upon annealing of these films were investigated using Raman spectroscopy. The analysis of the development of the different peaks upon annealing temperature reveals the transition from the amorphous structure to the more graphitic-like structure. This transition temperature increases by as much as 100°C when silicon is incorporated into the DLC film. However, when Si-O is incorporated instead of only silicon the same increase in temperature stability is observed .

INTRODUCTION

Amorphous hydrogenated carbon (a-C:H) or also named diamond-like carbon (DLC) films are widely used as protection coatings due to their many useful properties as high hardness, low friction coefficient, low wear and chemical inertness [1]. However, above 300°C the film structure changes irreversibly, resulting in a deterioration of the above mentioned properties [2]. Changing the thermal stability by introducing additional elements has been investigated by several authors [3, 4, 5], studying F, Si and N, respectively. Especially Dorfman and co-authors claimed that their so called diamond-like nanocomposites (DLN) deposited from siloxanes have a very high thermal stability in an inert gas atmosphere [6, 7, 8]. Furthermore it has been argued that the incorporation of silicon increases the probability of carbon atoms to be sp^3, rather than sp^2 hybridized [9]. Coincidentally it has been reported that silicon containing DLC coatings have an extremely low coefficient of friction even at a relative humidity as high as 70% [10], which is not the case for pure DLC.

Raman spectroscopy is used to investigate structural transformations of thermally annealed diamond-like carbon films since the original work by Dillon et al. [11]. In their paper it was shown that the intensity ratio I(D)/I(G) of the Raman D-band to the G-band can be used as an indication of structural transformations. The nucleation of graphitic-like clusters results in a first increase of this ratio, whereas the subsequent growth of these clusters is seen as a saturation followed by a decrease of the ratio. The influence of silicon alone on the thermal stability has been investigated earlier [12]. In this paper Raman spectroscopy was used to investigate the change of these structural transformations when silicon and oxygen are simultaneously incorporated into diamond-like carbon films.

Mat. Res. Soc. Symp. Proc. Vol. 434 © 1996 Materials Research Society

EXPERIMENT

Hydrogenated amorphous carbon films containing silicon and oxygen (Si-O-a-C:H) were prepared by rf plasma activated chemical vapor deposition from a mixture of acetylene and hexamethyldisiloxane (HMDSO, $[Si(CH_3)_3]_2O$). The deposition was carried out in an all stainless-steel high-vacuum system with a base pressure better than $2 \cdot 10^{-6}$Pa. The films were deposited onto polished p-type Si(100) wafers, 2.54cm diameter, with 1Ωcm resistivity which were mounted on the capacitively coupled rf powered electrode. The rf generator (13.56MHz) was regulated to yield a constant self-bias of -400V. Just prior to introduction into the vacuum system the silicon wafers were chemically etched in a 27% HF solution for 90s and then rinsed with pure water. In-situ cleaning was done using Ar sputtering at 3.4Pa for 60s. The etch rate for a Si(100) wafer had been determined to be 2nm/min. After this a Si-a-C:H layer was grown for 5min from tetramethylsilane (TMS) at a pressure reading of 2.0Pa, the Pirani pressure gauge having been calibrated for air. Then the plasma was turned off and the HMDSO was introduced until the desired background pressure was reached. Acetylene was then added up to a total pressure setting of 2.0Pa. Following that the plasma was ignited again and deposition was performed for 15min. The power dissipated in the plasma was 0.4W/cm^2 and the growth rate under these working conditions was 21nm/min for pure a-C:H films.

The chemical composition of the films was measured using X-ray photoelectron spectroscopy (XPS). Concentration values are always given in units of atomic% and normalized to a total of 100at%. The hydrogen concentration cannot be measured with this method. After coating, each wafer was cut into several pieces which were used for the annealing experiments. Thermal annealing was done in an argon atmosphere at 3.5kPa pressure. The heating rate was 20°C/min, holding time at the annealing temperature itself was 200 min and the cooling rate was 10°C/min. Raman spectroscopy with an excitation wavelength of 514.5nm was performed on a DILOR XY micro-Raman system.

RESULTS

DLC-films with four different silicon-oxygen concentrations were produced. The concentrations measured with XPS were a) 0.2at% silicon and 0.4at% oxygen, b) 11.1at% silicon and 4.3at% oxygen, c) 19.5at% silicon and 8.5at% oxygen and d) 34.3at% silicon and 14.3at% oxygen. The average silicon to oxygen ratio is 2.4 when the film a) is excluded. This corresponds to an average incorporation probability for oxygen atoms which is lower than the one of silicon considering that the ratio in the precursor is 2. A similar lower incorporation probability for oxygen was also observed in

Table 1: Raman Peak Data

Position [cm^{-1}]	Width [cm^{-1}]	Name	Description
1585	110	G_1	crystalline graphite (E_{2g})
1530	100	Si-C	Si-C compound (unknown)
1490	125	G_2	amorphous sp^2-carbon
1350	240	D_1	polycrystalline graphite (A_{1g})
1130	210	D_2	amorphous sp^3-carbon

the case of Ti-O containing a-C:H films [13].

For the annealing experiments sets of pieces from the silicon wafers with different Si-O containing a-C:H films were heat treated in the same run. Optical inspection of the

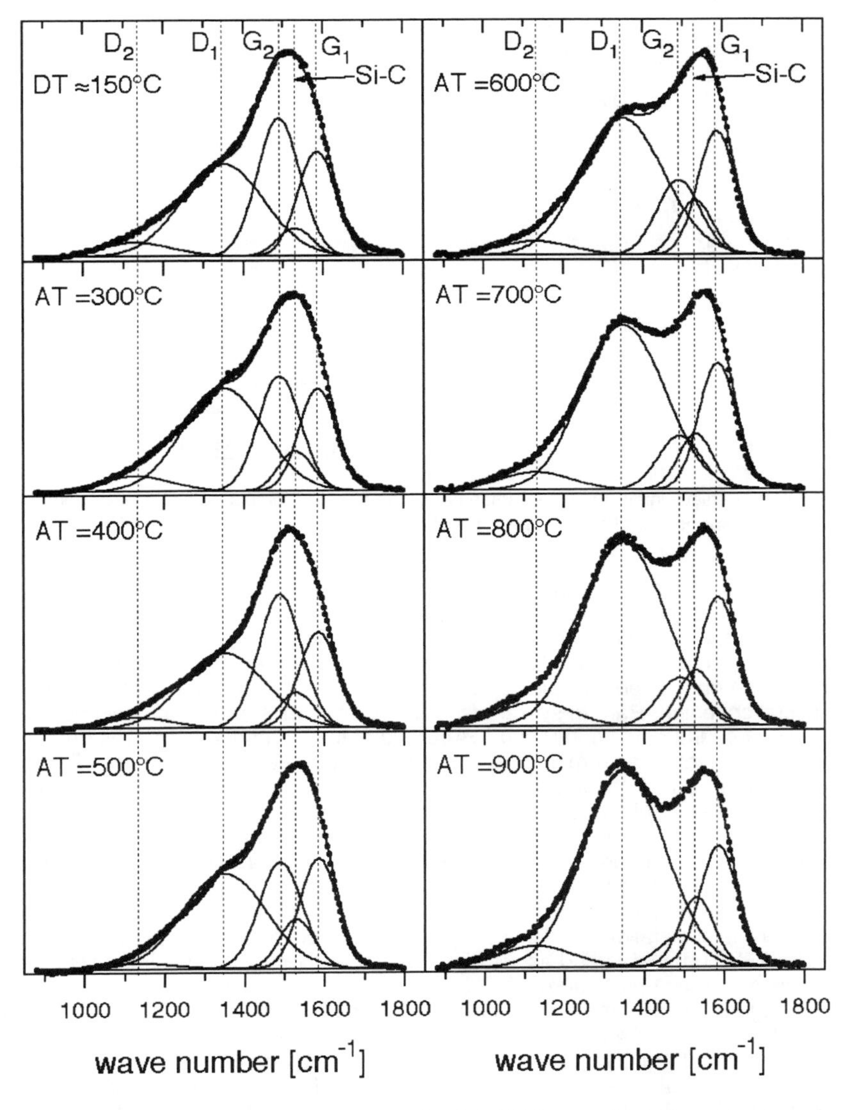

Figure 1: Set of Raman spectra of a Si-O - a-C:H film with 11.1at% silicon and 4.3at% oxygen content for the as deposited film (DT) and for annealing temperatures (AT) from 300°C to 900°C. Dots represent the measured data and the lines show the decomposition into five Gaussian peaks and the sum of these.

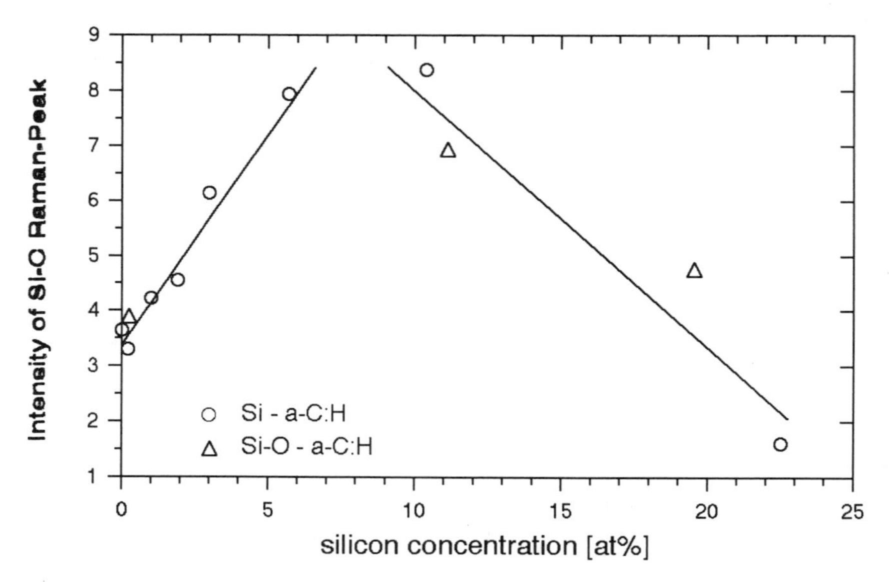

Figure 2: Relative intensity of the Raman peak from the Si-C compound for Si-a-C:H and Si-O-a-C:H films versus silicon content in the films. The lines serve only as guides for the eye.

coatings after this procedure showed no ablation or blistering of the films. Only the coating with the highest Si content, 34.3at%, had a few crack lines, when annealed at 800°C and a high density of crack lines when annealed to 900°C. This shows that the layered system Si-O-a-C:H/Si-a-C:H has an excellent adhesion to the silicon substrate. This is the case even for coatings with a very high Si-O content but they have a lower internal cohesion. Raman spectra were taken from all samples of coatings a), b) and c) in the range 880cm^{-1} to 1800cm^{-1}. Analysis of the Raman spectra was done by decomposition of each spectrum into five components of Gaussian line shape. Earlier analysis were performed by decomposition into two peaks, namely the D-peak and the G-peak [11]. However, at higher annealing temperatures this scheme does not describe the experimental data [5,11]. To obtain a more accurate fitting of the data at least four peaks have to be used for pure and fluorine containing amorphous carbon films [3]. In the case of silicon containing DLC films five peaks had to be used to obtain a good fit [12]. This is also the case for Si-O - a-C:H films where under these conditions a fit with an average correlation coefficient of 99.5% was obtained. Fixed peak positions and widths (FWHM) were used as listed in Table 1 and only the peak intensities were adjusted by the fit procedure.

Figure 1 shows the set of normalized Raman spectra from the DLC film containing 11.1at% silicon and 4.3at% oxygen annealed at the temperatures indicated. The relative intensity of the G_2 peak decreases whereas the one of the D_1 peak increases with increasing annealing temperature. The intensity of the additional peak at 1530cm^{-1} relative to the total intensity is shown in Figure 2 versus the silicon content in the films. For comparison also the data from silicon containing DLC films are included. As can clearly be seen for concentrations below 5at% this intensity depends linearly on the silicon concentration with a certain zero offset which is due to the mathematical dependence of these fits on the data. As the Si content increases, the intensity decreases. This behavior reflects presumably the transition from the regime where the silicon is incorporated into the carbon network by substituting carbon atoms to the regime where the formation of a more silicon carbide-like compound starts.

From all the spectra, the intensity ratio I(D)/I(G), where I(D) is the sum of the D-band intensities, $I(D_1)+I(D_2)$, and I(G) is the sum of the G-bands including the Si-C component, $I(G_1)+I(G_2)+I(Si\text{-}C)$, can be calculated. These curves exhibit a turning point defining a transition temperature from one state to the other [3]. This transition temperature is a measure for the stability temperature of these films. In figure 3 the transition temperature is shown as a function of the silicon content. Again the two types of silicon containing a-C:H, with and without oxygen, are displayed together. The overall trend is an increase of the transition temperature when the silicon content is increased. However, within the error limits little difference is observed between Si- and Si-O-DLC films. Dorfman has claimed that DLN films are stable under long-term annealing at T>1200°C in an oxygen-free atmosphere [7]. This claim is not supported by the results of this investigation where a maximum increase of the transition temperature of approximately 100°C is found. Unfortunately, he does not specify the exact composition and deposition methods nor the method of determining this stability temperature [7]. But our results clearly show that diamond-like carbon films containing silicon and oxygen in such concentrations that these films are still diamond-like carbon have a higher structural stability but not as high as Dorfman claims. This increased structural stability however does not have to be accompanied by an increased stability against oxidation in an oxygen containing atmosphere. Tamor [4] measured the weight loss of silicon containing amorphous hydrogenated carbon films when baked in air. The onset of weight loss was shifted to higher temperatures by roughly 100°C, in accordance with our results.

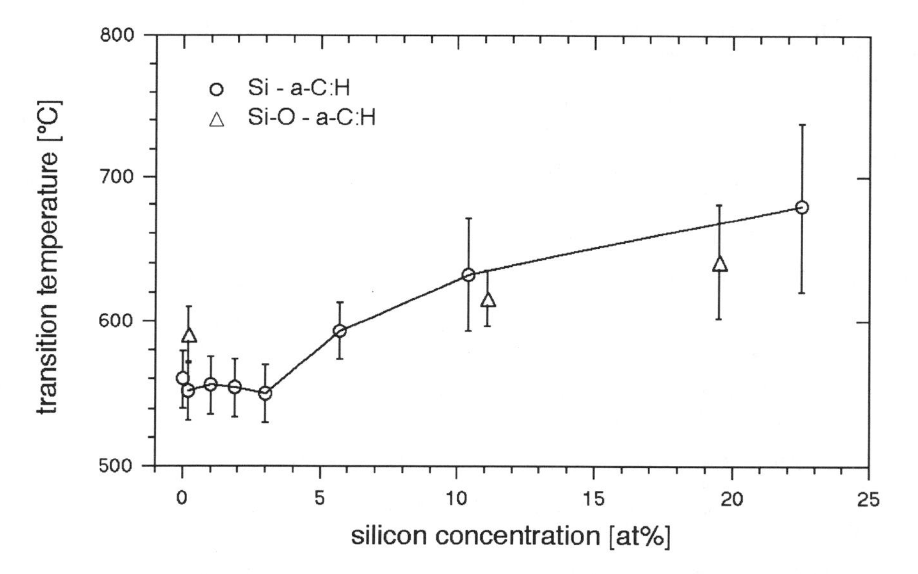

Figure 3: Transition temperature where the carbon film transforms from the amorphous state to the more graphitic-like state.

CONCLUSIONS

Silicon and oxygen containing diamond-like carbon films were produced by RF-PACVD. Raman investigations have shown that silicon incorporation increases the transition temperature and hence the thermal structural stability by as much as 100°C. No additional influence has been seen when also oxygen was added to silicon containing DLC films. However, the internal cohesion strength of the DLC film was reduced when too much silicon and oxygen are incorporated, whereas the adhesion was still excellent.

ACKNOWLEDGMENTS

We thank E. Blank and Y. von Kaenel for the kind permission to use their Raman equipment. Financial support by the Swiss Priority Program on Materials Research (PPM) is gratefully acknowledged.

REFERENCES

1. J. Robertson, Prog. Solid State Chem. **21**(4/1991), 199-333

2. D.R. Tallant, J.E. Parmeter, M.P. Siegal, R.L. Simpson, Diamond Related Mater. **4**(03/1995), 191-199

3. U. Müller, R. Hauert, B. Oral, M. Tobler, Surf. Coat. Tech. **76-77**(1995), 367-371

4. M.A. Tamor in "Applications of diamond films and related materials", Eds: A.Feldman, Y.Tzeng, W.A.Yarbrough, M.Yoshikawa, M.Murakawa, NIST Special Publication **885**(1995), 691-702

5. F.L. Freire, C.A. Achete, G. Mariotto, R. Canteri, J. Vac. Sci. Technol. A **12**(6/1994), 3048-3053

6. P. Asoka-Kumar, B.F. Dorfman, M.G. Abraizov, D. Yan, F.H. Pollak, J. Vac. Sci. Technol. A **13**(3Pt1/1995), 1044-1047

7. V.F. Dorfman, Thin Solid Films **212**(1992), 267-273

8. V.F. Dorfman, B.N. Pypkin, Surf. Coat. Tech. **48**(1991), 193-198

9. A. Chehaidar, R. Carles, A. Zwick, C. Meunier, B. Cros, J. Durand, J. Non-Cryst. Solids **169**(1994), 37-46

10. T. Hioki, Y. Itoh, A. Itoh, S. Hibi, J. Kawamoto, Surf. Coat. Tech. **46**(1991), 233-243

11. R.O. Dillon, J.A. Woollam, V. Katkanant, Phys. Rev. B **29**(06/1984), 3482-3489

12. U. Müller, R. Hauert, M. Tobler, in "Proceedings of ECASIA '95", Wiley&Sons, in press (1996)

13. U. Müller, R. Hauert, ICMCTF '96 in San Diego, April 22-26, 1996 to be published in the Proceedings

MECHANICAL AND THERMAL STABILITY OF HEAVILY DRAWN PEARLITIC STEEL WIRE

ETIENNE AERNOUDT*, JAVIER GIL SEVILLANO***, HILDE DELRUE*, JAN VAN HUMBEECK*, PIET WATTÉ**, IGNACE LEFEVER****
* Departement Metaalkunde en Toegepaste Materiaalkunde (MTM), K.U.Leuven, Belgium
** NV Philips Industries, Turnhout, Belgium
*** Centro de Estudios e Investigaciones Tecnicas de Guipuzcoa, San Sebastian, Spain
**** Bekaert Steel Wire Corporation, Kortrijk, Belgium

ABSTRACT

Having interlamellar spacings on the nanometer scale, there is no doubt about considering heavily drawn pearlitic steel wire as a nano-layered material. This extremely fine structure is of great technical importance: indeed, as the interlamellar distance determines the onset of plastic flow, the wire can be brought to a tensile strength beyond 4000 MPa and is therefore one of the strongest materials on the market nowadays.

At extremely large strains (well beyond $\varepsilon = 4$) and/or at moderate temperatures, the pearlitic steel loses its strength. Several possible failure mechanisms, like fragmentation of the cementite or thermal and strain-induced cementite dissolution, are put forward, but until now, there is no definite understanding of the really active mechanism.

In the present work, the calorimetric differential scanning technique, in combination with thermopower measurements and the high-resolution atomic force microscopy, have turned out to be most promising tools to reveal some of the mechanisms that are responsible for the degradation of the lamellar aggregate.

INTRODUCTION

At high strains ($\varepsilon = 4$), pearlitic steel wire exhibits an extremely fine structure with interlamellar cementite distances of 10 nm and therefore it merits to be considered as a nano-structured material. It is generally assumed that at strains well beyond $\varepsilon = 4$, cementite instabilisation might induce not only a collapse of the wire strength, but might also deliver an explanation for a number of ageing phenomena which occur during wire processing. This paper presents the investigation of the phenomena that occur during the static ageing of a severely deformed lamellar pearlite. Though such ageing is an intentional post-drawing treatment, it is almost sure that some of the same phenomena occur during drawing, as a consequence of the deformation and friction induced heating in the successive wire drawing dies.

Strain ageing in hard drawn pearlitic wire is already investigated by means of various experimental techniques ranging from tensile testing [6-8] and electrical measurements [9-11] to ion probe microscopy [3,12] and Mössbauer spectroscopy [1,2,4]. In all studies, as-drawn wire was submitted to artificial ageing treatments at moderate temperatures (< 350 °C) during relative short periods (< 30 minutes). According to the authors, one can adopt at least three different stages of ageing, occurring in consecutive order with increasing temperature:

Mat. Res. Soc. Symp. Proc. Vol. 434

stage I

Stage I of strain ageing is attributed to the migration of interstitially dissolved carbon atoms to ferrite dislocations to pin them down. Here, thermal activation is brought about by subsequent intentional ageing. This first stage of ageing occurs at temperatures below 100 °C.
This diffusion mechanism is most probably accompanied by a strain-induced dissolution of cementite [1,2,4].

stage II

Arguments have been given that at higher temperatures (between 100 °C and 300 °C approximately) stage I is followed by a stage II, which is characterised by post-straining cementite dissolution [7,13]. Here, a transfer of carbon atoms furnished by the cementite takes place from the cementite towards the ferrite phase. This thermally activated dissolution enables further locking of the dislocations by the supplied carbon atoms. Although already discussed by several authors, using different techniques, this stage might still be considered as rather speculative.

stage III

At temperatures above 300 °C, phenomena such as carbon clustering, carbide precipitation and recovery can be considered.

More profound investigation concerning the cementite stability in this higher-temperature range has been carried out by Languillaume [5]. In his study, the ageing treatments were performed during a longer time period (1 hour) at temperatures ranging from 200 °C to 700 °C. Observations with neutron diffraction techniques and high-resolution microscopes have pointed out that, next to a release of internal stresses and the appearance of a strong texture in the cementite phase, there is a gradual precipitation and development of cementite globules at the carbon enriched ferrite-cementite interfaces, which is a clear indication of the cementite dissolution.

It can be expected that the dissolution of cementite will have a direct influence on mechanical properties through changes in morphology and size of the cementite lamellae, but equally well an indirect influence, because of the increased carbon content in the ferrite lattice available for ageing and reprecipitation of cementite globules, as confirmed by Goes et al. [14]. In this study, the work hardening rate has been observed to change as a consequence of thermal treatments at temperatures of at least 150 °C. As can be seen from figure 1, ageing of hard drawn eutectoid wire during 300 minutes at 150 °C produces a shift in the gradient of the hardening rate versus tensile stress curve.

The advent of the powerful Differential Scanning Calorimetry (the DSC-technique) - which is applicable to fine wires- and the high-resolution Atomic Force Microscope (AFM) - emerging as one of the most promising scanning probe microscopes developed so far-, opened new perspectives to assess static strain ageing. In this paper, it is shown that DSC, in combination with thermopower measurements, can be easily adopted to discriminate between the stages of ageing. Special attention is given to the cementite decomposition as a mechanism of strain ageing. In the same perspective, the AFM-technique was applied as a tool to disclose this ageing mechanism.

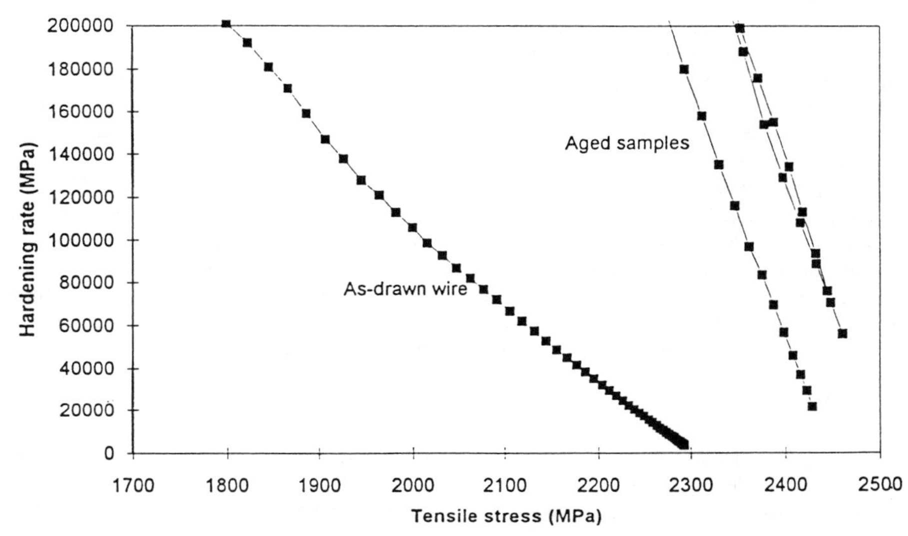

Figure 1: Hardening rate versus tensile stress curves of as-drawn and aged (150 °C, 300 minutes) eutectoid steel wire. Note the change in the gradient of the curve as a consequence of the ageing treatment [14].

EXPERIMENT

The wire under investigation was a plain carbon eutectoid pearlitic steel wire, manufactured by N.V. Bekaert S.A., with a diameter of 0.175 mm, a total strain of 3.9 and stored under frozen conditions to prevent any spontaneous ageing.

The DSC-experiments were performed on a commercially available calorimeter of TA-Instruments type DSC-10. Samples were prepared by filling aluminium pans with pieces of the investigated wire. Heat flow signals were recorded relative to a reference sample. The latter was cut from a batch of wire and subjected to a heating cycle from room temperature to 600 °C in the DSC-cell. As the exothermic heat exchange is irreversible, no exothermic events were recorded in the second or following runs with the same sample; the DSC-spectrum showed a flat baseline. Consequently, such a sample can be considered as fully chemically inert and used as a reference, leading to a better resolution of the recorded signals. The measurements were performed in a protective atmosphere of Argon gas, led through the DSC-cell.

Besides, the thermopower of samples that were artificially aged in a silicone oil bath was measured. The samples were isothermally aged at different temperatures between 120 °C and 180 °C (this is the temperature range of the first peak in the DSC-spectra) and for different ageing times (between 10 seconds and 30 minutes). The measurements were performed at room temperature and relative to a calibration sample with an absolute thermopower of 12.2 μV/K. During the measurements, a stable temperature gradient was applied. The sample ends were held at a temperature T_1 =15 °C and T_2=25°C, respectively.

In order to visualise cementite dissolution as a possible ageing mechanism, AFM-investigations of the pearlite structure were performed at longitudinal sections of hard drawn wire before and after an intentional ageing treatment for 5 minutes at 170 °C, the maximum peak temperature acquired from the DSC-experiments.
All samples were embedded in a resin and were ground in order to obtain longitudinal sections. The bare surface of each section was polished with two diamond pasts of which the grain size was 4 and 1 μm respectively. The wires were removed from the resin and glued onto a substrate which is placed on top of the piezotube of the microscope.
The measurements were performed on an atomic force microscope Digital Nanoscope III. Commercially available silicon nitride levers have been used. The levers are triangular and have 2 μm wide legs. The spring constant of the lever is about 0.6 N/m. In the experiments, the scanning area varied from 200 nm x 200 nm to 1000 nm x 1000 nm. The force microscope was operated under the equiforce mode. In this mode, the force induced by the sample surface atoms on the probing tip of the microscope, is controlled by adapting the sample position under the cantilever in order to keep the deflection of the cantilever constant.

RESULTS

Figure 2 displays a typical DSC-spectrum of the wire, drawn to a total strain of 3.9. This spectrum can be decomposed in four different peaks, which can be brought in connection with different stages of strain ageing. It can also be seen that there is a considerable overlapping of the peaks.

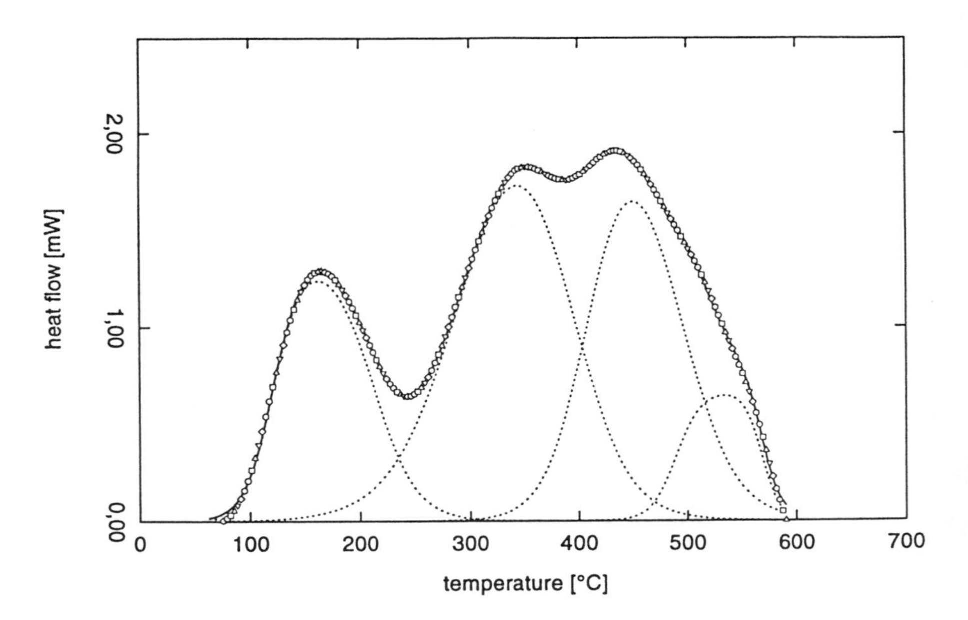

Figure 2: DSC-spectrum of the investigated wire, drawn to a strain of 3.9. This spectrum can be decomposed in four different peaks, each of them referring to different stages of ageing.

Regarding the wire properties during processing and in use, our interest concerns especially stage I and stage II.
During stage I carbon (and nitrogen) atoms in interstitial solid solution migrate to ferrite dislocations and pin them down. The carbon concentration in the ferrite is not sufficient enough to lock all ferrite dislocations despite its increase after plastic deformation (by deformation-induced cementite dissolution). This led several investigators [7,8] to conclude that some other mechanism than interstitial solute diffusion must operate. This opinion is shared by the authors since interstitial carbon diffusion is unlikely to yield the extremely high heat fluxes encountered during the first peak in the DSC-experiments.

The authors believe that in the temperature range covered by the first peak in the DSC-spectrum, stage I interferes with stage II. As mentioned earlier, this second stage of strain ageing is characterised by a temperature-induced partial decomposition of the cementite phase.
The thermopower of wire artificially aged at the corresponding temperatures of the first peak was measured in order to justify this statement. Figure 3 shows the TEP-measurements of the investigated samples as a function of time for the different ageing temperatures.

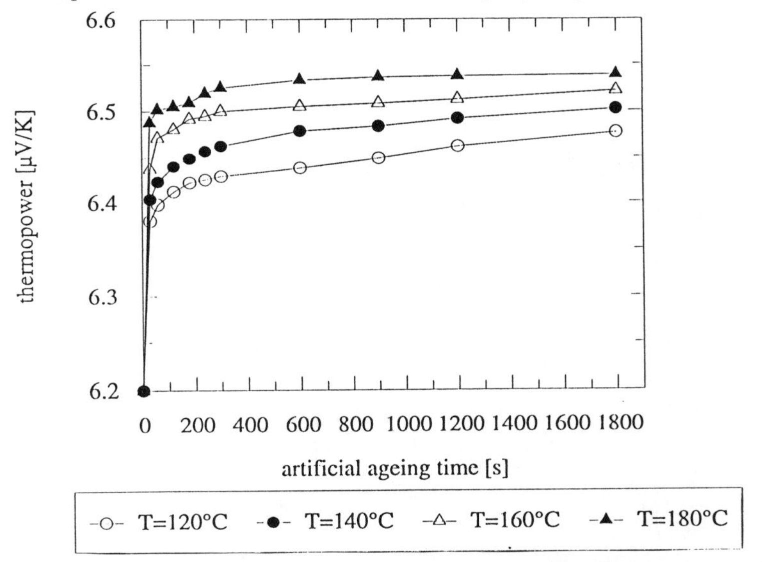

Figure 3: Thermo-electric power of the investigated wire as a function of ageing time and ageing temperature.

The rise in thermoelectric power after artificial ageing was plotted to a Johnson-Avrami-Mehl equation. According to this theory, the fraction of the solid state reaction, f(t), which evolves during isothermal heat treatments can be described as:

$$f(t)=1-\exp(-(kt)^n) \quad (1)$$

with $f(t) = \dfrac{S(t) - S_0}{S_{sat} - S_0}$, in which S(t) is the thermopower after an artificial ageing time t, S_{sat} is the thermopower at saturation and S_0 is the thermopower of the as-drawn wire. The time exponent n is characteristic for the intrinsic nature of the ageing process and is was estimated by means of linear regression. n is temperature independent and is found to be of the order of 1/3. This implies that the dominant mechanism of strain ageing below 250°C differs from matrix diffusion of carbon interstitials which is characterised by a $t^{1/2}$ dependency [15,16]. Time exponents of the order of 1/2 are expected when carbon interstitials are captured in cylindrical zones around dislocations, which was used as the basis for the models of Cottrell [17] and Harper [18]. According to the model of Lement and Cohen [19] a time exponent of 1/3 is in favour of a decomposition mechanism of one phase. In their model to describe the kinetics of such a decomposition, they assumed a planar flow of interstitials from one phase to another. This is applicable to eutectoid pearlite where this flow takes place from the cementite perpendicular towards the ferrite-cementite interface.

In figure 4 and 5 the topographic equiforce AFM-image of the pearlite structure probed at longitudinal sections of as-drawn wire and artificially aged wire (during 5 minutes at 170 °C) respectively, is labelled. A substantial difference between both images lies in the interface sharpness. The interfaces in the as-drawn structure are clearly more sharply defined than in the aged structure. This result is very interesting in the framework of the proposed mechanism of temperature-induced dissolution of the cementite, where ferrite interfacial dislocations are believed to attract carbon atoms from the cementite.

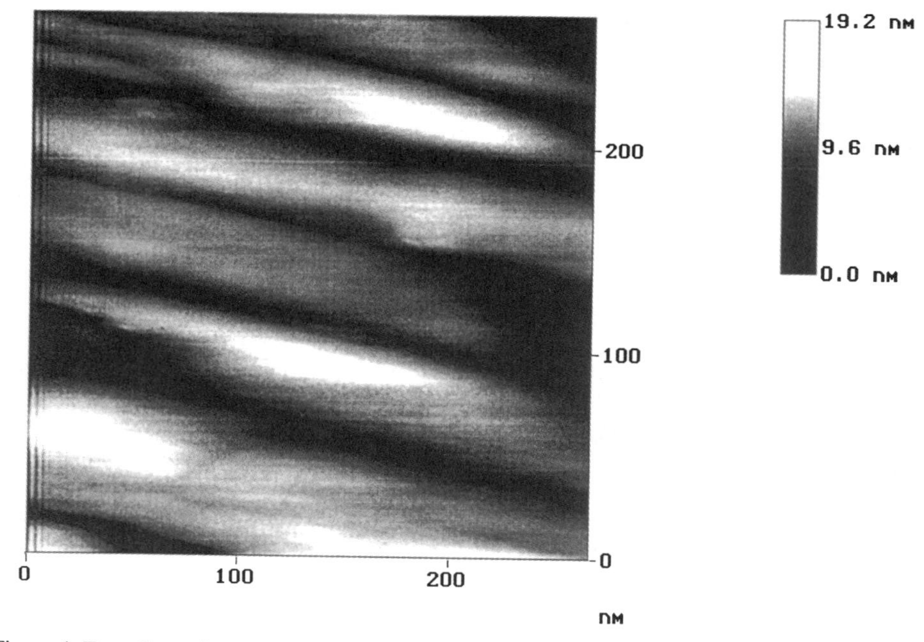

Figure 4: Two-dimensional topographic equiforce AFM-image of the pearlite structure probed at a longitudinal section of the investigated as-drawn sample. The scanning area was 267 nm x 267 nm.

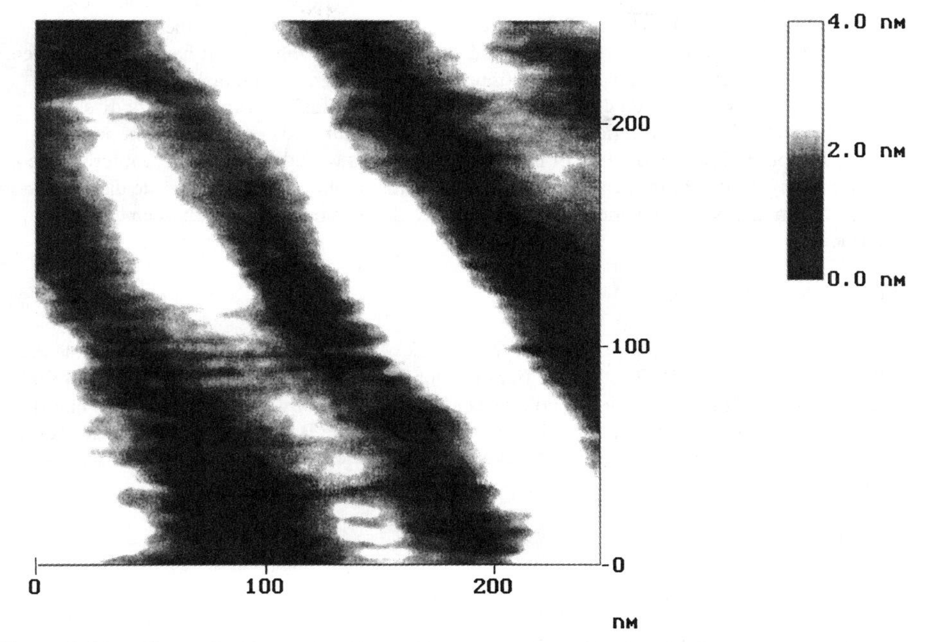

Figure 5: Two-dimensional topographic equiforce AFM-image of the pearlite structure probed at a longitudinal section of the artificially aged sample (170 °C, 5 minutes). The scanning area was 247 nm x 247 nm. Note the loss in interfacial sharpness compared to figure 4.

CONCLUSIONS

The extremely large strains that are reached in wire drawing, connected with a nearly homogeneous deformation behaviour of the two phases, points to an extensive dislocation activity in cementite as well as in ferrite. The deformation behaviour of cementite, including dislocation activity and substructure patterning, has been thoroughly investigated by Inoue et al. [20]. Slip in cementite is localised in parallel slip bands, 0.1 to 0.5 μm apart, at smaller strains, but doubtless more closely spaced at higher strains, dividing the cementite lamellae in a series of undeformed blocks moving over mentioned bands.

Decomposition during wire drawing is believed to take place by a dislocation attraction mechanism, in which ferrite interfacial dislocations attract carbon atoms from the cementite. It is indeed a well-established fact that *in undeformed pearlite,* the binding energy of carbon to cementite is only about 60 percent of the interaction energy of carbon to dislocations (0.5 eV and 0.8 eV respectively) [21,22]. In the *deformed* pearlite under consideration, the increasing interface energy between ferrite and cementite, as well as the increasing disorder induced in the cementite lattice by moving dislocations along the boundaries of the cementite "blocks", make that phase increasingly unstable. The "dynamic" mechanism of dissolution of cementite, occurring in the drawing die hence is to some extent mechanically activated. Additional thermal activation by the deformation and friction heat in wire drawing will enhance the process as well as the diffusion of carbon to the dislocation sites.

The decomposition process of cementite seems to continue during the second stage of "static" ageing. By combining the DSC- and TEP-data of hard drawn wire, the proposed mechanism of thermally activated cementite decomposition during stage II has been confirmed. This partial dissolution takes place to continue the decoration of ferrite dislocations. The decoration process has started during stage I of ageing by diffusion of interstitially dissolved carbon in the ferrite to dislocations, combined with strain-induced cementite decomposition.

The observation of the loss of interfacial soundness with the atomic force microprobe is another strong argument in favour of the proposed temperature-induced cementite dissolution in which carbon atoms of the cementite are attracted to dislocations in the ferrite-cementite interface.

ACKNOWLEDGEMENTS

Part of the present work took place in the framework of the "IUAP" (Inter University Poles of Attraction) action of "DWTC" (federal Services for Scientific, Technical and Cultural affairs) financed by the Ministry of Science Policy. The authors gratefully acknowledge that support. The first author thanks especially the Research Council of the Bask Government (Spain) for the financial support during his stay in San Sebastian where another part of the work presented in the paper tooks place.

REFERENCES

1. V.N. Gridnev and V.G. Gavrilyuk, Phys. Metals, **4**, (3), p. 531, (1982).
2. V.G. Gavrilyuk, V.G. Prokopenko, O.N. Razumov, Phys. Stat. Sol., **53**, p. 147, (1979).
3. T. Tarui, T. Takahashi, S. Ohashi, R. Uemori, Wire Industry International, p. 25, (1995).
4. F.G.S. Araujo, B.M. Gonzalez, P.R. Cetlin, A.R.Z. Coelho, R.A. Mansur, Wire Industry International, p. 1, (1992).
5. J. Languillaume, Ph.D. thesis, Institut National Polytechnique de Grenoble, (1995).
6. J.D. Baird, Metall. Rev., **149**, (16), p. 1, (1971).
7. I.P. Kemp, G. Pollard, A.N. Bramley, Mat. Sc. & Techn., **6**, p. 331, (1990).
8. Y. Yamada, Trans. ISIJ, **16**, p. 417, (1976).
9. A.H. Cottrell, A.T. Churchman, Journ. of the Iron and Steel Inst., p. 271, (1949).
10. H. Abe, Scandinavian Journ. of Metallurgy, **13**, p. 226, (1984).
11. J. Campbell, H. Conrad, Scripta Metall., **31**, (1), p. 69, (1994).
12. A.R. Waugh, S. Paetke, D.V. Edmonds, Metallography, **14**, p. 237, (1981).
13. M.L. Rudee, R.A. Huggins, Acta Metall., **12**, p. 501, (1964).
14. B. Goes, E. Aernoudt, J. Gil Sevillano, A. Martin Meizoso, confidential.
15. H.B. Aaron, G.R. Kotler, Metall. Trans. A, **2**, p. 393, (1971).
16. C. Zener, J. Appl. Phys., **21**, p. 5, (1950).
17. A.H. Cottrell, B.A. Bilby, Proc. Phys. Soc., **62**, p. 49, (1949).
18. S. Harper, Phys. Rev., **83**, p. 709, (1951).
19. B.S. Lement and M. Cohen, Acta Metall., **4**, p. 469, (1956).
20. A. Inoue, T. Ogura, T. Masumoto, Trans. Japan Inst. Metals, **17**, p. 149, (1976).
21. A.H. Cottrell, Dislocations and Plastic flow in Crystals, Oxford University Press, p. 134, (1953).
22. L.J. Dijkstra, J. Metals, **1**, p. 252, (1949).

INTERFACIAL DIFFUSION EFFECTS AND NON-STABILITY OF DISPERSE LAYERED STRUCTURES

L. N. PARITSKAYA AND V. V. BOGDANOV
Department of Crystal Physics, Kharkov State University, 310077 Kharkov, Ukraine

ABSTRACT

The kinetics and atomic mechanism of the low temperature phase formation and accompanying effects in disperse multilayered structures of the binary systems $Ni - Cr$ and $Ni - Al$ have been studied under the conditions when volume diffusion is practically "frozen".

INTRODUCTION

One of the basic problems that has to be solved for practical applications of disperse layered materials is the thermal stability of their layered structure. The high energy stored in such structures provides the driving forces for recovery and recrystallization processes. If layered object consists of two or more components the important factors determining the structure stability are evolution of phase composition and distribution of components and phases. These factors depend on the regularities of diffusion and diffusional processes which are governed by the high density of structural defects especially interfaces.
Our purpose in this paper is the investigation of the role of interfaces in diffusion processes that result in formation of new phases, changes in phase composition , phase redistribution and structure evolution which accompanies the phase formation process.

EXPERIMENT

The objects under investigation have been important for practical purposes binary metallic systems with different types of phase diagrams: $Ni - Cr$ with unipolar limited solubility of Cr in Ni and $Ni - Al$ with intermetallics. The studied samples were mechanically alloyed powder mixtures ($Ni - 18at\%Cr$ and $Ni - 50at\%Al$) subjected to short ($t \simeq 1h$) pulsed mechanical treatment. As a result we have obtained crystallites of sizes of the order of hundred microns with a disperse lamellar structure (Fig. 1).

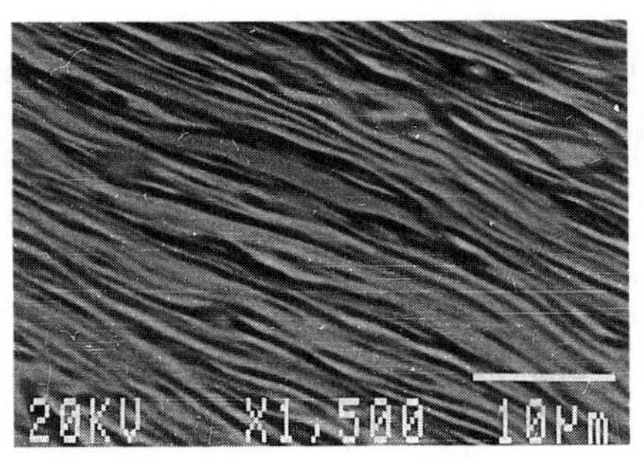

Fig. 1 Typical microphotograph of the disperse layered structure of mechanically alloyed $Ni - 18at\%Cr$ powder mixture.

The thickness h of alternating (A-B-A-B...) layers was $(1-5) \cdot 10^{-7}m$. The layers consisted of the structural elements of sizes $\ell \simeq 10^{-8}m$, which were determined by X-ray analysis

Mat. Res. Soc. Symp. Proc. Vol. 434 © 1996 Materials Research Society

from widening of the diffraction maxima. The analysis of the diffraction maxima has shown that the used regime of mechanical treatment did not lead to amorphization and new phase formation.

Using the method of optical and scanning electron microscopy, X-ray microprobe and phase analysis we investigated the evolution of phase composition and structure as well as the changes in the local elementary composition during isothermal annealings at $T \simeq (0.3-0.5)T_m$ (T_m - melting temperature of Ni for system $Ni-Cr$ or of intermetallics $NiAl$ which is the main phase that has been formed in the system $Ni-Al$). From the evolution of the diffraction maxima from the multilayered $Ni-Cr$ crystallites after the successive stages of isothermal annealing we determined the kinetic dependencies of the average Cr concentration $\bar{c}_{Cr}(t)$ in the formed solutions and their volume part $\kappa(t)$. In the multilayered samples of the system $Ni-Al$ the separate intermetallics $NiAl$ inclusions elongated along the layers are formed during the isothermal annealings (Fig.2)

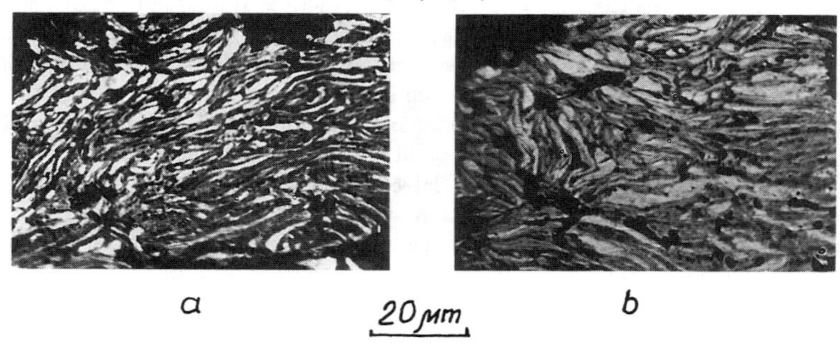

Fig. 2 Typical microphotographs of the disperse multilayered mechanically alloyed mixture $Ni-50at\%Al$ after successive stages of isothermal annealing at T=623 K; a - t=9h, b - t=25 h.

In this case from the evolution of the diffraction maxima we determined the phase composition, calculated the volume part $\kappa(t)$ of the intermetallics using integral intensities of close diffraction lines of the phase $NiAl$ (112) and pure Ni (311) and measured mean transverse and longitudinal sizes of inclusions $\ell_{\perp}$ and $\ell_{\parallel}$ and their density ΣN per m^2. Averaging was made over 200 inclusions measured on the area of 10^{-8} m^2. Typical kinetic dependencies $\kappa(t)$ and ΣN are shown in Fig. 3

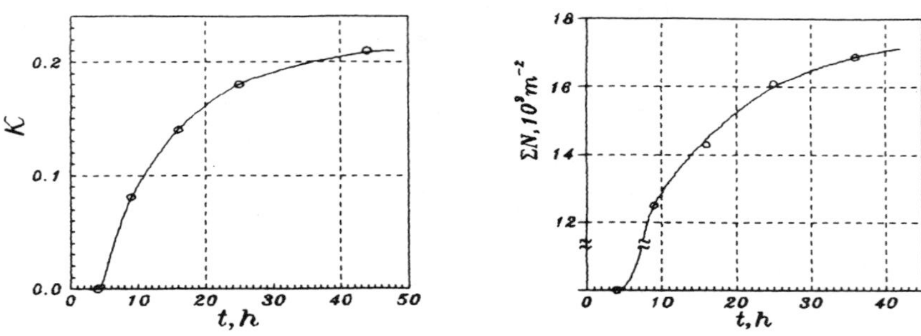

Fig. 3 Typical kinetic dependencies of $\kappa(t)$ (a) and $\Sigma N(t)$ for multilayered structures of $Ni-50at\%Al$ at T=623 K.

RESULTS OF THE EXPERIMENTS AND DISCUSSION

The main peculiarities of the diffusion effects obtained with the disperse multilayered samples are as follows:

1. New phase formation is occurring at the temperatures several hundreds degrees less than in massive objects, i.e. under the conditions when lattice diffusion is practically "frozen".
2. In the case of the solid formation (system $Ni-Cr$) we observed the kinetic dependencies $\bar{c}_{Cr}(t)$ of different types (Fig. 4) up to anomalous one when $\bar{c}_{Cr}(t)$decreases with time (Fig. 4b). This is on the contrary to the usual case with the formation of solutions under the diffusion along the stationary boundaries with a penetration into the volume when concentration increases.
3. On the kinetic curves $\kappa(t)$ (Fig. 4 a,b) two stages of phase formation process are clearly seen which differ in the rate of κ increase in the course of time: in the first stage up to $t \simeq 1h$ the volume of new phases grows rapidly; in the second stage ($t > 1h$) $d\kappa/dt$ decreases with time. For a long time $\kappa < 1$; it means that solutions like intermetallics are formed in "spots".
4. In multilayered objects the formation of solid solutions is accompanied by a decay of lamellar structure and its transformation into a disperse isomeric one.

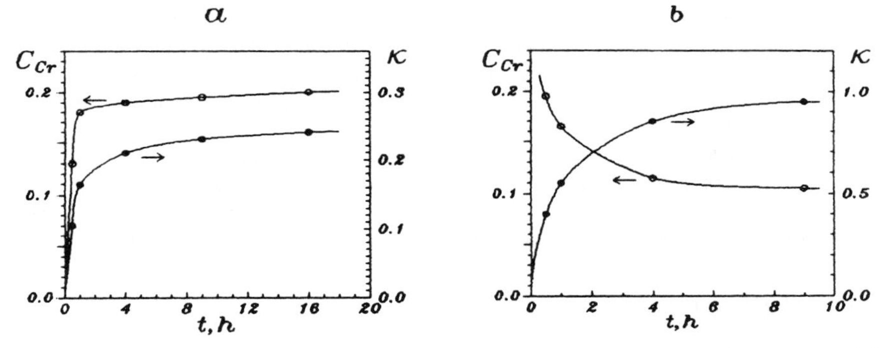

Fig. 4 Kinetic dependencies of $\bar{c}_{Cr}(t)$ and $\kappa(t)$ for multilayered structures of $Ni - 18at\%Cr$ at different temperatures: a- T=773 K, b- T=873 K.

The observed peculiarities of phase formation are typical for low temperature diffusion process governed by migrating boundaries according to the mechanism of diffusion induced grain boundary migration (DIGM) [1]. Let us discuss low temperature phase formation (LTPF) process with the example of the disperse layered structure of $Ni-Cr$ system as in the $Ni-Al$ system like in any system with intermetallics phase formation process could be limited by nucleation of a new phase but not by diffusion process itself. At $T \simeq 0.4T_m$ (773K) the volume diffusion coefficient D_V of Cr in Ni is $D_V \simeq 10^{-23}\mathrm{m^2s^{-1}}$ could provide the penetration of diffusing atoms during the time $t \simeq 10^4$s at a distance $(D_V t)^{1/2} \simeq 3 \cdot 10^{-10}$m, i.e. interatomic distance. From Fig. 4 it follows that under these conditions the solutions with a high mean concentration $\bar{c}_{Cr} \simeq 0.22$ are formed which volume part is $\kappa > 0.1$. It is clear that this could not be a result of an accelerated diffusion along the stationary boundaries with a penetration into the volume. Therefore we can suppose that the observed solution formation is carried out by migrating boundaries in the following scheme (Fig. 5): atoms A

(Cr) and B(Ni) diffuse along the grain boundary in opposite directions which induces the migration of the boundary with a normal velocity v_n such that migrating boundary leaves atoms A in the lattice B behind itself.

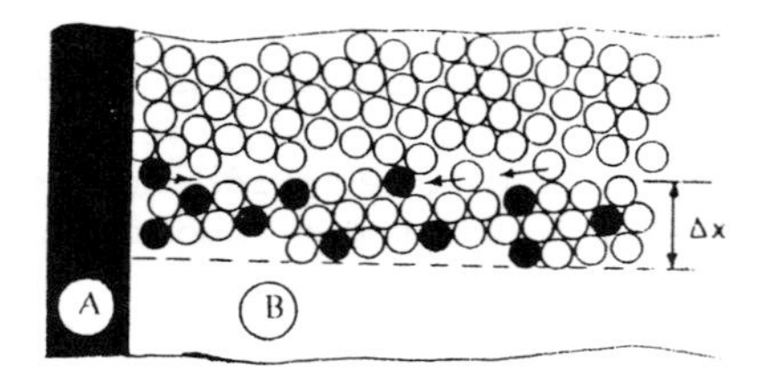

Fig. 5 a- The scheme of solution formation by DIGM mechanism (Δx - boundary shift induced by interdiffusion of atom A and B); b- variation of step density at the boundary sides changing the migration direction in different regions of the boundary.

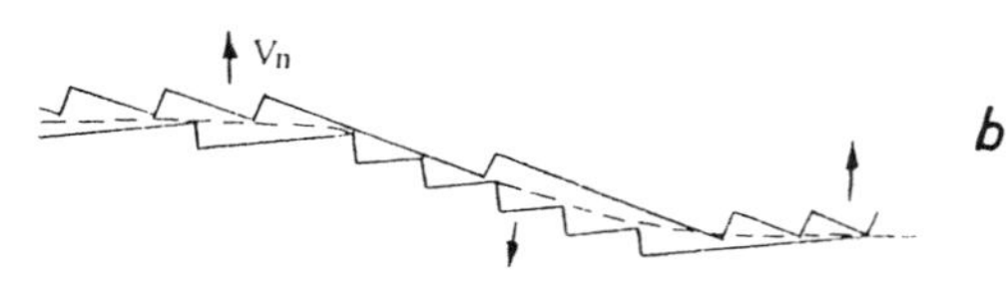

In order to estimate the conditions necessary for the phase formation by migrating boundaries we consider in [2] interdiffusion process of the atoms A and B along a semiinfinite grain boundary of a general type with different step densities ρ_i on its sides (Fig. 5). At low temperatures, when the penetration of atoms A into the volume B is neglected atoms A sink at the boundary steps of B and preferably are captured by them on one of the boundary side because of different step densities. The normal velocity v_n of boundary migration is determined by the flux of atoms B from the opposite side across the boundary. As the chemical potentials of atoms A and B in the solution are lower than in the pure components, atoms A and B mix at the steps and form the layers of solution on one side of the boundary. As a result the boundary migrates to the opposite side leaving the new layers of the solution behind itself. For such atomic mechanism of DIGM we obtained v_n in the following form [2]:

$$v_n = \beta_B N_0 \omega_B [\rho_2 - \rho_1(1-c)], \tag{1}$$

where β_B is the kinetic coefficient of the steps characterizing the embedding rate of atom B in steps; N_0 - two dimensional boundary density of atomic positions; ω_B - atomic volume of B; ρ_1 and ρ_2 - step densities on both boundary sides; $c = c(y)$ - concentration distribution of atoms A in the solution formed in the vicinity of the boundary; y - coordinate along the boundary. According to [2] $c(y)$ has the following form:

$$c(y) = c_0 \exp(-\frac{y}{\lambda_b}), \tag{2}$$

where c_0 is the solution concentration at $y = 0$, $\lambda_b = [(D_A - D_B)/(\beta_A - \beta_B)\rho_1)]^{1/2}$ - characteristic length determining the extent of the diffusion zone where solid solution is formed by the migrating boundary; D_A and D_B - partial grain boundary diffusion coefficients of atoms A and B.

From Eqs. (1), (2) both the criteria and direction of boundary migration under DIGM and the reason why the different places of one boundary migrate in different directions are

clearified. The latter is caused by the dependence of DIGM velocity v_n on the density of steps at different sides of the boundary. The boundary migrates to those of its sides where the step density is higher as the expression in the square brackets in (1) should be positive. Correspondently, the solid solution is formed on the opposite side (Fig. 5b).
Due to (2) the necessary condition for the phase formation by migrating boundary is that the length h of the boundary must be small compared to the diffusion length λ_b. If $h \ll \lambda_b$ a constant concentration along the entire boundary is established and behind the migrating boundary a solution with a constant concentration will be formed in the volume (but different from that within the boundary). Thus if solutions are formed behind the migrating boundary without volume diffusion their concentration can be practically unchanged with time and the volume part of the solutions are less than 1 (Fig. 4a). However, with the increase of the temperature and annealing time and correspondently of the volume diffusion contribution (the value $(D_V t)^{(1/2)}$increases) the initially formed regions ("spots") of the solutions with sufficiently high concentrations can spread with time and the mean concentration of the solutions will somewhat decrease (Fig. 4 b).
Thus, the observed peculiarities of the kinetic dependencies $\bar{c}_{Cr}(t)$ and $\kappa(t)$ (Fig. 4) demonstrate that the solutions form by mechanism of diffusion along migrating boundaries. This mechanism is confirmed also by the observed phenomenon of the lamellar structure decay. We propose the following model of this phenomenon (Fig. 6).

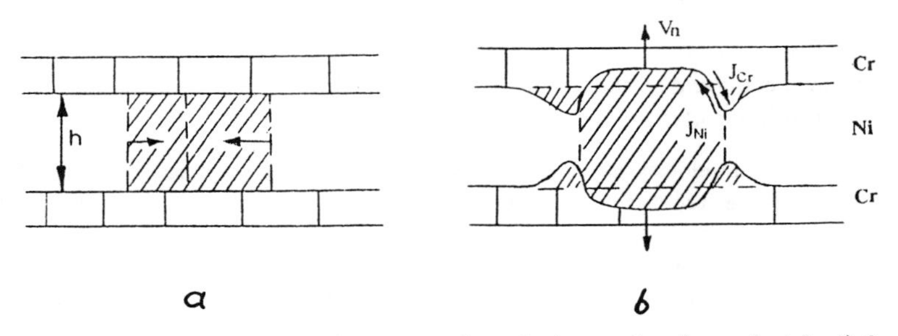

Fig.6 The scheme of layered structure decay: a - the solution region (cross-hatched) formed as a result of migration of two adjacent grain boundaries toward each other; b- evolution of the lamellar structure element leading to layer decay.

The migration of the two adjacent grain boundaries in Ni towards one another has resulted in a region of Cr solution formed in Ni by the above described mechanism of interdiffusion of Cr and Ni along the grain boundaries. If we consider an element of the structure where the region of the solution is located between two grains of Ni (Fig. 6a), we can see that further evolution of this element is possible only by interdiffusion of Ni and Cr along the interfaces between the layers: Ni diffuses from Ni-grains and Cr diffuses in the opposite direction. This leads to the migration of convex regions of the interface "Cr-solution" towards Cr and concave ones "Ni-solution" towards each other (Fig. 6b). In both cases solutions of Cr in Ni are formed. The components are redistributed near the interface "Ni-solution" at a distance about λ_b. This distance determines the characteristic dimension of the structure element being formed.
The observed decay of the layered structure could be caused by two different forces: capillary $\Delta G_{cap} \simeq \alpha_i/h$ (α_i - specific interface free energy), i.e. Laplace pressure; and chemical

$\Delta G_{ch} \simeq kT(c_{0s} - \bar{c})^2/\omega c_{0s}^2$ (c_{0s} - initial diffusant concentration in the diffusion source, $\bar{c}$ - mean solution concentration), i.e. the specific free energy decrease due to solution formation from the initial components. Comparison of their values (at $c_{0s} \simeq 1$, $\bar{c} \simeq 0.2$, $\alpha_i \simeq 1$ J m^{-2} and $h \simeq 5 \cdot 10^{-7}$m $\Delta G_{ch} \simeq 10^8$ J m^{-3} and $\Delta G_{cap} \simeq 10^6$J m^{-3}) shows that the observed decay of the layered structure is caused by the chemical force as $\Delta G_{ch}/\Delta G_{cap} \simeq 10^2$. As the decay of the layered structure takes place in the process of solution formation and is induced by the diffusion along the interfaces it can be called diffusion induced decay. This phenomenon leads to the loss of stability or degradation of the lamellar structure at low temperature solution formation.

CONCLUSIONS

The following conclusions can be drawn from the investigations of the interfacial diffusion effects in the disperse layered structures of $Ni - Cr$ and $Ni - Al$ systems.

1. We discovered the low temperature ($T \simeq (0.3-0.5)T_m$) formation of the solid solutions Cr in Ni and intermetallics $NiAl$ in the material volume when the volume diffusion is practically "frozen".
2. The low temperature solution formation is accompanied by the decay of the disperse layered structure and its transformation into an isomeric one.
3. On the basis of analysis of LTPF kinetics and peculiarities of the layered structure decay we suggested the atomic mechanism of this process consisting of interdiffusion along migrating interfaces.
4. The normal velocity v_n of migrating interfaces has been calculated and it depends on the interface structure determined by the densities of the atomic steps on the different sides of the interface. The proposed model clarifies the criteria and the main peculiarities of LTPF and also the mechanism of the lamellar structure decay caused by the interdiffusion along the interfaces and thus called "diffusion induced decay".

We emphasize the practical importance of this phenomenon leading to non-stability of layered structure at LTPF and the necessity of accounting for it when creating multilayered functional materials.

ACKNOWLEDGEMENTS

This work is funded by the International Association Project No. 93-2617.

REFERENCES

1. J. W. Cahn, J. D. Pan, and R. W. Balluffi, *Scripta Met.* **13**, 503 (1979).
2. Yu. S. Kaganovskii, L. N. Paritskaya, and A. O.Grengo, *Functional Materials* **1**, 30 (1994).

Part IV

Mechanical Behavior

YIELD STRESS OF NANO- AND MICRO- MULTILAYERS

P. M. HAZZLEDINE and S. I. RAO
UES Inc., 4401 Dayton-Xenia Road, Dayton. OH 45432

ABSTRACT

An outline theory is given for the strengthening in polycrystalline and 'single crystal' multilayers. The model is based on the Hall-Petch theory applied to both the soft mode (in plane) and hard mode (cross plane) of deformation. In this theory the parameters to be evaluated are a Taylor factor M, the shear stress τ_0 to move a dislocation within a multilayer and τ^*, the shear stress needed to push a dislocation over a grain or interphase boundary. All three parameters are material-specific and attention is focussed on coherent multilayers of γ TiAl with micron thick layers and Cu-Ni with nanometer thick layers. M and some components of τ^* are estimated classically. The remaining components of τ^* and some components of τ_0 are estimated from embedded atom simulations. The model captures the main experimental facts, that γ TiAl is plastically very anisotropic with a rising yield stress as the lamellar thickness is refined and that Cu-Ni displays a peak in the yield stress at a layer thickness of approximately 10nm.

INTRODUCTION

The Hall-Petch equation between the yield stress σ and the grain size d is generally written $\sigma = \sigma_o + Kd^{-1/2}$ with σ_o and K taken to be constants. In multilayers, however, K is a function of d and both d and σ_o are directional. These dependancies may give rise to strong plastic anisotropy.

GEOMETRY OF DEFORMATION IN MULTILAYERS

Fig. 1 illustrates the geometrical constraint imposed by multilayers on the shapes of the slip planes. In Fig. 1(a) a single crystal multilayer is being compressed perpendicular to the layers ($\phi = 90^o$). Slip or twinning must occur in extremely elongated slip zones, with widths close to h, the layer thickness. Dislocations must overcome closely spaced interface barriers before general yield can occur. Most coherent multilayers deform in this 'hard mode' but, if the slip plane happens to coincide with the layer plane, Fig. 1(b) a 'soft mode' operates when ϕ is close to neither 0° nor 90°. In γ TiAl the slip plane does coincide with the layer plane and in CuNi, it does in certain circumstances. In the soft mode the size of the slip zones and hence the yield strength is determined by the grain size (called 'domain size' in TiAl).

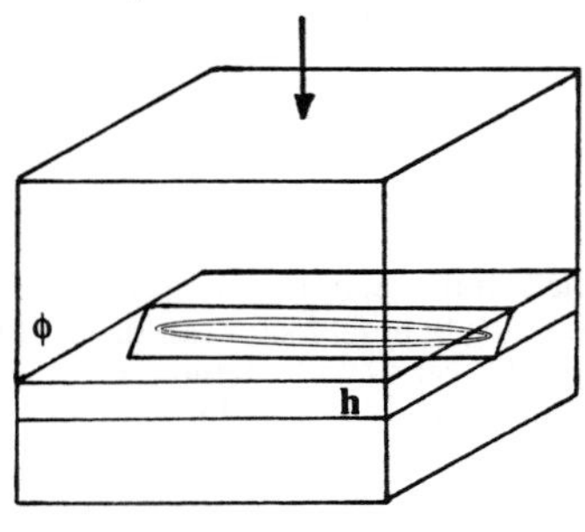

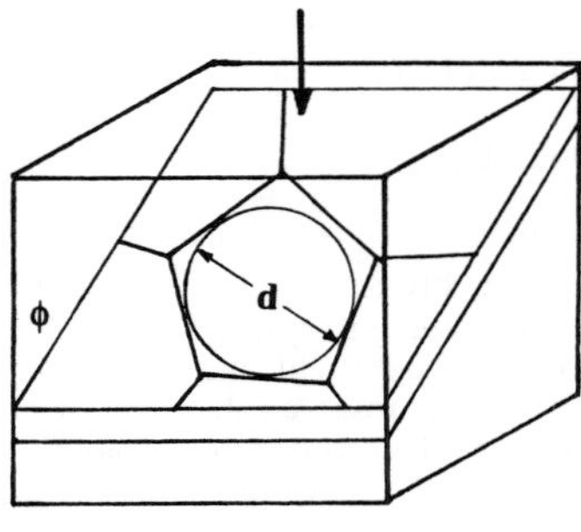

(a) (b)

Figure 1 (a) Elongated pile ups in the hard mode (b) Equiaxed pile ups in the soft mode

Mat. Res. Soc. Symp. Proc. Vol. 434

HALL-PETCH THEORY

When the soft mode is operating, the slip zones are equiaxed and, assuming that sources operate in the centers of the zones, the dislocations pile up at the boundaries. The dislocations in the pile ups are approximately circular. The leading dislocation experiences a concentrated stress τ^* and yield occurs when τ^* is sufficient to move the dislocation across the interfacial barrier. The tensile yield stress σ is then [1,2,3]

$$\sigma = M\{\tau_o + [(2-\nu)\pi\tau^*Gb/(2(1-\nu)d)]^{1/2} \qquad (1)$$

Where M is a Taylor factor, τ_o is the shear stress to move a dislocation through a single crystal of the layer material, ν is Poisson's ratio, G is the shear modulus and b the Burgers vector.
When the hard mode is operating, the slip zones are such elongated shapes that the elliptical dislocations become almost like straight dislocations piled up at the interfaces. In this limit, the yield stress is controlled by the smaller dimension of the slip zone h and it may be written:

$$\sigma = M\{\tau_o + [4\tau^*Gb/\alpha\pi h]^{1/2}\} \qquad (2)$$

Where α is 1 for screw dislocations and $(1 - \nu)$ for edge dislocations.
As pointed out by Yamaguchi and Umakoshi [4] the main origin of the plastic anisotropy in γ TiAl lies in the difference between the yield stresses of the soft mode (eq.1) and the hard mode (eq.2), in particular that M in eq.1 is generally smaller than M in eq. 2 and that d >> h. More subtle effects are caused by the barrier strengths τ^* not being the same in eqs. 1 and 2 and by the fact that the τ_o values need not be the same when different types of dislocations (super, perfect, twin) operate in soft and hard modes [5].
Because of the geometry of the specimens available, nanomultilayers are invariably tested in the hard mode (either in tension with $\phi = 0^\circ$ or by indentation with $\phi = 90^\circ$) and eq. (2) is appropriate. A soft mode, in which shear displacements occur in the plane of the layers, may exist but it has not been reported.

APPLICATION TO PST Ti-Al

Polysynthetically twinned TiAl is a semi-coherent multilayer in which the layer thickness is typically 1µm. Within each layer, the γ ($L1_0$ structure) grains are also separated by semi-coherent interfaces, the grain (or domain) size being typically 40µm. A fraction of the layers retains the α_2 (hexagonal) structure and this fraction depends on the processing conditions. In the theory we assume, for simplicity, that the PST multilayer is entirely composed of γ grains.

Soft mode yield.

The three parameters to be calculated and inserted into eq.1 are the Taylor factor M, the critical resolved shear stress τ_o and the barrier strength τ^*. In the soft mode, individual layers can shear without constraining their neighbors so M is the inverse of the Schmid factor for the most favored deformation mode.Its value may be calculated precisely and it varies with orientation in the range 2.0 to 3.0. Critical resolved shear stresses have been calculated [6](by embedded atom simulation methods) for perfect <110]/2 dislocations and for superdislocations <101] at 0K but not for twinning Shockley dislocations. For perfect dislocations, τ_o depends on dislocation character, in the range 40MPa to 250MPa and for superdislocations, the τ_o values range from 25MPa to 120MPa. These may be compared with experimental measurements at 300K [5,7] which show that twinning, perfect and super dislocations operate in that order but with approximately the same CRSS of about 50MPa.

The calculation of τ^* presents much greater difficulty than the calculations of either M or τ_o because there is a spectrum of values of τ^*. Since the domains are rotated relative to each other by multiples of 60°, there are three physically distinct domain boundaries: 60°, 120° and 180°. Within each domain the strain may be carried by twin dislocations, perfect dislocations or superdislocations. If the strain is carried by one kind of dislocation in one domain it is often carried by another kind of dislocation in the next domain. In principle, τ^* should be calculated for each pair of dislocations interacting at each type of boundary. In addition to this proliferation of τ^* values, there are at least three physically different contributions to τ^* : the presence of mismatch dislocations at the interfaces, caused by the tetragonality of $L1_o$, the small change in Burgers vector, also caused by tetragonality and the gross change in Burgers vector across an interface caused by the crystallography of $L1_o$.

The three contributions to τ^* may be estimated as follows: At 60° and 120° (but not 180°) interfaces mismatch dislocations, spaced by s, exist to relax the internal stresses. These act as forest obstacles to dislocations passing the interface and an estimate of their contribution to τ^* is Gb/2s. Since s is $\sqrt{2}a/(c/a - 1)$ where a and c are the unit cell dimensions, τ^* is approximately 0.003G for twin dislocations, 0.005G for perfect dislocations and 0.01G for superdislocations. The second contribution to τ^* comes from the fact that a perfect dislocation has Burgers vector $a/\sqrt{2}$ whereas a superpartial has Burgers vector $(a^2 + c^2)^{1/2}/2$. When a perfect dislocation crosses either a 60° or a 120° interface to become a superpartial, it must increase its energy and this makes a contribution to τ^* of G(b/w)(c/a - 1) where w is the interface width. Putting w equal to 20b and c/a to 1.02, τ^* is found to be small, 0.001G. The third contribution to τ^* , from gross changes in b across an interface has only been calculated in the case of a perfect dislocation transforming into a superpartial dislocation at 60° and 120° interfaces and consequently dragging an antiphase boundary (APB). The value for τ^* is then γ/b or 0.03G when the APB energy γ is given the calculated value of 0.51 Jm^{-2} [8]. Atomistic simulations of this interaction show that an intrinsic stacking fault rather than an APB may form and that τ^* may be higher than 0.03G. The dislocations most affected by large changes in Burgers vector as they cross an interface are twinning dislocations because there is no interface across which the twin Burgers vector is continuous or nearly continuous. Consequently a twin in one domain must always trigger either a different twin or a glide dislocation in the next domain. The contribution to τ^* from this effect has not been calculated, although it is clearly large. 180° interfaces appear to be transparent to both perfect and super dislocations, for these dislocations at twin boundaries, $\tau^* = 0$.

For any combination of dislocation and boundary type, τ^* may, as a first approximation, be taken as the sum of the individual contributions because a mobile dislocation (the leader in the pile up) has to overcome all the contributions to the barrier simultaneously. The overall response of the solid is then controlled by some weighted average of the total τ^* values. This may be estimated, for example, by assuming that half the strain is carried by superdislocations, and a quarter each by perfect dislocations and by twins (there are two super, one perfect and one twin Burgers vector on each slip plane). If the three types of boundary, 60°, 120° and 180° are equally abundant, then the weighted average value of τ^* is close to 0.01G which, using eq. 1, gives a Hall-Petch slope of about 0.39 MPa$\sqrt{m}$, which is a little higher than the experimental value [9], 0.25+- 0.1 MPa$\sqrt{m}$.

The best estimates which can be made of the parameters in eq. 1 are M = 2.0 to 3.0, depending on orientation, τ_o = 50 MPa, τ^* = 0.01G and values close to these provide a good fit [10] with experimental data. However, twinning has been almost ignored in the theory. One further effect which should be included in a full theory is that there are two additional potentially very soft modes, basal slip in α_2 layers and interface sliding between the γ layers. Neither α_2 layers nor interfacial sliding (the 'supersoft mode') should suffer much Hall-Petch hardening because, for them, d>>50μm.

Hard mode yield

In the hard mode, the values of τ_o in γ TiAl are identical to those in the soft mode but both M and τ^* are higher. On the face of it, the Hall-Petch slope in the hard mode (eq. 2) is smaller than that in the soft mode (eq. 1) by a factor of $[(2-\nu)\alpha\pi^2/8(1-\nu)]^{1/2}$ = 1.4 to 1.8 but the increases in M and τ^* more than compensate for this factor with the result that the Hall-Petch slope is higher in the hard mode than in the soft mode.

The Taylor factor M should take account of the fact that individual layers must deform in a compatible manner in the hard mode. No calculation has been made of this effect but an estimate for M can be made by assuming that they deform identically. For example, if a specimen were deformed perpendicular to the layers ($\phi = 90^0$) the Schmid factors for all the glide systems are the same, $2/3\sqrt{6}$, and an estimate for the Taylor factor is then $3\sqrt{6}/2 = 3.7$.

Calculations of τ^* in the hard mode include the fact that all the interfacial obstacles which are present in the soft mode (interface dislocations, changes in Burgers vectors) are also present in the hard mode but there are also additional obstacles caused by the non-continuity of the slip planes in the hard mode. It is also found in EAM simulations that dislocations tend to spread their cores into the interfaces, thereby lowering their energy there. The work required to reconstrict them makes a further contribution to τ^* of uncertain magnitude [11].

The discontinuities in the slip planes are of two kinds: across 60° and 180° layer interfaces, the slip planes are twin related and consequently they bend through an angle of 39°. Across a 120° interface there is a small tilt caused by the tetragonality. The latter effect provides a weak obstacle, equivalent to placing jogs spaced at intervals of 2b/(c/a - 1) on the crossing dislocations. The estimated contribution to τ^* is 0.001G. The bend in the slip plane at 60° and 180° interfaces would appear to be a strong obstacle to most dislocations since it requires a large change in the Burgers vector at the interface. The single exception is a screw perfect dislocation crossing a 180° interface which is merely required to cross slip, a comparatively easy process in TiAl.

Since every interface is a barrier to every kind of dislocation in the hard mode and since τ^* contains all the soft mode contributions as well as some more in the hard mode, its value must be higher in the hard mode than in the soft. Averaging in the same way as in the soft mode, and ignoring those contributions which have not been quantified, an estimate for τ^* of 0.015G is reached. Using this value and an M value of 3.7, the hard mode Hall-Petch slope is found to be 0.55MPa$\sqrt{m}$, again a little higher than the experimental value of 0.45+-0.1 MPa$\sqrt{m}$ [9]. A possible reason why both the theoretical Hall-Petch slopes are higher than the experimental values could be that the deformation mode is selected which has a low, rather than an average, yield stress.

Using values of M = 3.7, τ_o = 50MPa and τ^* = 0.015G, a reasonable fit is achieved with experimental hard mode yield stresses in PST TiAl. However a full theory should take account of the presence of α_2 layers which deform by prism slip when ϕ is close to 0° and by pyramidal slip when ϕ is close to 90° [5].

APPLICATION TO Cu-Ni MULTILAYERS

Nano multilayers, such as Cu-Ni, are always tested in the hard mode, so eq. 2 is appropriate. In that equation, M varies between 2 and 4 depending on the type of test and the crystallography of the epitaxial layers, τ_o is very low since both Cu and Ni are soft metals and τ^* is a function of layer thickness. The interface strength has three components: τ^*_l caused by the difference in Burgers vector between Cu and Ni, τ^*_d caused by the bowing between the interface dislocations and τ^*_K the Koehler stress, caused by the fact that Cu and Ni have different shear moduli. τ^*_K is positive at Cu to Ni interfaces and negative at Ni to Cu interfaces and does not vary strongly with the layer spacing (expressed here as a wavelength λ, the sum of the Cu and Ni layer thicknesses).

The Koehler stress was originally calculated by elasticity theory [12] and has recently been calculated by the embedded atom method [13]. Both calculations give a result close to 0.01G in Cu-Ni. In the coherent (low λ) region τ^*_K is the only component in τ^* but in the semicoherent region, the interface barriers become stronger as the density of interface dislocations rises. The components τ^*_l and τ^*_d depend on λ in the same way which may be calculated as follows: Since the lattice parameters of Cu and Ni differ (by δa), both metals strain (Cu in biaxial compression, Ni in biaxial tension) and since their elastic constants are similar, the two strains are similar,$\pm\ \varepsilon/2$. The remaining mismatch is taken up by interface dislocations with spacing s such that $b/s = \delta a/a - \varepsilon$. As Frank and van der Merwe [14] showed, the energy is minimised when the elastic strain ε is approximately b/λ so long as λ is above the coherence limit λ_c ($\lambda_c = ab/\delta a$ and is close to 10nm in Cu-Ni). In very thin layers, the mismatch is all taken up elastically. In the semicoherent region, therefore, the barrier strength τ^* may be approximated by

$$\tau^* = \tau^*_K + \tau^*_d + \tau^*_l = 0.01G + G(\delta a/a - b/\lambda)(1/2 + 2b/w) \tag{3}$$

and this may be inserted into eq. 2 to give the yield stress. A similar expression could be used for σ when $\lambda < \lambda_c$ except that, in Cu-Ni, λ reaches another critical length first, the length at which a layer contains only one dislocation in its pile up. In this case ($\lambda < 20$nm) the yield stress when single dislocations cross Cu to Ni interfaces is

$$\sigma = M[\ \tau_o + \tau^*\ ] \tag{4}$$

The calculated values of the yield stress as a function of wavelength [13] from eqs. 2 or 4 are compared with experiment in Fig. 2. The model captures the main features of the experimental results: at $\lambda > 100$nm the yield stress follows the normal Hall-Petch equation, with slope K = $0.25\text{MPa}\sqrt{\text{m}}$. When $\lambda = 1\mu$m there are 10 dislocations in a pile up. As λ decreases, the number of dislocations N drops and the discrete nature of the dislocations in the pile up becomes increasingly important until, at $\lambda = 20$nm, the yield stress peaks when N = 1. Below this wavelength dislocations cross interfaces singly. The yield stress drops sharply between $\lambda = 20$nm and $\lambda = \lambda_c = 10$nm as the density of the mismatch dislocations falls. At $\lambda = 10$nm the interfaces become coherent and for smaller wavelengths than this, according to the current theory, the yield stress is constant at the tensile Koehler stress. The behavior in the coherent region has not been much explored theoretically but it seems that the main obstacles for the dislocations to overcome are the operation of the source (probably in a Ni layer) and the first Cu to Ni barrier encountered as the dislocation loop expands. The yield stress must drop sharply as λ drops below 10nm because, when λ becomes very small, say < 1nm, the multilayer has become a superlattice and a typical value for the yield stress of such an ordered alloy is low, 0.1GPa.

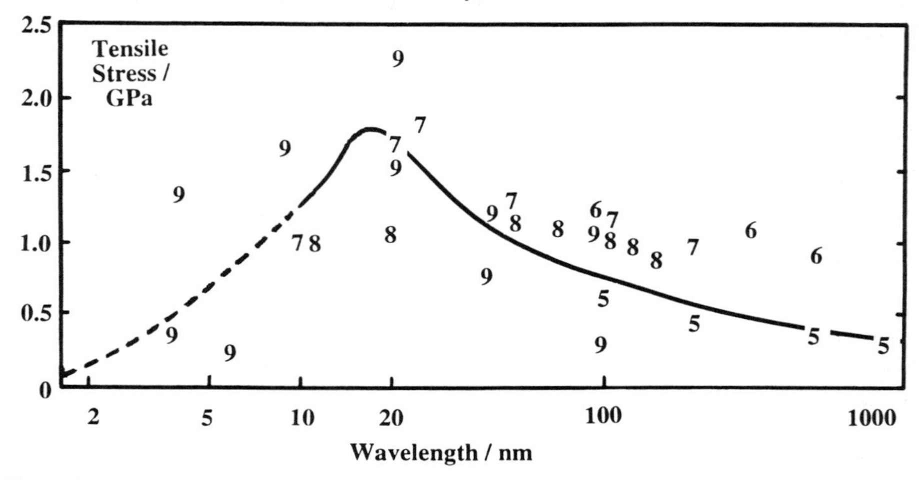

Figure 2. Yield stress versus log wavelength for Cu-Ni. Full line is the theory (eqs. 2,3,4). Numbers are experimental measurements taken from the literature, see [13] for details.

CONCLUSIONS

The Hall-Petch theory can be used to describe the yield stresses of both micro- and nano-multilayers. Because of the geometry of the layers, the strength is anisotropic. The theory does not predict a simple $d^{-1/2}$ dependence on grain size when the strengths of the interfaces depend on the layer thicknesses. The theory is far from being fully quantitative; particular candidates for future work are (1) to replace classical estimates of barrier strengths by atomistic calculations and (2) to simulate the operation of dislocation sources in fully coherent multilayers.

ACKNOWLEDGMENTS

The authors acknowledge support from Air Force contract # F 33615-91-C-5663 with Wright Laboratory Materials Directorate, WL/MLLM, Wright-Patterson Air Force Base, OH 45433

REFERENCES

1. J. D. Eshelby, Phys. Stat. Sol. **3**, 2057 (1963).
2. J. C. M. Li and G. C. T. Liu, Phil. Mag. **15**, 1059 (1967).
3. J. C. M. Li and Y. T. Chou, Met. Trans. **1**, 1145 (1970).
4. M. Yamaguchi and Y. Umakoshi, Prog. Mater. Sci. **34**, 1 (1990).
5. M. Yamaguchi, H. Inui, K.Kishida, M.Matsumuro and Y. Shirai, MRS Proc. **364**, 3 (1995).
6. S. I. Rao, C. Woodward, J. Simmons and D. M. Dimiduk, MRS Proc. **364**, 129 (1995).
7. H. Inui, M. H. Oh, A. Nakamura and M. Yamaguchi, Acta Metall. Mater. **40**, 3095 (1992).
8. C. L. Fu and M. H. Yoo, MRS Proc. **186**, 265 (1990).
9. T. Nakano, A. Yokoyama and Y. Umakoshi, Scripta Metall. Mater., **27**, 1253 (1992).
10. P. M. Hazzledine, B. K. Kad and M. G. Mendiratta, MRS Proc. **308**, 725 (1993).
11. S. I. Rao, C. Woodward and P. M. Hazzledine, MRS Proc. **319**, 285 (1994).
12. J. S. Koehler, Phys. Rev. **B2**, 547 (1970).
13. S. I. Rao, P. M. Hazzledine and D. M. Dimiduk, MRS Proc. **362**, 67 (1995).
14. F. C. Frank and J. H. van der Merwe, Proc. Roy. Soc. London **A198**, 216 (1953).

MICROMECHANICS OF DEFORMATION AND FRACTURE IN LOW SYMMETRY LAYERED MATERIALS

Bimal K. Kad, Ming Dao and Robert J. Asaro
Department of Applied Mechanics and Engineering Sciences
University of California-San Diego, La Jolla, CA 92093-0411

Abstract

Deformation microstructures in γ-TiAl + α_2Ti_3Al based low symmetry layered materials, with fully lamellar (FL) microstructures, have been simulated using micro-mechanical methods. In this particular effort we embed the specific contributions of scale and temperature dependent plastic anisotropies of individual colonies of Poly Synthetically Twinned (PST) lamellar TiAl and demonstrate their effect on overall deformation and fracture response.

1. Introduction

Recently, an effort has been made to apply finite element procedures [1], incorporating physically based crystal plasticity models [1-6], to study the evolution of non-uniform deformation in TiAl based polycrystalline alloys of engineering development interest [3]. The impetus for such efforts is to gather fundamental insight into microstructure sensitive deformation mechanisms, and to extract additional information, that is nominally not obtainable from traditional mechanical property measurements. In lamellar TiAl alloys, such an effort is particularly desirable to help track various aspects of plastic anisotropy of single crystals (i.e., strength, failure strain etc.), and the contribution of micro-constituents (γ-grain vs lamellar colony volume fraction) as implicit in polycrystalline aggregates.

The microstructural deformation inhomogeneity in polycrystal lamellar TiAl, for a fixed level of soft vs hard-mode plastic anisotropy, was explored earlier [3]. However, this anisotropy is variable, and extremely processing dependent. In this paper, we illustrate the contribution of this single PST crystal anisotropy to the deformation response of the polycrystal aggregate.

2. Mechanical & Microstructural Details

The lamellar geometry is comprised of flat slabs of α_2-Ti3Al and γ-TiAl with an HCP-FCC type orientation relationship as $(0001)\alpha_2 \parallel \{111\}\gamma$ and $<11\bar{2}0>\alpha_2 \parallel <1\bar{1}0>\gamma$ in a given colony. Ti-(48-50)at%Al alloys initially solidify as disordered α-hcp, which upon cooling undergo solid state transformations α->γ and α_2->γ yielding the laminate morphology. The unique orientation of the basal plane normal in the parent α-hcp phase determines the eventual orientation of the flat slabs in the polycrystalline aggregate.

In the single PST laminate form, γ-TiAl is the softer phase and an orientation dependent flow anisotropy is derived particularly from the laminate geometry where shear deformation parallel to the slabs (soft mode) is easier than across the slabs (hard mode) owing to the difficulty of propagating slip through the harder α_2-phase [7], Figure 1. In lamellar single crystals, the most prominent deformation modes are: i) Soft-mode: lamella oriented about 45° to the loading axis. Deformation occurs in the soft γ-phase with deformation vectors parallel to the lamellar interface, and ii) Hard-mode: lamella oriented perpendicular to the loading axis. Maximum constraint exerted by the α_2 phase is the stress required (τ_{crss}=910MPa, [7]) to activate the $<11\bar{2}6>\{11\bar{2}1\}$

Mat. Res. Soc. Symp. Proc. Vol. 434 © 1996 Materials Research Society

pyramidal slip. In particular, the hard-mode strengths are strongly scale (i.e. lamellar thickness) and composition dependent, figure 1(b), and can be varied by processing or imposing different cooling rates on alloys cooled from the α-phase field, and is the specific problem addressed here.

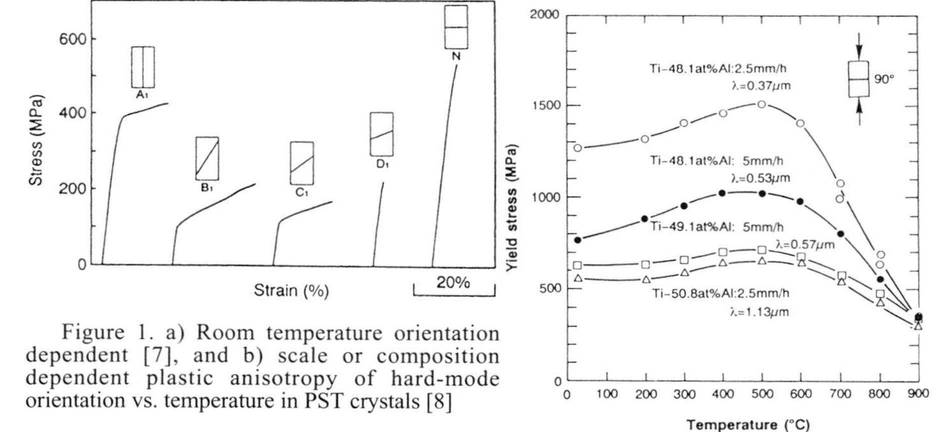

Figure 1. a) Room temperature orientation dependent [7], and b) scale or composition dependent plastic anisotropy of hard-mode orientation vs. temperature in PST crystals [8]

2. Theoretical & Microstructural Details

Single crystal constitutive law: The single crystal constitutive law, in its current form, is developed by Asaro and his coworkers [1-6]. The total deformation gradient is decomposed into plastic ($\mathbf{F}^P$), thermal ($\mathbf{F}^\theta$) and lattice parts ($\mathbf{F}^*$), i.e., if $\mathbf{u}$ is the displacement vector and $\mathbf{X}$ is the material position vector with respect to the reference (undeformed state) then $\mathbf{F} = \mathbf{I} + \partial\mathbf{u}/\partial\mathbf{X}$, and

$$\mathbf{F} = \mathbf{F}^* \cdot \mathbf{F}^\theta \cdot \mathbf{F}^P. \tag{1}$$

Plastic deformation occurs by the flow of the material through the lattice, via simple shearing, across planes with unit normals $\mathbf{m}_\alpha$ and in directions $\mathbf{s}_\alpha$; $\mathbf{m}_\alpha$ and $\mathbf{s}_\alpha$ are crystallographic slip plane normal and slip direction, respectively and the symbol α designates the specific slip system. If $\dot{\gamma}_\alpha$ is the shear rate on the αth slip system and $\mathbf{F}^P$ is the plastic part of the deformation gradient, the value of $\mathbf{F}^P$ is given by the path dependent integration of equation (2):

$$\dot{\mathbf{F}}^P \cdot \mathbf{F}^{P\text{-}1} = \sum_\alpha \dot{\gamma}_\alpha \, \mathbf{s}_\alpha \mathbf{m}_\alpha \tag{2}$$

The kinetic laws governing the rate of shear on a particular slip system have the general form:

$$\dot{\gamma}_\alpha = f_\alpha(\textit{stress state, material state, temperature}) \tag{3}$$

The kinetic description of plasticity on each slip system, as a particular case of (3), is cast in terms of the *loading parameter* τ_α^D and the slip rate on that system as

$$\dot{\gamma}_\alpha = \sum_\alpha \dot{a} sgn\{\tau_\alpha\}\left\{\left|\frac{\tau_\alpha^D}{g_\alpha}\right|\right\}^{\frac{1}{m}} \tag{4}$$

where τ_α is the current value of the resolved shear stress, τ_α^D is the *loading parameter* for slip, $g_\alpha > 0$ is the current value of the *slip system hardness*, and m is the material rate sensitivity exponent (which will, in the examples described herein, be taken the same for each slip system), and $\dot{a}$ is the reference shear rate and sgn( ) is the sign of the quantity. τ_α^D, is derived primarily from the *resolved shear stress* on that slip system along with minor contributions from *non-Schmid effects*. This generalized stress which acts to load a slip system is given as

$$\tau_\alpha^D = \tau_\alpha + \eta_\alpha : \tau = \mathbf{m}_\alpha^* \cdot \tau \cdot \mathbf{s}_\alpha^* + \eta_\alpha : \tau \tag{5}$$

where τ is the Kirchhoff stress tensor, and η_α is the tensor of *non-Schmid effects* for slip system α. $\mathbf{s}^*_\alpha$ is along the αth slip direction in the current configuration and $\mathbf{m}^*_\alpha$ is normal to the αth slip plane.

The slip system hardness g_α is obtained by the path dependent integration of the evolution equation

$$\dot{g}_\alpha = \sum_\beta h_{\alpha\beta}(\gamma_a)\left|\dot{\gamma}_\beta\right| + g^\theta_\alpha\dot{\theta}, \qquad \gamma_a = \int_o^t \sum_\alpha \left|\dot{\gamma}_\alpha\right| dt \tag{6}$$

where $h_{\alpha\beta}$ is a matrix of hardening moduli, g^θ_α is the rate of change of slip system hardness with respect to temperature alone, and γ_a is the accumulated sum of slips. The initial condition for this evolution is given by $g_\alpha(\gamma_a = 0, \theta = \theta_o) = g_\alpha(\theta_o)$ where θ_o is the initial temperature.

This constitutive theory has been implemented into our finite element code, specific details of which are available elsewhere [1,3]. In the present paper we only deal with isothermal deformations without non-Schmid effects.

Input configuration for 2-D FEM idealization: Complete three dimensional finite element polycrystal analyses require large amounts of computer memory and CPU time. Fortunately much progress can be made using two dimensional models, as demonstrated earlier [1]. In the 2-dimensional $\gamma+\alpha_2$ composite formulation, the ascribed slip geometry is based on the $\gamma+\alpha_2$ composite mixture, figure 2(a). Thus, the soft mode ($\mathbf{s}_1$, $\mathbf{m}_1$), is ascribed to the γ-TiAl phase, the hard modes ($\mathbf{s}_2$, $\mathbf{m}_2$) and ($\mathbf{s}_3$, $\mathbf{m}_3$), are ascribed to the α_2-Ti_3Al phase. In both cases of two dimensional configurations, the lamellar normals are configured to lie in-plane and only in-plane slip vectors are considered. These are arranged in an non-equilateral triangle, figure 2(b), and the reference crystal base vectors $\mathbf{a}_i$ are aligned with the crystal lattice as shown in figure 2(c).

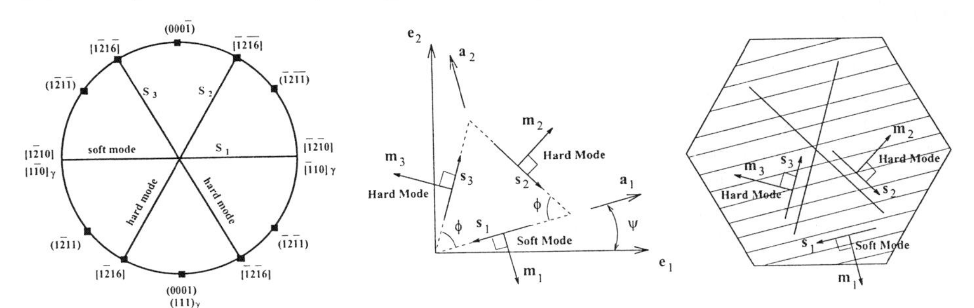

Figure 2. a) $\langle 10\bar{1}0\rangle\alpha_2 \parallel \langle 11\bar{2}\rangle\gamma$ projection indicating the two dimensional input slip systems; soft mode derived from shear in γ-TiAl parallel to the laminates and hard mode from the $\langle 11\bar{2}6\rangle\{11\bar{2}1\}$ pyramidal slip system constraint imposed by the harder α_2-Ti_3Al phase, b) Final construction of the slip-system morphology for the 2-D model, and c) Embedding the 2-D slip geometry into the single crystal lamellar morphology.

The reference configuration of the two dimensional polycrystal microstructure to be analyzed is shown in figure 3, where the Cartesian base vectors describe the orientation of the polycrystal's reference configuration with respect to the laboratory. Each of the 27 grains (or single PST crystals) is defined by an orthogonal transformation $\mathbf{a}_i^n = \psi_{ij}(\psi^n)\mathbf{e}_j$, n =1 to 27, where ψ^n is one of the angles ψ shown in figure 2(b). The finite element mesh used in the polycrystal calculations consists initially of rectangular "crossed-triangle" quadrilateral elements in a uniform grid 40 rectangles wide by 56 rectangles high, which is 8960 constant strain triangles totally. All the "continuum grain boundaries" in this polycrystal model coincide with either an edge or a diagonal of a quadrilateral element, i.e. with an edge of a constant strain triangle. The polycrystal's initial reference configuration is assumed to be stress free and without any lattice perturbations. Therefore, no lattice distortion and residual stress caused by lattice distortion near grain boundary are included. Each grain boundary is a line across which the initial lattice orientation ψ^n has a jump. All simulation runs are loaded in e_2 direction, figure 3.

The PST single crystal (colony) data for two-phase $\alpha_2+\gamma$ fully lamellar alloys is obtained for tension [7], and at varying lamella thickness [9].

τ_{crss} (soft) = g_1^o = 50 MPa,

$\delta\tau/\delta\gamma$ (soft) = h_{11} = 150MPa

τ_{crss} (hard) = $g_2^o = g_3^o$ = 600, 900, 1200 MPa

$\delta\tau/\delta\gamma$ (hard) = $h_{22} = h_{33}$ = 1500MPa,

The tensile deformation is simulated at 10^{-3}sec^{-1} strain rate, consistent with the strain rates applied in quasi static tests. The reference shearing rate is 10^{-3}sec^{-1}, and the material strain rate sensitivity exponent m=0.005. Isotropic elasticity is assumed with λ=69.7GPa and μ=71.4GPa. A simple linear hardening is taken with $h_{ij}=0$ ($i \neq j$).

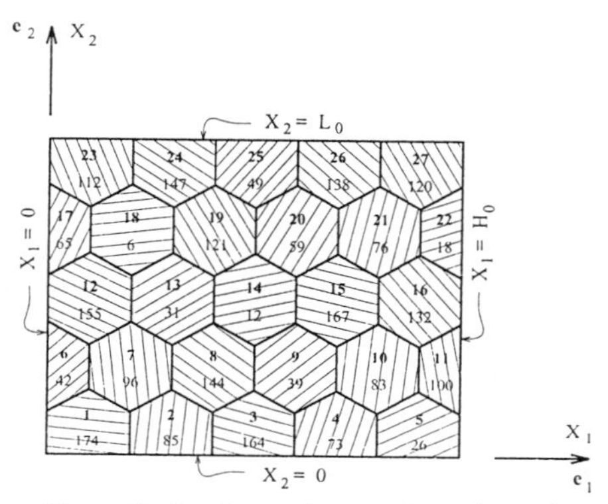

Figure 3. Random polycrystal configuration

3. Results and Discussion

Flow Stress Behavior: Figure 4 shows the computed stress strain behavior for the two-dimensional model microstructure, for an input soft vs. hard-mode plastic anisotropy of 12, 18, and 24. and is typical of the experimentally observed plastic behavior in lamellar microstructures, as shown in figure 1(b). Matching in the elastic regime is poor on account of the limited initial constraints in the two dimensional polycrystal, but despite the two dimensional idealization, the single crystal properties input parameters predict the polycrystalline response with reasonable accuracy, and the underlying deformation mechanisms are well captured. Note that while the model is scale independent, these input parameters can be scaled to account for scale effects of grain size and lamella thickness changes as reported elsewhere [3]

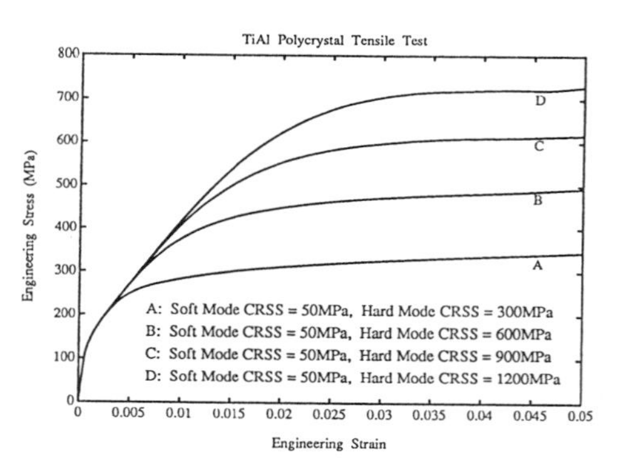

Figure 4. Computed stress-strain curves for the idealized FL microstructure of figure 3.

Strain Accumulations: The accumulated strains at aggregate tensile strains of 2.5% are shown for a plastic anisotropy of 12 ($\tau_{soft\text{-}mode}$=50MPa, $\tau_{hard\text{-}mode}$ =600MPa), figure 5(a), and 24

($\tau_{soft\text{-}mode}$=50MPa, $\tau_{hard\text{-}mode}$=1200MPa), figure 5(b), respectively. The magnitude, and distribution pattern, of accumulated strain localizations are essentially identical in the two microstructures. This derives from the fact that the deformation strains are largely contributed by the soft-mode operative in this 2-D microstructure. Since $\tau_{soft\text{-}mode}$=50MPa is kept fixed (analogous to a fixed colony/grain size) we have essentially identical strain distributions. In a given polycrystal microstructure, regions of gross inhomogenous deformation, both between and within grains are immediately obvious. Maximum localized strains on the order of 40% are computed at or near the grain boundaries as illustrated between crystal pairs 1-6, 7-13, 16-21, 19-25 in figure 5(a), and experimentally observed [3]. The triple point between crystals 12, 13, and 18 constitutes strain localization of ≈25%. Significant inhomogeneities are also observed within the bulk of crystals 3, 4, 5, 12, 13, 16, 26, 27, with transition zones from roughly 0% strain to >20% strain. Within a given crystal, the strain anisotropies are both parallel (in crystals 6, 8, 24) and perpendicular (in crystals 12, 13, 25, 26) to the laminates. As discussed in the next section, strain localizations parallel to the soft mode yield lower hydrostatic stress components, whereas strain localizations across the plate require activation of hard mode slip systems, or cutting of the α_2 plates and are considered as being additional constraining events requiring higher stress buildups, and presumably affecting fracture response.

A peculiar characteristics of the crystals is their response to the varying nature of constraint at its boundaries. For example, in crystals 12, 13 (both oriented in the soft mode) large portion of the crystals do not deform as they are constrained by their neighbors. Additionally, in crystal 7, while the bulk of the crystal does not deform, some small deformations are indeed observed very close to the boundaries to comply with its more readily deformable neighbors. The orientation dependent single crystal deformation behavior is thus collectively affected by its neighbors, which in essence raises the flow stress.

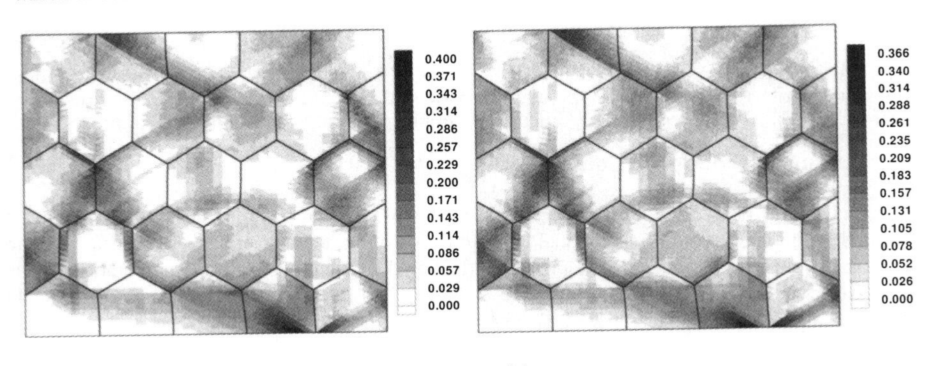

a) **b)**

Figure 5. Computed strain accumulations (at 2.5% macroscopic strain) in FL microstructure for a) $\tau_{soft\text{-}mode}$=50MPa, $\tau_{hard\text{-}mode}$=600MPa, b) $\tau_{soft\text{-}mode}$=50MPa, $\tau_{hard\text{-}mode}$=1200MPa

Hydrostatic Stresses: Hydrostatic stresses developed in the fully lamellar microstructures (corresponding to strain accumulations shown in figure 5) at 2.5% aggregate strain are shown for plastic anisotropy of 12 ($\tau_{soft\text{-}mode}$=50MPa, $\tau_{hard\text{-}mode}$=600MPa), figure 6(a), and 24 ($\tau_{soft\text{-}mode}$=50MPa, $\tau_{hard\text{-}mode}$=1200MPa), figure 6(b) respectively. For ease of viewing, only hydrostatic stresses > 1000MPa are plotted and, the distributions are scaled with respect to peak stresses computed for the $\tau_{soft\text{-}mode}$=50MPa, $\tau_{hard\text{-}mode}$=1200MPa input data-set (i.e., plastic anisotropy of 24). As described previously [3], the largest values of hydrostatic stresses, far exceeding the experimentally observed fracture stress, are encountered at the boundaries and triple points. The hydrostatic stresses arise from a large accumulation of strain, especially those accumulations that activate hard mode slip. For $\tau_{hard\text{-}mode}/\tau_{soft\text{-}mode}=12$, figure 6(a), the peak stress is 1815MPa, and only a minute volume fraction of the microstructure experiences stresses > 1000MPa, whereas for $\tau_{hard\text{-}mode}/\tau_{soft\text{-}mode}=24$, figure 6(b), the peak stress is ≈2800MPa, and a significant volume fraction of the microstructure experiences stresses > 1000MPa. It was shown earlier [3] that the presence of large hydrostatic stresses at these boundaries provides nucleation sites for fracture of debonding/decohesion type. In the present case, increasing the level of plastic anisotropy (while holding $\tau_{soft\text{-}mode}$ fixed at 50MPa) puts the microstructure at greater risk of possible crack initiation and failure. These trends are expected to account for both scale and positive temperature dependent plastic anisotropies as shown in figure 1.

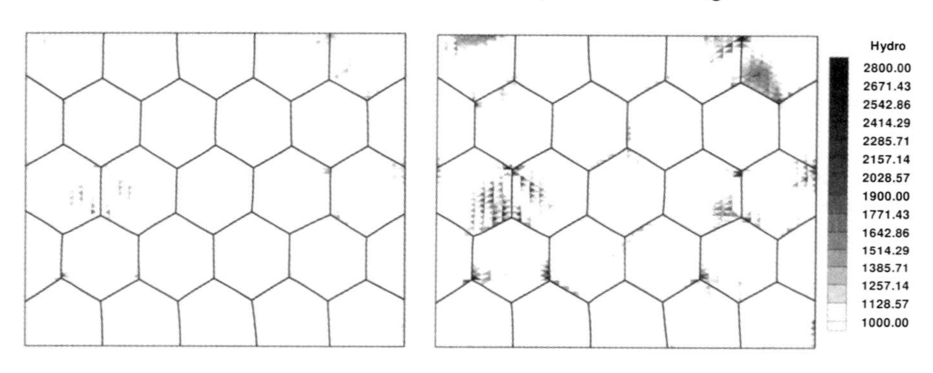

Figure 6. Localization of hydrostatic stresses >1000MPa in FL alloys with soft vs hard-mode anisotropy of a) 12, and b) 24. Both (a) and (b) are scaled with respect to the peak hydrostatic stress (≈2800MPa) generated in (b), thereby yielding comparative site and distribution details.

4. Summary and Conclusions

Increasing the hard-mode plastic anisotropy (analogous to decreasing lamella spacings), while keeping the soft mode fixed (analogous to fixed colony/grain size) has a small effect on the ensuing strain distributions, but a significant effect on the magnitude of peak hydrostatic stresses and stress-distributions. Fracture initiation events are expected to be enhanced with this scale and/or temperature dependent plastic anisotropy.

References

1 S.V. Harren H.E. Deve and R.J. Asaro, *Acta Metall.*, **36** (1988) 2435.
2 R.J. Asaro and A. Needleman, *Acta. Metall.*, **33** (1985) 923.
3 B.K. Kad, M. Dao and R. J. Asaro, *Philos. Mag.*, **71** (1995) 567; *Mat.Sci. & Eng.*, **A192/193** (1995) 97; *MRS Symp. Proc.*, **364**, 169.
4 R.J. Asaro, Acta Metall., **27** (1979) 445.
5 D. Pierce, R.J. Asaro and A. Needleman, Acta Met., **31** (1982) 1951.
6 P.E. McHugh, R.J. Asaro and C.F. Shih, *Acta Metall.*, **41** (1993) 1461
7 H. Inui, M.H. Oh, A. Nakamura and M. Yamaguchi, *Acta Metall.*, **40** (1992) 3095.
8 Y. Umakoshi and T. Nakano, *ISIJ Int.*, **32** (1992) 1339
9 T. Nakano, A. Yokoyama and Y. Umakoshi, *Scripta Metall.*, **27** (1992) 1253

COHERENCY STRAIN AND HIGH STRENGTH AT HIGH TEMPERATURE

M.E. BRENCHLEY [*], D.J. DUNSTAN [**], P. KIDD [***], A. KELLY [****]

[*]Department of Physics, University of Surrey, Guildford, Surrey GU2 5XH, England, M.Brenchley@surrey.ac.uk

[**]Department of Physics, Queen Mary and Westfield College, London E1 4NS, England,

[***]Department of Materials Science and Engineering, University of Surrey, Guildford, Surrey GU2 5XH, England,

[****]Department of Materials Science and Metallurgy, University of Cambridge, Cambridge CB2 3QZ, England,

ABSTRACT

We propose an athermal strengthening mechanism for high-temperature structural materials in which large coherency strains are built in to a layered structure in order to prevent dislocation mulitplication mechanism from functioning. A practical model system is provided by semiconductor strained-layer superlattices of InGaAs grown on InP. We report results from high-resolution X-ray diffraction and from direct tensile testing which provide evidence for athermal strengthening. A discussion of methods of micro-mechanical testing is also included.

INTRODUCTION

High elastic yield strength together with plasticity to absorb energy before fracture are the objects of research on strong structural materials or composites. Strategies include the control of dislocation movement in a relatively soft phase by the introduction of obstacles which may be small (as in precipitation hardening alloys) or extended (as in the platelets of cementite in steel) [1]. It has also been proposed that strengthening could be achieved by the introduction of interfaces between soft phases with a significant change in elastic constants [2]. If, additionally, high strength is required at high service temperature, it is necessary that the blocking mechanism should continue to operate at high fractions of the melting points of the constituent phases [3]. In this paper, we propose a new method for athermal strengthening of a microstructure, consisting of interfaces between layers which have high values of coherency strain. This approach could operate right up to the melting point. We propose that the very highly developed crystal growth and processing techniques for semiconductors make them an ideal model system in which to test concepts such as this, and we present preliminary data which show that athermal strengthening due to coherency strain is a real effect.

It is straightforward to grow by molecular-beam epitaxy layers of alloy semiconductor crystal from 1nm thick up to several microns with layer thicknesses and interface flatness accurate to a few ångstroms. Coherency strains up to about 1.5% are readily obtained by changing the alloy composition so as to change the misfit between the lattice constant of the alloy layer and the lattice constant of the crystal wafer used as a substrate. Such layers are thermodynamically stable and retain the full misfit strain if they are grown with less than a critical strain, related to the thickness by critical thickness theory (reviewed by Fitzgerald [4]). For our purposes it is sufficient to note that the critical strain is given approximately by

$$\varepsilon_c = \frac{0.1\ (\text{nm})}{h} \qquad (1)$$

Mat. Res. Soc. Symp. Proc. Vol. 434 © 1996 Materials Research Society

where h is the thickness of the layer. We previously studied the behaviour of layers grown above critical thickness [5, 6]. The key points to retain here are that high-quality layers grown under compression do not relax—they do not undergo plastic deformation—until a thickness several times critical thickness, and then they follow the relationship

$$\varepsilon(h) = \frac{0.8\ (\mathrm{nm})}{h} \tag{2}$$

This behaviour was quite surprising, and indeed it still remains controversial.

The semiconductor layers obeying Eqn.2 were grown at temperatures between 350°C and 650°C, the latter being well over half their melting-points [6]. Even at much higher temperatures they retain a large fraction of the strain of Eqn.2 [7]. Bulk semiconductors at these temperatures support strains only of the order of 10^{-4}, becoming weaker rapidly with temperature, while observations of dislocation mobility in strained layers showed the expected thermal activation (see, e.g. [8]). Thus both the absolute magnitude of the strain-thickness product (of 0.8nm) of Eqn.2, and its independence of temperature were unexpected.

An explanation was put forward by Beanland [9] in terms of the thickness needed for dislocation multiplication mechanisms such as spiral or Frank-Read sources to operate. If the dislocation loops required for these sources cannot extend into the substrate nor beyond the free surface, the radii of the loops are required to be small (a fraction of the layer thickness) and the stresses required to drive the source are then high. This model will apply as well to tensile layers as to compressive, and this suggests that a multilayer structure, rather than a single layer, could be constructed in which dislocation multiplication in each layer would be confined to that layer and therefore blocked at any temperature for strains below the value of Eqn.2. If the independence of each layer could be maintained, such a structure could be made indefinitely thick and become the basis of the microstructure of a high-temperature high-strength structural material with a designed elastic yield strength.

EXPERIMENT

Sample Structures

We have investigated these ideas using superlattices 2.5µm thick grown by molecular beam epitaxy (MBE) on InP of alternate compressive and tensile 50nm layers of $In_xGa_{1-x}As$. The misfit in each layer to the InP substrate is given in terms of the layer composition as

$$\varepsilon(x) = 0.07 \times (0.53 - x) \tag{3}$$

(tensile for x less than the lattice-match indium content of 0.53). Such structures can be made with individual layer strains and net strains (averaged over the layers) freely chosen from 0 to 0.015 tensile or compressive. Detection of effects of coherency strain on the structural properties are made by comparison with 2.5µm thick layers of homogeneous alloy of the same average composition as the superlattices.

High-Resolution X-Ray Diffraction

We started with structures designed to demonstrate the effect in the simplest possible way, using high-resolution X-ray diffraction (HRXRD). A superlattice was grown with (nominally)

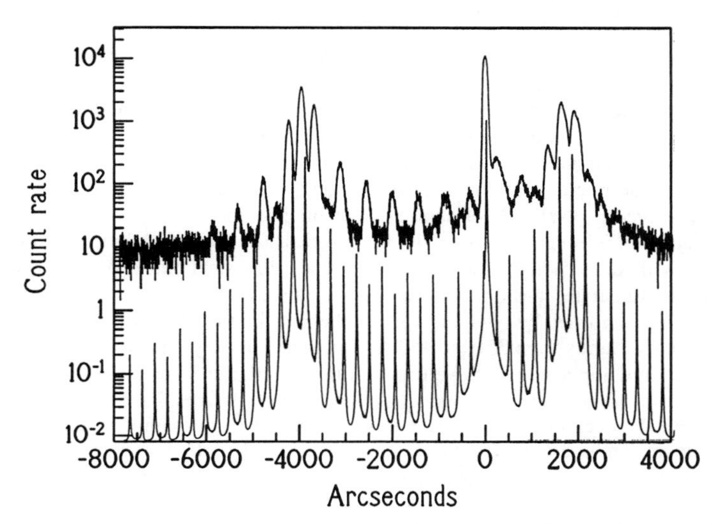

Figure 1. An HRXRD rocking curve for the superlattice with -0.0026 net (compressive) strain is shown, together with the predicted rocking curve for the structure (shifted down one decade for clarity). The fine structure on the experimental curve shows that no significant relaxation has occurred in this structure. The good agreement between the positions of the principal peaks in the theoretical and experimental curves and the period of the fringes confirms this, while the slight differences in fringe peak positions and intensities correspond to trivial differences between the sample and the model.

$\varepsilon = -0.008$ compressive strained 50nm layers alternating with (nominally) $\varepsilon = +0.004$ tensile strained 50nm layers, to a total thickness of 2.5μm. The superlattice has an average strain of -0.002 (compressive), and a homogeneous layer, also 2.5μm thick, was grown with this strain for comparison. Both the superlattice and the homogeneous layer have a (nominal) strain-thickness product of 5nm, well in excess of that allowed by Eqn.2. HRXRD rocking curves of these samples were recorded. The rocking curve of the superlattice shows a wealth of detail (Fig.1) which demonstrates the high quality of its growth and which is analysed by simulating the curve with a software package (Bede RADS) which calculates what the structure should give according to dynamical X-ray scattering theory. The result shows that the structure actually has a net strain of –0.0026 (more than the nominal value but within growth tolerances), and is completely unrelaxed. This means that it is actually supporting a strain-thickness product of 6.5nm. In contrast, the homogeneous layer gave broadened rocking curves characteristic of samples that have undergone plastic deformation, or relaxation. A full analysis of a set of rocking curves [10]. gave a composition of $x = 56.9$ and a strain of only -0.0021, significantly relaxed from its misfit value of -0.0026. Thus the coherency strain has had the effect of increasing the elastic yield strength of the superlattice, at the growth temperature of 500°C, to at least 320MPa (for a biaxial modulus of 123GPa). This may be compared with 10MPa for bulk material [12], or with 39MPa for samples falling on the curve of Eqn.2. The alloy sample lies above the curve of Eqn.2 and supports 260MPa which may be due to work-hardening during relaxation [5]. But the superlattice is not relaxed at all and therefore cannot gain its increased strength from work-hardening.

Micromechanical Structural Testing

To explore the effects of coherency strain further, we have designed two sets of experiments to measure tensile yield strength as a function of time and temperature, strain and superlattice period, in superlattices with (nominal) zero net strain. These experiments are in an early stage and it is largely the techniques which we wish to discuss here. We exploit the selective etches available in semiconductor technology [11] to remove the InP substrate and leave free-standing epitaxial layer for micro-mechanical measurements. We have attempted to measure the yield strength by inducing plastic deformation at elevated temperature (with a nitrogen ambient to avoid oxidation of the specimens). Two geometries are used. For bulge-testing, a sample is cemented growth-side down over a 2mm hole in a stainless steel block. A hole about 3mm square is etched in the substrate; masking with wax leaves a solid frame around the hole. Biaxial tensile strain can then be imposed by applying vacuum behind the sample. Preliminary results have been obtained at temperatures from 350° to 450°C. Samples normally fail by fracture or tearing (Fig.2) at widely scattered applied strains; we find that the superlattices sometimes support much higher strains than the comparison homogeneous layers, but the results are not yet reproducible. Problems with this technique include the mode of failure of the samples, and the lifetime of the cement at elevated temperature (a silicone cement lasts well at 450°C). Furthermore, bulge testing is liable to introduce artefacts if the initial state of strain of the sample is not accurately known; while we do measure this routinely by HRXRD the corrections are only approximate and comparisons between samples not completely rigorous.

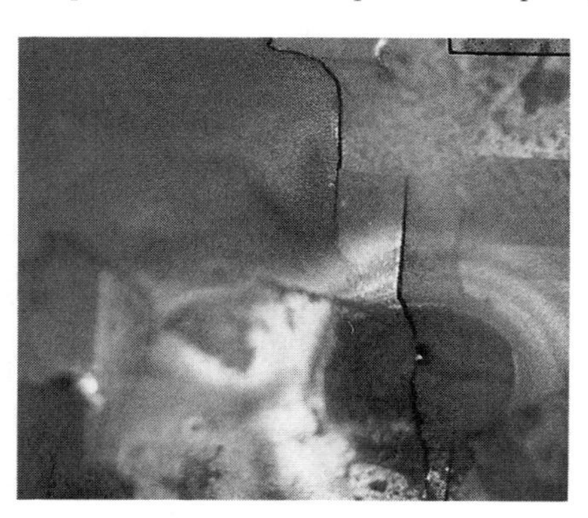

Figure 2. A typical sample after failure in bulge-testing. Note that there are regions in which brittle fracture (cleaving) appears to have occured as well as regions where the fracture is curved.

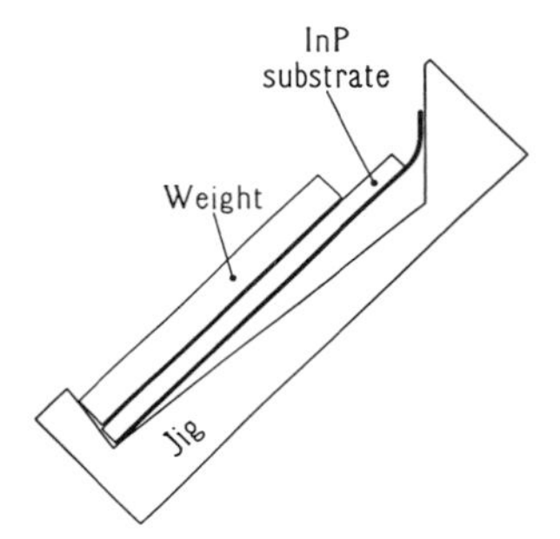

Figure 3. A specimen prepared for beam-bending experiments by etching away the substrate at one end, leaving a straight free-standing beam approximately 1.5mm long by 1mm wide and 2.5μm thick. The epitaxial layer is represented by the heavy line. The sample is placed in the jig with a stainless steel weight and the stress in the beam is increased by rotating the jig towards horizontal.

The other geometry we are using is a cantilever beam. It is easy to etch away the substrate so as to leave a free-standing beam about 1.5mm long by 1mm wide, cantilevered out from the substrate (Fig.3). It is necessary to symmetrise the superlattice structure to avoid the beam curling up; this has proved feasible (growth tolerances are sufficiently tight) and a typical beam is quite straight. The beam is then placed on a stainless steel jig in the furnace with weights behind it (Fig.3), and a controlled stress can then be induced by rotating the jig to bring the beam towards horizontal. Remarkably, we have been unable to induce plastic deformation even in the homogeneous alloy beams, at temperatures as high as 600°C. The largest loads we have applied give a maximum stress of about 200MPa, an order of magnitude above the yield stress of about 10MPa at 600°C [12], yet even after this load has been applied overnight the samples spring back to their original shape on unloading. We believe that this behaviour may be understood in terms of the critical strain-thickness product of 0.8nm of Eqn.2. We have shown elsewhere that in a layer with a varying strain profile $\varepsilon(h)$, such as a multi-layer or a graded layer, that it is the average strain times the thickness which is to be used in Eqn.2 [13, 14]. For a radius of curvature of R and a thickness d the strain at the surface of the beam is

$$\varepsilon_s = \frac{d}{2R} \tag{4}$$

and the strain-thickness product integrated from the neutral plane to the surface is

$$S = \int \varepsilon(h)\, dh = \frac{1}{4}\varepsilon_s d = \frac{1}{8}\frac{d^2}{R} = 0.78 \text{ nm} \tag{5}$$

where the numerical value is for d = 2.5μm and R = 1mm corresponding to a surface stress of about 125MPa.

DISCUSSION

Preliminary results provides three data in support of our proposed mechanism of strengthening and the theory of Eqn.1 and 2: (i) The compressive net strain superlattice which did not relax, (ii) the higher maximum pressures required to fracture the superlattice in the bulge testing; and (iii) the completely unexpected failure to deform the alloy beams plastically in beam-bending. Further development of the micro-mechanical testing is required, firstly to obtain a quantitative measure of the increase in yield stress in the superlattices, and then to find how it varies with superlattice layer thicknesses and strains.

CONCLUSIONS

The net compressive strain superlattice sample has clearly demonstrated the existence of an strengthening mechanism due to coherency strain in alternating-strain superlattices. With further developments in experimental technique for studying stretching and beam bending in epitaxial films at elevated temperatures, this phenomenon and perhaps other problems in metallurgy can usefully be studied in semiconductor structures.

ACKNOWLEDGEMENTS

We are grateful to Dr M. Hopkinson of the Sheffield Central Growth Facility for the MBE growth of the structures, and to the Engineering and Physical Sciences Research Council (UK) for financial support.

REFERENCES

1. A. Kelly and N.J. Macmillan, Strong Solids, Oxford University Press, 1986.

2. S.L. Lehoczky, Phys. Rev. Lett. **41**, p. 1814 (1978); J. Appl. Phys. **49**, p. 5479 (1978).

3. A. Kelly, Design of a High Temperature Structural Material in 2nd International Conference on Advanced Materials and Technology. New Compo '91 Hyogo. Hyogo Japan 205 (1991).

4. E.A. Fitzgerald, Mat. Sci. Reports **7**, p. 87 (1991).

5. D.J. Dunstan, P. Kidd, L.K. Howard and R.H. Dixon, Appl. Phys. Lett. **59**, p. 3,390 (1991).

6. D.J. Dunstan, P. Kidd, R. Beanland, A. Sacedón, E. Calleja, L. González, Y. González, and F.J. Pacheco, Mat. Sci. and Technol. **12**, p. 181 (1996).

7. M. Lourenço, K.P. Homewood and L. Considine, Mat. Sci. and Eng. B**28**, p. 507 (1994).

8. R. Hull, J.C. Bean, D. Bahnck, L.J. Peticolas, K.T. Short and F.C. Unterwald, J. Appl. Phys. **70**, p. 2052 (1991).

9. R. Beanland, J. Appl. Phys. **72**, p. 4031 (1992).

10. H.-J. Herzog and E. Kasper, J. Cryst. Growth **144**, 177 (1994).

11. See e.g. I. Adesida in Properties of Lattice-Matched and Strained Indium Gallium Arsenide, edited by P. Bhattacharya, INSPEC IEE, London, 1993, p.250, and K. Matsushita, S. Adachi and H.L. Hartnagel in Properties of Indium Phosphide, INSPEC IEE, London, 1991, p.333.

12. V. Swaminathan and S.M. Copley, J. Am. Ceram. Soc. **58**, 482 (1975).

13. D.J. Dunstan, P. Kidd, P.F. Fewster, N.L. Andrew, R. Grey, J.P.R. David, L. González, Y. González, A. Sacedón and F. González-Sanz, Appl. Phys. Lett. **65**, p. 839 (1994).

14. A. Sacedón, F. González-Sanz, E. Calleja, E. Muñoz, S.I. Molina, F.J. Pacheco, D. Araújo, R. Garcia, M. Lourenço, Z. Yang, P. Kidd and D.J. Dunstan, Appl. Phys. Lett. **66**, p. 3334 (1995).

SYNTHESIS AND MECHANICAL PROPERTIES OF NIOBIUM FILMS BY ION BEAM ASSISTED DEPOSITION

H. Ji, G. S. Was, and J. W. Jones, University of Michigan, Ann Arbor, MI 48109

ABSTRACT

Mechanical properties of niobium thin films are studied by controlling the microstructure, texture and residual stress of the films using ion beam assisted deposition (IBAD). Niobium films were deposited onto (100) Si substrates and their microstructure, texture and residual stress were measured as a function of ion energy and R ratio (ion to atom arrival rate ratio). The grain sizes of these films ranged from 20 nm to 40 nm and no effect of ion bombardment was observed. All the films have strong (110) fiber texture, but the in-plane texture is a strong function of the incident angle, energy and flux of the ion beam. Results show that while the degree of the texture increases with increasing ion energy and flux, it is also a strong linear function of the product of the two. The residual stress of the films was measured by a scanning-laser reflection technique. As a function of normalized energy, the stress is tensile for En < 30 eV/atom with a maximum of 400 MPa at about 15 eV/atom. It becomes compressive with increasing normalized energy and saturates at - 400 MPa for En > 50 eV/atom. Both PVD (physical vapor deposition) and IBAD films have a hardness of about 6 GPa at shallow depth measured by nanoindentation. The different stress state may be responsible for the 15% difference on hardness observed between the PVD and IBAD films.

INTRODUCTION

Recently, there has been significant interest in the origins of the interfacial strength of metal/ceramic systems because of the importance of such interfaces in composites, microelectronics and multilayer structural materials. The ultimate goal of this project is to control the interfacial strength of the niobium/aluminum oxide system by controlling the orientation relationships between the two constituents at the interface. This is achieved by controlling the in-plane texture of the niobium films by ion bombardment during deposition. Unavoidably, microstructure, residual stress state and mechanical properties of these films will also be modified along with the texture modification and need to be fully characterized and understood. The present work was conducted to serve this purpose.

EXPERIMENT

Niobium films were synthesized in an ultra high vacuum (UHV) chamber by vapor deposition with simultaneous ion bombardment (IBAD). The base pressure of the chamber is $2x10^{-10}$ torr. The system consists of two 6 kw electron gun vapor sources with 15 cc hearth and a 3 cm Kaufman ion gun. The rate of the deposition is monitored and controlled by quartz crystal thickness monitors and the ion beam flux is measured by a Faraday cup. Argon ions were used for all IBAD films. All substrates were sputter cleaned with a 500 eV Ar^+ ion beam prior to deposition. The deposition rate for all samples was 0.5 nm/sec.

The actual thickness was measured using a Dektak profilometer. Rutherford backscattering spectrometry (RBS) was used to determine the area density and composition of the films. The oxygen content of these films was detected by nuclear reaction analysis (NRA) using the $^{16}O(d, p)^{17}O$ reaction. The residual stress in the films was measured by a scanning laser deflection technique. The stress-induced curvature change of the silicon substrate was determined by the deflection of the laser beam from the sample surface, and the residual stress was then calculated using Stoney theory [1]. Microstructure of the niobium films was studied by transmission electron microscopy (TEM) using a JOEL 2000FX. The TEM samples were prepared by depositing 50 nm niobium films on single crystal sodium chloride substrates. The film was then removed by dissolving the substrate in water. Texture of the niobium films was

Mat. Res. Soc. Symp. Proc. Vol. 434 © 1996 Materials Research Society

characterized by Schulz's x-ray pole figure method. Using $Cu_{K\alpha}$ radiation, intensity of (110) and (200) Bragg diffraction peaks was measured at $2\theta = 38.2^\circ$ and $2\theta = 55.1^\circ$, respectively, over a range from substrate normal to 70° from normal. Hardness was measured by nanoindentation using the Nano Indenter® II (Nano Instruments, Inc.). The experiments were conducted in the displacement control mode from 50 nm to 1 μm.

RESULTS AND DISCUSSION

Microstructure and Composition

The niobium films were crystalline under all deposition conditions. The PVD film exhibited a fine grain structure with grain sizes ranging from 20 nm to 40 nm. Similar microstructure and grain size distribution were found for the IBAD film deposited at E = 1000 eV and R = 0.4 (Fig. 1). The extra energy introduced by ion bombardment was not sufficient to move the microstructure of the films out of zone I in the structure diagram [2] because of the high melting temperature for niobium (2468 °C). The grain size of the films was in good agreement with the work of Grovenor [2], who found that the grain sizes for various metals deposited at low homologous temperatures ranged from 5 nm to 20 nm and were independent of the temperature. The PVD sample had azimuthally symmetric diffraction pattern indicating a random distribution of grain orientations. The IBAD sample, on the other hand, showed preferred orientations in the film with intensity minimum and maximum.

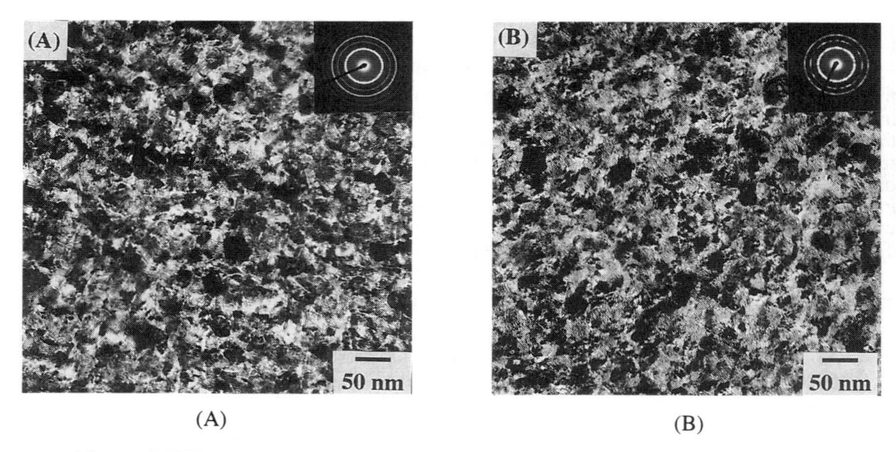

(A) (B)

Figure 1. TEM micrographs and SAD patterns for (A) PVD Nb film, and (B) IBAD Nb film deposited at 1000 eV, R = 0.4.

The compositions of the films were measured by RBS and oxygen content was determined by NRA. No impurity other than oxygen and argon was found in the films. The oxygen level for films deposited under all conditions was less than 1 at% and no significant difference was observed between PVD film and IBAD films. The "sputter cleaning" effect due to the preferential sputtering of oxygen is only effective for higher oxygen level. The argon in the IBAD films was incorporated by the Ar+ ion bombardment and was found to increase with increasing normalized energy of the ion beam. The maximum argon level was about 6 at% and occurred in IBAD films with normalized energy equal to 400 eV/atom.

Texture

Table I summarizes the deposition conditions for the samples used for texture study. The ion beam incident angle for all the IBAD samples was 50° relative to the substrate surface normal. Fig. 2 shows the (110) and (200) pole figures for niobium films deposited on (100) silicon substrates under two different conditions: (A) PVD, and (B) IBAD at E = 1000 eV, R = 0.4. It should be noted that the projection of the ion beam direction is at 0° azimuthal angle. Both samples have (110) fiber texture indicated by the intensity maximum at the center of the (110) pole figures. As expected, a second maximum should appear at a tilt angle of 60° (30° ring). For Sample (A), the uniform distribution of the intensity on this 30° ring indicates an azimuthally random distribution of grain orientation . The (200) pole figure of sample (A) confirmed the absence of an in-plane texture. On the other hand, both (110) and (200) pole figures for sample (B) show strong in-plane texture in the film as noted by the distinct intensity maxima at preferred azimuthal directions. The angular relationship of these poles matches with the crystallographic structure of niobium. Furthermore, the (200) pole is parallel to the ion beam incident direction. The same result was also found in niobium films deposited on amorphous glass under the same condition (1000 eV, R=0.4) implying that there is no substrate effect in this case.

Table I. IBAD Deposition Conditions of Various Samples Used for Pole Figure

Sample ID	Substrate	Beam Energy (eV)	R ratio	Normalized Energy (eV/atom)	Film Thickness (nm)
1	(100) Silicon	0	0	0	490
2	(100) Silicon	250	0.1	25	691
3	(100) Silicon	500	0.1	50	737
4	(100) Silicon	500	0.4	200	562
5	(100) Silicon	1000	0.4	400	828
6	Glass	1000	0.4	400	828

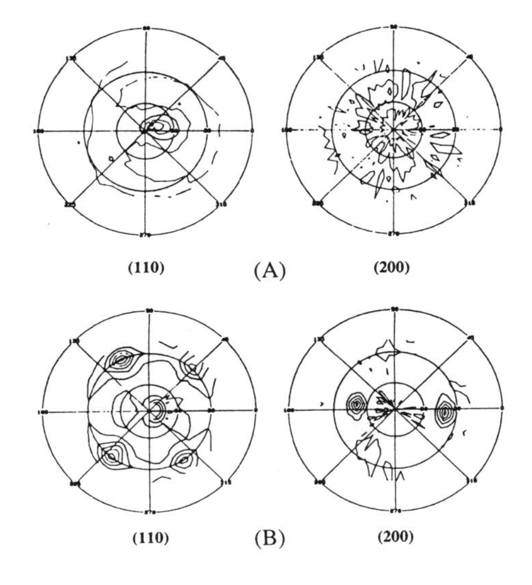

Figure 2. (110) and (200) pole figures for (A) PVD Nb films, and (B) IBAD Nb films deposited at 1000 eV, R = 0.4 on (100) Si substrates.

For b.c.c. films like niobium, (110) fiber texture is generally observed with the close packed planes parallel to the surface of the substrate to minimize the surface energy. However, the in-plane texture modification is solely due to the ion beam bombardment during deposition. Among the different possible mechanisms [3,4,5], we believe that the process is controlled by sputtering. Grains with the channeling directions aligned with the ion beam direction have a lower sputtering rate and will grow preferentially compared to the other grains. In order to obtain the desired in-plane texture without disturbing the (110) fiber texture, the ion beam incident angle has to be carefully chosen. We chosed 50° for the beam incident angle based on the niobium crystal structure and our system configuration. With the (100) channeling direction aligned with the beam, the (110) in-plane direction is parallel to the ion beam projection on the substrate. To measure the degree of in-plane texture, we introduce a quantity D (intensity ratio):

$$D = I_{avg}^{30^\circ} / I^{0^\circ}$$

where $I_{avg}^{30^\circ}$ is the average intensity of the four (110) poles on the 30° ring and I^{0° is the intensity of the center (110) pole. Fig. 3 shows a linear relationship between intensity ratio and the normalized energy. Since sputtering rate is linearly proportional to both the R ratio and the ion energy within the range used in this work, this result is consistent with the sputtering mechanism.

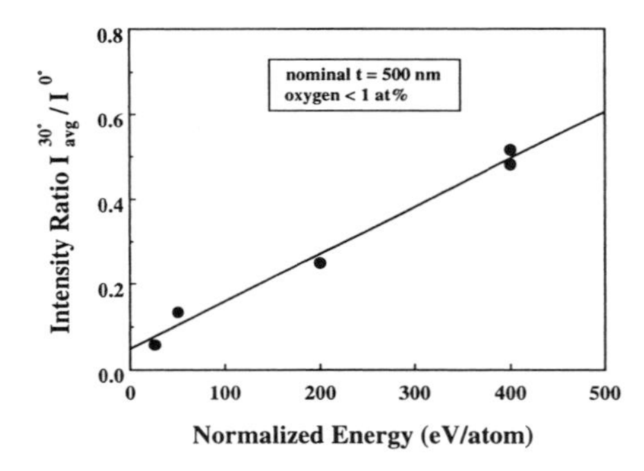

Figure 3. Intensity ratio $I_{avg}^{30^\circ} / I^{0^\circ}$ vs. nomalized energy, indicating an increase of degree of texture with increasing ion bombardment.

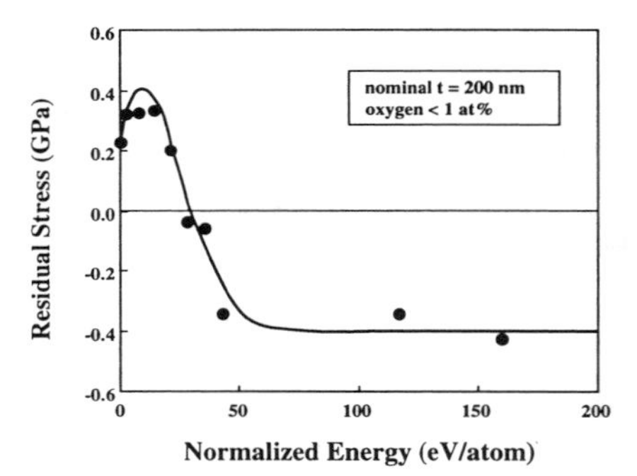

Figure 4. Residual stress in Nb films as function of normalized energy.

Residual Stress

While the in-plane texture of the films is effectively controlled by ion bombardment during deposition, the development of residual stress in the films is also dependent on deposition conditions. Residual stress was measured in the niobium films of thickness 200 nm by a laser deflection technique. As shown in Fig. 4, the residual stress in the niobium films is consistent with the behavior commonly observed in other b.c.c. metal films [6,7]. As a function of the normalized energy ($E_n = E \times R$) the stress was tensile (224 MPa) at $E_n = 0$ and increased to a maximum of about 400 MPa at $E_n = 15$ eV/atom. The stress then decreased in the compressive direction with increasing normalized energy, and eventually became compressive at about 30 eV/atom, saturating at about -400 MPa for $E_n > 50$ eV/atom.

Several mechanisms have been proposed to explain the intrinsic residual stress formation in the films with ion bombardment. Among these are structural relaxation, densification, grain growth and phase changes creating a tensile stress, and thermal spike, atom peening, recoil implantation and trapped bombardment gas for compressive stress formation[8]. However, none of these models individually can explain this common behavior of the b.c.c. metal films. So far, the only explanation is based on a combination of film purification and defect formation by ion bombardment [6]. The impurity (mainly oxygen) level for the previous works in these metals (Nb, W and Mo) are more than 10 at% and decrease significantly with normalized energy. The oxygen level in the niobium films in this work is less than 1 at% and is independent of ion bombardment rate. This is in contradiction with the film purification model for residual stress control. Further work is needed to understand the mechanism of residual stress formation in these metals.

Mechanical Properties

Hardness of PVD and IBAD (1000 eV, R = 0.4) niobium films deposited on (100) Si was measured by nanoindentation using a NanoII with a Berkovich indenter. The same measurement was also performed on the Si (100) substrate and bulk polycrystalline niobium with a grain size of 30 μm (Fig. 5). Both films have hardness value around 6 GPa at shallow depth. According to the Tabor relation [9], H = 3.2 Y, we estimate a yield strength of 1.87 GPa for both samples. The high yield strength in thin films is a result of both fine grain size and small thickness. The thickness of the PVD sample is 490 nm and 828 nm for the IBAD (1000 eV, R=0.4) film. The measured hardness increases and approaches the substrate hardness with increasing indent depth. Except for the result at 50 nm, the IBAD film has a higher hardness than the PVD film. This could not be caused by the different thickness of the films since the thinner PVD film should show a higher hardness. Neither is there strong evidence of microstructural difference that is responsible. However, the different residual stress in those films could in part be responsible for the difference in hardness. The presence of biaxial tension in the film should result in a measured hardness that is less than the true value. Likewise, biaxial compression should increase the measured hardness[10]. We observed a 15% increase in hardness from tensile PVD films to compressive IBAD films.

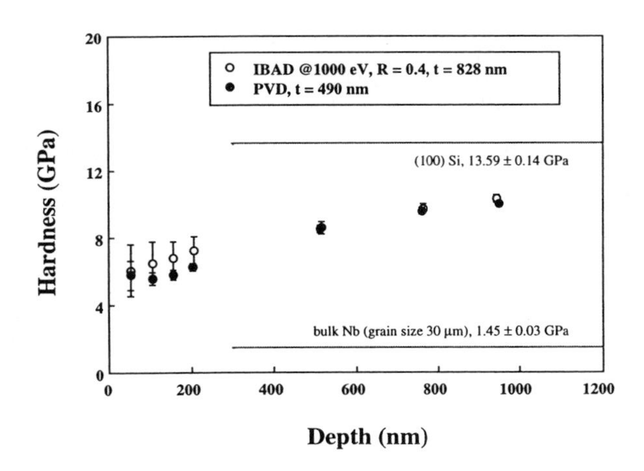

Figure 5. Measured hardness for both PVD and IBAD Nb films deposited on (100) Si substrates. Results for (100) Si and bulk Nb are also shown.

SUMMARY

Ion beam assisted deposition was used to control the texture of niobium films. Although all films have (110) fiber texture, the in-plane texture is controlled by the ion bombardment. The degree of texture is strongly dependent on the ion beam incident angle, ion energy and ion to atom arrival rate ratio (R ratio). The linear increase of degree of texture with increasing ion energy and R ratio supports the sputtering mechanism. Residual stress in these films were also modified by the ion bombardment and is strongly dependent on the ion beam conditions. As a function of normalized energy, the residual stress is tensile for $E_n < 30$ eV/atom with a maximum of 400 MPa at about 15 eV/atom. It then becomes compressive with increasing normalized energy and saturates at -400 MPa for $E_n > 50$ eV/atom. The mechanism(s) that controls the stress formation in these films under ion bombardment is still not clear. There is no effect of ion bombardment with ion energy up to 1000 eV and R ratio to 0.4 on the microstructure of the niobium films. Both the PVD and IBAD films have grain sizes ranging from 20 nm to 40 nm. Hardness measurement shows little effect of ion bombardment. Both PVD and IBAD have a hardness of about 6 GPa at shallow depth. The different stress state may be responsible for the 15% difference on hardness observed between the PVD and IBAD films.

ACKNOWLEDGMENT

The authors would like to thank the Michigan Ion Beam Laboratory for Surface Modification and Analysis and the Electron Microbeam Analysis Lab at the University of Michigan for the use of their facilities. This work is supported under NSF grant #DMR-9411141.

REFERENCES

1. P. A. Flinn, D. S. Gardner, and W. D. Nix, IEEE Transactions on Electron Devices, **ED-34(3)**, 689 (1987).
2. C. R. M. Grovenor, H. T. G. Hentzell, and D. A. Smith, Acta Metall., **32**, 773 (1984).
3. D. Dobrev, Thin Solid Films, **92**, 41 (1982).
4. G. N. Van Wyk and H. J. Smith, Nucl. Instrum. Meth., **170**, 433 (1980).
5. R. M. Bradley, J. M. E. Harper, and D. A. Smith, J. Appl. Phys., **60**, 4160 (1986).
6. J. J. Cuomo, J. M. E. Harper, C. R. Guarnieri, D. S. Yee, L. J. Ahanasio, J. Angilello, and C. T. Wu, J. Vac. Sci. Technol., **20(3)**, 349 (1982).
7. R. A. Roy, R. Petkie, and A. Boulding, J. Mater. Res., **6**, 80 (1991).
8. M. F. Doernwe, and W. D. Nix, Critical Reviews in Solid State and Materials Sciences, **14(3)**, 225 (1988).
9. D. Tabor, in The Hardness of Materials (Clarendon, Oxford, 1951) cited by M. F. Doerner, D. S. Gardner, and W. D. Nix, J. Mater. Res., **1**, 845 (1986).
10. T. Y. Tsui, W. C. Oliver, and G. M. Pharr, J. Mater. Res., **11(3)**, 752 (1996).

OROWAN-BASED DEFORMATION MODEL FOR LAYERED METALLIC MATERIALS

ERIC R. KREIDLER AND PETER M. ANDERSON
Department of Materials Science and Engineering, The Ohio State University, 116 W. 19th Ave., Columbus, OH 43210-1179

ABSTRACT

An Orowan-based deformation model for layered metallic materials is presented and used to calculate the stress-strain behavior for two deformation modes. This model assumes that layer thicknesses are sufficiently small so that single rather than multiple dislocation pile-ups form. Deformation then proceeds by increasing the density of single dislocation pile-ups. Furthermore, it is assumed that the controlling stress for plastic deformation is that to propagate a tunneling dislocation loop inside an embedded elastic-plastic layer. Initially, the resolved stress required to propagate an isolated tunneling loop does not depend on whether the loop shears the layer perpendicular to an interface or stretches it parallel to an interface. At larger strains, the tunneling arrays become sufficiently dense such that local dislocation interaction changes the line energy of a tunneling dislocation. As a result, the elastic-plastic layers may exhibit modest softening when sheared or substantial hardening when stretched. When the elastic-plastic layers are embedded into a multilayered specimen with alternating elastic-only layers, no macroscopic strain softening is observed. However, the predicted macroscopic stress-strain curves for stretching and shearing are significantly different in their dependence on layer thickness.

INTRODUCTION

Interest in the mechanical properties of multilayer thin films has been prompted by observed enhancements in yield and fracture stress [1-5]. Several models have been proposed to help understand the reported variation in strength, which increases to a peak or a plateau when the bilayer thickness of two-phase samples is reduced to the order of tens of nanometers. These models focus on the stress required for a critical deformation event in order to understand strengths at different bilayer thicknesses. However, our investigation of the tensile response of a silver/nickel multilayered sample with a nominal bilayer wavelength of 930 Å suggests that there are regimes of deformation which are not understood or modeled simply by a single critical event. In particular, the same fracture surface of a silver/nickel specimen shows some regions (Figure 1a) that still have a distinct layer morphology, while other regions (Figure 1b) reveal no individual layers with an appearance similar to cavity rupture in a ductile monolithic matrix. Clearly, Figure 1(b) is consistent with extensive co-deformation of layers while Figure 1(a) is the result of deformation that is confined much more to individual layers.

In order to understand the evolution of deformation and the resulting stress-strain behavior of multilayered materials, this work adopts an approach by Embury and Hirth [9] in which deformation proceeds by the propagation of tunneling dislocation loops within a layer. The layers are considered thin enough so that extended pile-ups do not form. Thus, continued plastic strain is produced by tunneling of successive loops on other planes. This approach allows the prediction of stress-strain behavior leading up to failure.

Mat. Res. Soc. Symp. Proc. Vol. 434

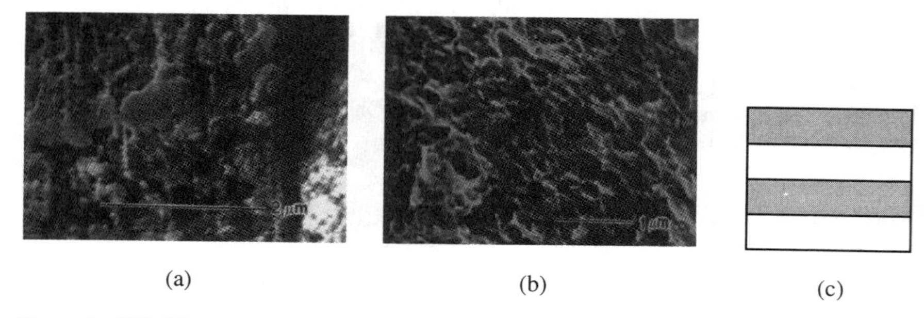

(a) (b) (c)

Figure 1. SEM Fractography (a) Distinct layers visible indicating independent layer deformation. (b) Layers indistinct with void sizes ranging to several bilayer dimensions. (c) Layer orientation in micrographs.

Other models, such as those by Lehoczky [1], Rao *et al.* [6], and Chu and Barnett [7], have estimated the stress for dislocation transmission across an interface based on the Koehler solution for a dislocation near an interface between materials of different moduli. These models suggest that a change in interface resistance to transmission, either from interdiffusion or an incoherent-to-coherent transition, may produce the observed peak or plateau in strength as layer thickness is decreased. Li and Anderson [8] observe a peak in critical resolved stress for transmission in Cu-Ni layered materials based on a model of small pile-ups generated from discrete Frank-Read sources in alternating Cu and Ni layers. The peak occurs because differences in elastic moduli, and hence image effects, cause the direction of transmission to change from Cu→Ni at larger layer thicknesses to Ni→Cu at smaller layer thicknesses. Finally, Chu and Barnett [7] predict a peak in tensile stress to occur if the critical event changes from transmission of dislocations at smaller layer thicknesses to dislocation tunneling at larger bilayer thicknesses. With the exception of the Embury and Hirth model, the approaches described consider only an isolated pile-up or tunneling loop corresponding to initial yielding.

It is clear from the observed fracture surfaces in Fig. 1 that substantial plastic deformation has occurred so that tunneling loops or pile-ups interact with other loops or pile-ups in the same or nearby layers. This work extends the approach by Embury and Hirth [9] by comparing stress-strain curves for in-plane stretching to those for shearing of the layers. The curves assume that tunneling loops of shear or prismatic character increase in density and thus, interact more strongly with continued plastic strain. The results predict a large plastic anisotropy between the stretching and shearing modes due to differences in dislocation line energies produced by each type of deformation. Future extensions to the model are also presented.

MODEL CONFIGURATION

Assumed Deformation Mechanism

The proposed model considers a multilayered material with an elastic-plastic phase (α) and an elastic-only phase (β) of equal volume fraction where the thickness of each layer type is h. The elastic-plastic phase is assumed to deform plastically by the propagation of dislocation loops in a tunneling mode as displayed in Figure 2. Tunneling loops are expected to propagate by glide on an inclined plane, with the dislocation geometry defined by the angle θ subtended by the slip direction, $\underline{s}$, and the x-axis normal to the interface. In general, a glide dislocation may be separated into shearing and stretching components $(b_{shear}, b_{stretch}) = (b\cos\theta, b\sin\theta)$. The limiting cases of pure shearing and pure stretching loops are depicted in Figs. 3a and 3b, respectively, where successive loops have expanded on planes with vertical spacing, λ, between them. The pure stretching case shown in Figure 3(b) corresponds to an array of prismatic dislocation loops.

The plastic strain generated in α is inversely proportional to the spacing, λ, of tunneling dislocations according to

$$\gamma_{\alpha\beta}{}^{P} = 2\varepsilon_{\alpha\beta}{}^{P}=(b_{\alpha}n_{\beta}+b_{\beta}n_{\alpha})/\lambda \tag{1}$$

where n_{α} (α = x, y) are components of the normal to the loop plane. Accordingly, the shear array has $(b_x, b_y) = (b, 0)$ and $(n_x, n_y) = (0, 1)$ so that only the shear strain components $\gamma_{xy}{}^{P} = \gamma_{yx}{}^{P} = b/\lambda$ are produced. For the stretching array, $(b_x, b_y) = (0, b)$ and $(n_x, n_y) = (0, 1)$ so that $\varepsilon_{yy}{}^{P} = b/\lambda$ is the only component of strain produced.

The corresponding driving force for tunneling is the resolved stress,

$$\tau_{ns} = \sigma_{\alpha\beta}n_{\alpha}s_{\beta} \tag{2}$$

Hence, $\tau_{ns} = \sigma_{xy} = \sigma_{yx}$ for the shear case and $\tau_{ns} = \sigma_{yy}$ for the stretching (prismatic) case.

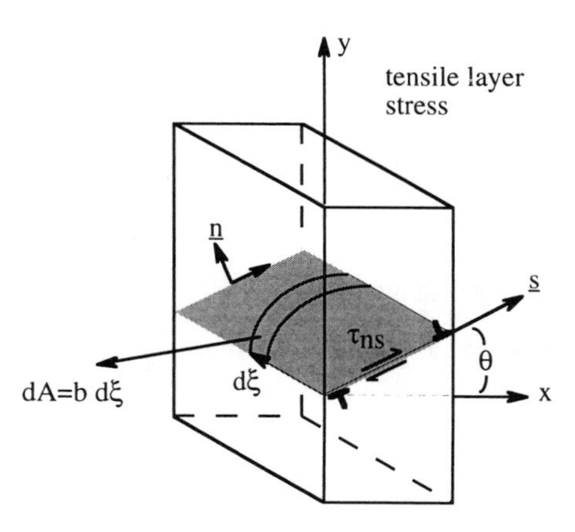

Figure 2. Tunneling mode of dislocation motion where slip occurs on an inclined plane defined by θ. The Orowan stress for propagation is calculated by equating the energy expended and released by propagating the loop through a distance $d\xi$.

The Orowan Stress and Dislocation Line Energy

The Orowan stress is calculated by equating the energy expended and released by the tunneling of a dislocation loop as depicted in Figure 2. The energy expended when the loop advances by an incremental distance, $d\xi$, is $2(W/L)d\xi$, where W/L is the line energy of each dislocation created by the advance. The energy released during the advance is $\tau_{ns}bhd\xi$. Equating the energy expended and released yields the familiar Orowan stress for propagation of the loop,

$$\tau_{ns} = 2(W/L)/h. \tag{3}$$

Clearly, the explicit dependence on 1/h has the effect of dramatically increasing the tunneling stress as layer thickness is decreased. However, W/L depends on both h and λ, and the remaining task is to determine this dependence.

The line energy is calculated from the work, W_{sep}, to separate two vertical arrays of oppositely signed dislocations from an initial separation distance of b to a final separation distance of h. The Cartesian coordinate system (x, y) is attached to the left-hand array, (L), and the shear or stretch configurations in Fig. 3 are produced by moving the right-hand array from x = b to x = h. The work expended per dislocation in the right-hand array is written by arbitrarily choosing an expansion path along y = 0 and equating this to 2(W/L),

Shearing case:

$$W_{sep} = 2\frac{W}{L}\bigg|_{Shear} = \int_b^h \sigma_{xy}^{(L)}(x, y=0)bdx \tag{4}$$

Stretching case:

$$W_{sep} = 2\frac{W}{L}\bigg|_{Stretch} = \int_b^h \sigma_{yy}^{(L)}(x, y=0)bdx \tag{5}$$

The stress components, $\sigma_{xy}^{(L)}$ and $\sigma_{yy}^{(L)}$, are those generated by the left-hand array at the site of a dislocation in the right-hand array. There are no contributions from dislocations in the right-hand array since all dislocations in that array are displaced simultaneously.

Stress Generated by Dislocation Arrays

The left-hand arrays in the shearing and stretching configurations generate stress components according to [10]:

Shearing:

$$\sigma_{xy}^{(L)} = -\sigma_o 2\pi X(\cosh 2\pi X \cos 2\pi Y - 1) \tag{6}$$

Stretching:

$$\sigma_{yy}^{(L)} = -\sigma_o[2\sinh 2\pi X(\cosh 2\pi X - \cos 2\pi Y) - 2\pi X(\cosh 2\pi X \cos 2\pi Y - 1)] \tag{7}$$

The normalized positions are defined by $X = x/\lambda$ and $Y = y/\lambda$.

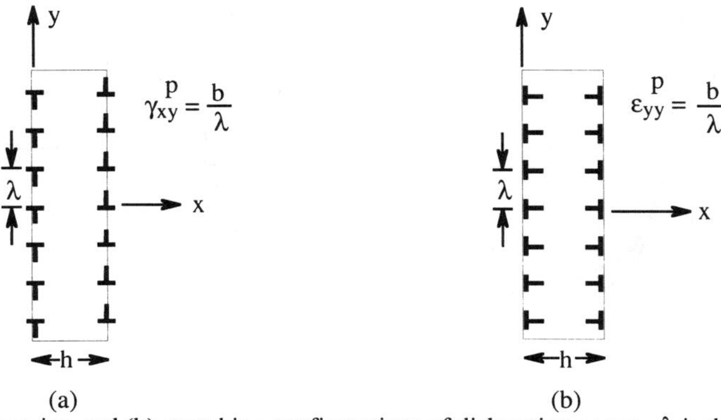

Figure 3. (a) Shearing and (b) stretching configurations of dislocation arrays. λ is the dislocation spacing in an array.

Further,

$$\sigma_o = \frac{\mu' b}{2\lambda(\cosh 2\pi X - \cos 2\pi Y)^2} \tag{8}$$

where $\mu' = \mu/(1-\nu)$ depends on the elastic shear modulus μ and Poisson's ratio ν, and b is the magnitude of the Burgers vector. For the specific case of Y = 0 considered in eqns. (4) and (5), the relevant components of stress are given by

Shearing: $$\sigma_{xy}^{(L)}(X, Y = 0) = \frac{\mu' b}{2\lambda} \frac{2\pi X}{\cosh 2\pi X - 1} \tag{9}$$

Stretching: $$\sigma_{yy}^{(L)}(X, Y = 0) = \frac{\mu' b}{2\lambda} \frac{2\sinh 2\pi X - 2\pi X}{\cosh 2\pi X - 1} \tag{10}$$

Figure 4 displays important distinguishing features of the stress distribution produced by the shearing versus stretching configurations. The shearing array produces a short-range shear stress which decays to nearly zero ($\mu' b/85\lambda$) at a distance $x = \lambda$ away from the array. Alternately, the direct stress σ_{yy} produced by the stretching array decays to a long-range value of $-\mu' b/\lambda$ at a distance $x > \lambda$. This difference in the long-range stress produces two important differences between the arrays. Since W_{sep} in each case is proportional to the area under the σ(x) curves in Fig. 4, the results suggest that W_{sep} for the shearing case increases with λ and that the opposite is true for the stretching case. Further, W_{sep} for the shearing case is expected to reach an asymptotic value for $h > \lambda$, but no asymptotic value is observed for the stretching case. Instead, W_{sep} grows linearly with h for $h > \lambda$.

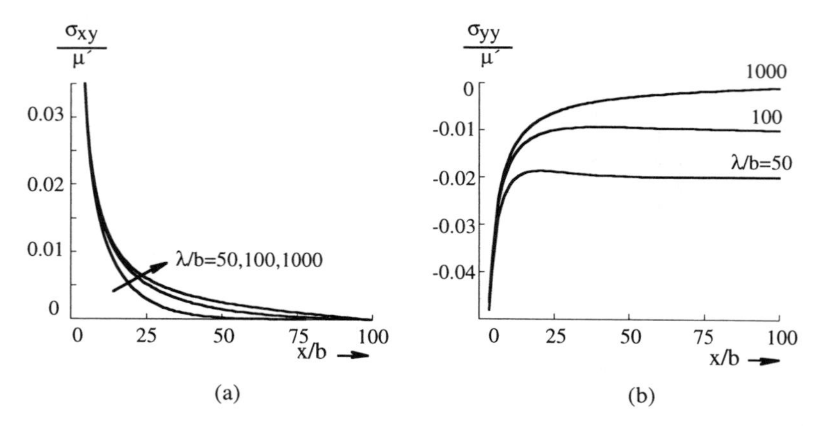

Figure 4. (a) Stress field produced by tilt array and (b) misfit array. A long-range stress field exists for the misfit array. Line energies are proportional to the area under the curve.

Dislocation Line Energies

The dislocation line energies for the shearing and stretching configurations are determined by substituting into eqns. (4) and (5) the expressions for $\sigma_{xy}^{(L)}$ and $\sigma_{yy}^{(L)}$ given in eqns. (6) and (7), respectively. The results are

Shearing:
$$\frac{W/L}{\mu' b^2/4\pi} = \ln\left(\frac{\sinh \pi h/\lambda}{\sinh \pi b/\lambda}\right) - \pi\left(\frac{h}{\lambda}\coth \pi h/\lambda - \frac{b}{\lambda}\coth \pi b/\lambda\right) \tag{11}$$

Stretching:
$$\frac{W/L}{\mu' b^2/4\pi} = \ln\left(\frac{\sinh \pi h/\lambda}{\sinh \pi b/\lambda}\right) + \pi\left(\frac{h}{\lambda}\coth \pi h/\lambda - \frac{b}{\lambda}\coth \pi b/\lambda\right) \tag{12}$$

Figure 5 displays the normalized line energies as a function of layer thickness for the shearing and stretching configurations. For $h \ll \lambda$, corresponding to small strains, the line energies appear to be independent of the type of configuration and the value of λ, but increase monotonically with layer thickness. For $h > \lambda$, the two types of configurations display quite different behavior. In particular, the line energy in the shearing configuration becomes independent of layer thickness and *increases* monotonically with λ. In contrast, the line energy in the stretching configuration increases linearly with layer thickness, and *decreases* monotonically with λ. The line energies in these two limits may be found analytically from eqns. (9) and (10).

$\lambda \gg h \gg b$:
$$\frac{W/L}{\mu' b^2/4\pi} \approx \ln\left(\frac{h}{b}\right) \tag{13}$$

$$h > \lambda >> b: \qquad \frac{W/L}{\mu' b^2/4\pi} \approx \begin{cases} 1 - \ln 2\pi b/\lambda & \text{(shearing)} \\ 2\pi h/\lambda - \ln 2\pi b/\lambda - 1 & \text{(stretching)} \end{cases} \tag{14}$$

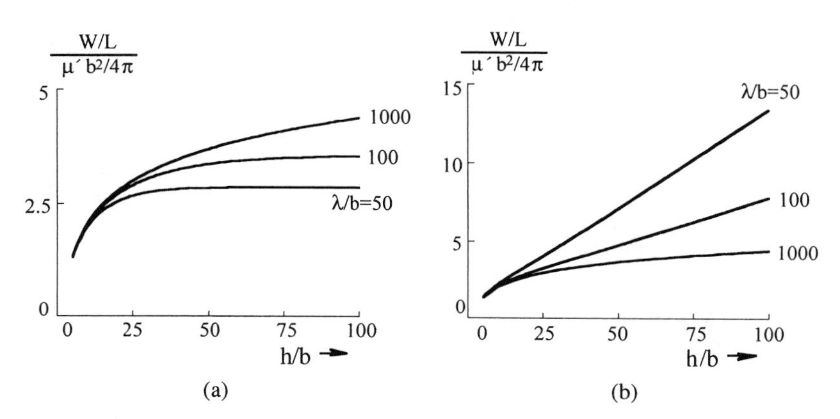

Figure 5. Normalized dislocation line energies vs. separation distance for the (a) shearing and (b) stretching configurations.

Extension from Isolated to Alternating Layer Deformation

The analysis presented may be extended from the isolated deforming layer of α presented to all alternating layers of α deforming plastically. Accordingly, eqns. (4) and (5) are still used to determine W/L, but with $\sigma_{xy}^{(L)}$ and $\sigma_{yy}^{(L)}$ in the integrands replaced by

$$\begin{aligned} \sigma_{xy}^{\text{multiple}}(x,y) &= \sum_{n=-\infty}^{\infty} \sigma_{xy}^{(L)}(x-2nh,0) - \sum_{\substack{n=-\infty \\ n\neq 0}}^{\infty} \sigma_{xy}^{(L)}(2nh,0) \quad \text{(shearing)} \\ \sigma_{yy}^{\text{multiple}}(x,y) &= \sum_{n=-\infty}^{\infty} \sigma_{yy}^{(L)}(x-2nh,0) - \sum_{\substack{n=-\infty \\ n\neq 0}}^{\infty} \sigma_{yy}^{(L)}(2nh,0) \quad \text{(stretching)} \end{aligned} \tag{15}$$

Figure 6 presents the line energies for the multiple layer cases as a continuous function of strain for discrete values of h/b. The isolated and multiple layer results are comparable for the stretching configuration over the range of h and λ shown, but in the shearing configuration, they are comparable only when $h >> \lambda$. For other ranges, the line energy in the shearing configuration is lowered significantly by multiple layer plasticity. The difference arises since the nearest tilt array from the additional layers generates a positive shear stress along the expansion path. For $h >> \lambda$, the short-range stress field from this nearby array decays too quickly to change the line energy significantly.

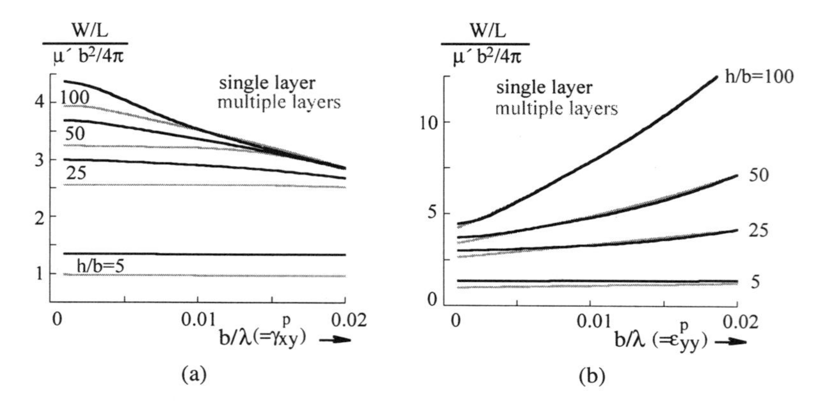

Figure 6. Comparison of line energy as a function of strain for the isolated (darker lines) and multiple (lighter lines) layer deformation modes, for (a) shearing and (b) stretching configurations.

The multiple layer results are considered to be more realistic and are used to generate subsequent plots in the manuscript. However, the approximate analytic results from the isolated layer case will continue to be developed in order to provide some insight to the analytic behavior. In particular, the line energies in eqns. (13) and (14) may be expressed in terms of strain by noting that from eqn. (1), $b/\lambda = \gamma_{xy}$ for the shearing case and that $b/\lambda = \varepsilon_{yy}$ for the stretching case. Accordingly,

$h\varepsilon^p_{yy}/b$ or $h\gamma^p_{xy}/b << 1$:
$$\frac{W/L}{\mu' b^2/4\pi} \approx \ln\left(\frac{h}{b}\right) \quad \text{(shearing or stretching)} \qquad (16)$$

$h\varepsilon^p_{yy}/b$ or $h\gamma^p_{xy}/b > 1$:
$$\frac{W/L}{\mu' b^2/4\pi} \approx \begin{cases} 1 - \ln 2\pi\gamma^p_{xy} & \text{(shearing)} \\ 2\pi\varepsilon^p_{yy} h/b - \ln 2\pi\varepsilon^p_{yy} - 1 & \text{(stretching)} \end{cases} \qquad (17)$$

An important feature displayed in Fig. 6 and in eqn. (17) is that line energy increases with strain in the stretching case, but decreases with strain in the shearing case.

Orowan Stress vs. Strain

The Orowan stress to propagate a tunneling dislocation is determined by substituting the multiple layer line energy results from Fig. 6 into eqn. (3). The most apparent feature in Fig. 7 is that the tunneling stress to initiate plastic deformation increases dramatically with decreasing layer thickness. Second, the shearing configuration displays modest strain softening while the stretching configuration displays substantial hardening. The softening and hardening in each case are more pronounced at larger layer thickness. The softening in shear has been reported earlier by Embury and Hirth [9].

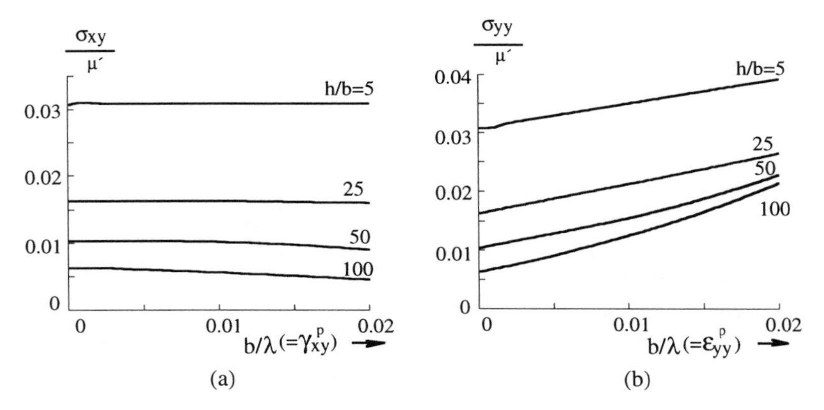

Figure 7. Orowan Stress as a function of plastic strain for the (a) shearing and (b) stretching configurations.

The principal observation made thus far is the dramatic anisotropy in plastic work hardening exhibited by this material. The corresponding analytic expressions generated from the isolated layer analysis are obtained by substituting eqns. (16) and (17) into eqn. (3),

$$h\varepsilon^p_{yy}/b \text{ or } h\gamma^p_{xy}/b << 1: \quad \frac{\sigma_{xy}}{\mu'}, \frac{\sigma_{yy}}{\mu'} \approx \frac{1}{2}\frac{\ln(h/b)}{\pi h/b} \quad \text{(shearing, stretching)} \tag{18}$$

$$h\varepsilon^p_{yy}/b \text{ or } h\gamma^p_{xy}/b > 1: \quad \begin{aligned} \frac{\sigma_{xy}}{\mu'} &\approx \frac{1-\ln 2\pi\gamma^p_{xy}}{2\pi h/b} && \text{(shearing)} \\ \frac{\sigma_{yy}}{\mu'} &\approx \frac{2\pi\varepsilon^p_{yy}h/b - \ln 2\pi\varepsilon^p_{yy} - 1}{2\pi h/b} && \text{(stretching)} \end{aligned} \tag{19}$$

The predictions in Fig. 7 indicate that shear or tensile stresses on the order of (0.005 to 0.03)μ' are required to initiate yield. Thus, the elastic strains produced are comparable to the magnitude of plastic strains shown. Accordingly, the Orowan stress is presented as a function of total strain by noting that

$$\begin{aligned} \gamma_{xy} &= \gamma^p_{xy} + \frac{\sigma_{xy}}{\mu} && \text{(shearing)} \\ \varepsilon_{yy} &= \varepsilon^p_{yy} + \frac{\sigma_{yy}}{2\mu(1+\nu)} && \text{(stretching)} \end{aligned} \tag{20}$$

and taking $\nu = 1/3$. The important features displayed in Fig. 8 are the large elastic strain predicted prior to yielding in each configuration and the dramatic hardening and softening observed in the shearing and stretching configurations, respectively.

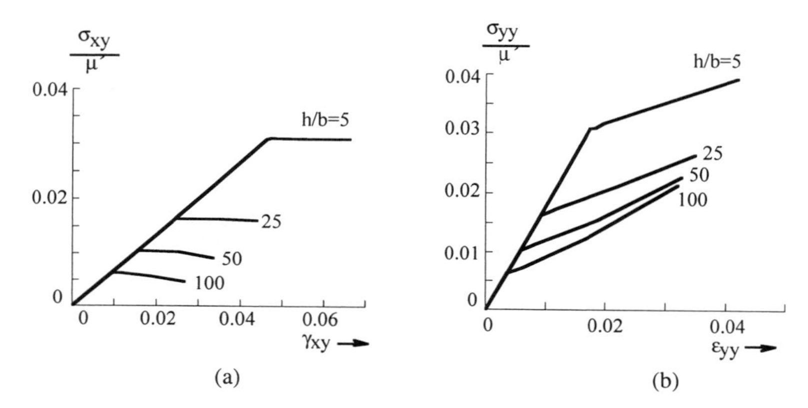

Figure 8. Orowan Stress as a function of total strain for (a) shear deformation perpendicular to the layers, and (b) tensile deformation parallel to the layers.

The elastic-plastic α layers may be embedded into a layered composite material of alternating α and elastic-only (β) layers, to determine macroscopic response of a sample. The volume fractions and elastic properties of α and β layers are taken to be equal. An isostrain approach is used so that the composite stress is simply the average of the stresses in the α and β layers and the total strain in each layer type is assumed to be equal. The predictions in Fig. 9 show that neither deformation mode exhibits any softening due to the addition of the elastic-only phase. However, the elastic-plastic slope for shear deformation is much smaller than for stretching. A consequence is that the shear stress-strain curves exhibit a much larger dependence on layer thickness than the stretching stress-strain curves.

DISCUSSION

Several models of mechanical behavior of multilayered thin films focus on a critical event to interpret hardness and fracture data as a function of bilayer thickness. In contrast, this work develops the entire stress-strain behavior of a multilayered material, for the special case that one phase is elastic-plastic while the other phase is impenetrable to dislocations below some stress level. The model does not account for observed peaks or plateaus in strength that have been observed as a function of layer thickness in some materials. In fact, the tunneling stress is predicted to increase monotonically with decreasing layer thickness as long as the elastic phase is impenetrable to dislocations. Such a peak or plateau can only occur in this model by permitting the stress for transmission across an interface to vary with layer thickness.

This stress-strain model extends an analysis of shear deformation perpendicular to the layers by Embury and Hirth [9] to include tensile (stretching) deformation parallel to the layers. The layers display strong plastic anisotropy primarily because the line energy of the trailing segments of a tunneling loop depends on the Burgers vector and local configuration of surrounding dislocations. A principal result is that the elastic-plastic layers exhibit strain hardening when stretched in tension, but strain softening when sheared. Even when the

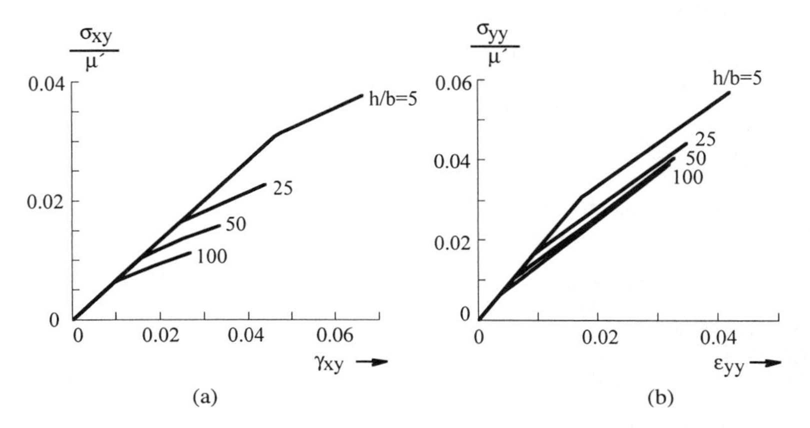

Figure 9. Multilayer Stress-Strain Predictions for (a) shear deformation perpendicular to the layer and (b) tensile deformation parallel to the layer. The multilayered sample has alternating elastic-plastic layers (α) and elastic-only layers (β) of equal thickness h.

elastic-plastic layer is embedded with alternating elastic-only layers, the stress-strain curves for shearing depend on layer thickness much more than the corresponding tension curves.

Future modeling efforts are to study the effects of coherency strains, source limited plasticity in α and β, and the regimes of stress and layer thickness for which tunneling motion of single pile-up arrays is preferred to multiple dislocation pile-ups.

CONCLUSIONS

The Orowan model presented suggests that multilayered materials have large plastic anisotropy as a result of differences in dislocation line energies for the arrays formed by shearing and stretching. This model assumes that the controlling stress is that to propagate a tunneling dislocation loop through an embedded layer, although it is uncertain whether other events such as loop nucleation require an even larger stress. Initially, the resolved stress required to propagate an isolated tunneling loop does not depend on whether the loop shears the layer perpendicular to an interface or stretches it parallel to an interface. At larger strains, the tunneling arrays become dense enough so that local dislocation interaction changes the line energy of a tunneling dislocation. This produces strain softening for shearing perpendicular to the interface and strain hardening for stretching parallel to the interface. When the elastic-plastic layers are embedded with alternating elastic-only layers, no macroscopic strain softening is observed. However, the predicted macroscopic stress-strain curves change substantially with deformation mode, so that tensile stress-strain curves are predicted to depend less on layer thickness than shear stress-strain curves.

ACKNOWLEDGMENTS

The authors wish to thank Dr. John P. Hirth of Washington State University for his useful insight, the Office of Naval Research, contract N00014-91-J-1998, for financial support of this work, and Dan Jossel of NIST for providing Ag-Ni multilayers used in this study.

REFERENCES

1. S.L. Lehoczky, J. Appl. Phys. 49 (1978).
2. K. Yoshii, K. Takagi, M. Umeno, and H. Kawabe, Metall. Trans. A 15A, 1273 (1984).
3. S. Menezes and D.P. Anderson, J. Electrochem. Soc. 137 (2), 440 (1990).
4. U. Helmersson, S. Todorova, S. A. Barnett, and J.E. Sundgren, J. Appl. Phys. 62 (2), 481 (1987).
5. S.A. Barnett and M. Shinn, Annu. Rev. Mater. Sci. 24, 481 (1994).
6. S.I. Rao, P.M. Hazzledine and D.M. Dimiduk, Mat. Res. Soc. Symp. Proc. 362, 67 (1995).
7. X. Chu and S.A. Barnett, J. Appl. Phys. 77, 4403 (1995).
8. P.M. Anderson and C. Li, Nanos. Mater. 5 (3), 349 (1995).
9. J.D. Embury and J.P. Hirth, Acta. Metall. Mater. 42, 2051 (1994).
10. J.P. Hirth and J. Lothe, *Theory of Dislocations*, 2nd Edn., Wiley, New York (1982) pp. 733-34.

AN ANALYSIS OF DUCTILE BRITTLE FRACTURE TRANSITION IN LAYERED COMPOSITES

S.B. BINER
Ames Laboratory, Iowa State University, Ames IA 50011, U.S.A

ABSTRACT

In this study the failure of the ductile layers in laminated composite systems was studied numerically. The results indicate that similar maximum stress values develop in the ductile layers as in the fracture test of the same ductile material if the crack tip in the brittle layer is already at the interface. For nondebonding interfaces brittle behavior of the ductile layers is dependent upon the extent of the cracks and the fracture characteristic of the brittle layers.

INTRODUCTION

Laminated composites containing ductile reinforcements are currently under development not only to improve the fracture toughness of inherently brittle intermetallics and ceramics [1-3] but also to increase the relatively low fracture toughness of metal matrix composite systems [4,5]. The changes in the fracture mode of the ductile layers have been observed in many of the high temperature composite systems currently under investigation consisting of b.c.c. ductile reinforcements [3,6]. In this study a detailed FEM analysis of the fracture behavior of the ductile layers is presented in order to elucidate the conditions leading to brittle fracture (cleavage or intergranular).

NUMERICAL ANALYSIS

In this study, the growth of colinear cracks perpendicular to the interface of the ductile layer is analyzed as schematically shown in Fig. 1a. The ductile layer was modeled using a constitutive relationship that accounts for strength degradation resulting from the nucleation and growth of micro-voids. The basis for the constitutive model is a flow potential introduced by Gurson [7,8], in which voids are represented in terms of a single internal variable, f, the void volume fraction :

$$\phi = \frac{\sigma_e^2}{\sigma_o^2} + 2 f^* q_1 \cosh\left(\frac{q_2 \sigma_h}{2 \sigma_o}\right) - 1 - q_1^2 f^{*2} = 0 \tag{1}$$

where σ_o is the flow strength of the ductile layer, σ_e is the equivalent stress and σ_h is the hydrostatic stress. The parameters q_1 and q_2 were introduced in order to provide a better relationship between unit-cell analysis and Eq. 1 [9,10]. Later, the function f^* was proposed to account for the effect of rapid void coalescence at failure [11]. Initially $f^* = f$, but at some critical void fraction, f_c, the dependence of f^* on f is changed. This function is expressed by

$$f^* = \begin{cases} f & f \le f_c \\ f_c + \dfrac{f_u^* - f_c}{f_f - f_c}(f - f_c) & f \ge f_c \end{cases} \tag{2}$$

The constant f_u^* is the value of f^* at zero stress in Eq. 1 (i.e., $f_u^* = 1/q_1$) and f_f is the void fraction at fracture. As $f \rightarrow f_f$, $f^* \rightarrow f_u^*$ and the material loses all stress carrying capacity.

Mat. Res. Soc. Symp. Proc. Vol. 434 © 1996 Materials Research Society

The increase in void volume fraction f arises from the growth of existing voids and from the nucleation of new voids. The increase in the void volume fraction due to the nucleation process is assumed to occur with strain controlled nucleation and follow a normal distribution. Thus, the rate of void nucleation is specified by

$$(\dot{f})_{nucleation} = \left[\frac{f_N}{s_N (2\pi)^{1/2}} \exp(-\frac{1}{2}(\frac{\varepsilon^p - \varepsilon_N}{s_N})^2) \right] \dot{\varepsilon}^p \tag{3}$$

The microscopic effective plastic strain rate $\dot{\varepsilon}^p$ is represented by the power law relation

$$\dot{\varepsilon}^p = \dot{\varepsilon}_o \left[\frac{\sigma_o}{g(\varepsilon^p)} \right]^{\frac{1}{m}} \tag{4}$$

where m is the strain rate hardening exponent, $\dot{\varepsilon}_o$ is a reference strain rate and ε^p is the current value of the effective plastic strain representing the actual microscopic strain state. The function $g(\varepsilon^p)$ represents the effective tensile flow stress in a tensile test carried out at strain rate that is equal to reference strain rate, $\dot{\varepsilon}_o$. For a power hardening ductile layer material the function $g(\varepsilon^p)$ is taken to be

$$g(\varepsilon^p) = \sigma_o \left[\frac{E\varepsilon^p}{\sigma_o} + 1 \right]^n \qquad g(0) = \sigma_o \tag{5}$$

with strain hardening exponent n, Young's modulus E and reference stress σ_o. The material parameters for the ductile layer appearing in Eqs. 4 and 5 were chosen as $E = 500\sigma_o$, $\nu = 0.3$, $n = 0.1$, $m = 0.01$ and the reference strain rate $\dot{\varepsilon}_o = 2\text{x}10^{-3}$. The parameters appearing in Eq. 3 for void nucleation were taken as $f_N = 0.04$, $s_N = 0.1$, and $\varepsilon_N = 0.3$. For accelerated void growth, the parameters appearing in Eq. 2 were chosen as $f_f = 0.25$, $f_c = 0.10$ and $f_u^* = 1/1.25$. Also, $q_1 = 1.25$ and $q_2 = q_1^2$ were selected for Eq. 1. In the analysis, the behavior of the crack containing brittle layers is assumed to be elastic. The thickness ratio of the ductile to the brittle layers was 0.2. The Young's modulus of the brittle layers was 1.5 times the modulus of the ductile layers, and both layers had the same Poisson's ratio.

Because of the symmetry, only one quarter of the geometry seen in Fig. 1a is analyzed in plane-strain condition with the FEM mesh shown in Fig. 1b. The crack tip is located at the interface, and it is modeled with a hole having a radius of 10^{-4} of the crack length as shown in Fig. 1c. The element size around the crack tip region was 1.8 x 10^{-4} of the crack length; to accommodate this large difference in element size and preserve the aspect ratio of the elements, in the finite element discretization the elements were scaled exponentially to the crack tip. There were 1004 nodes and 932 elements in the model. The axial displacement rate was 5 x $\dot{\varepsilon}_o$, and resulting load values were calculated from the reaction forces. The crack extension is implemented by element vanish technique, when the f values became greater than the f_f value at all integration points.

RESULTS AND DISCUSSION

First, the fracture behavior and void growth kinetics in the material of the ductile layer is elucidated. To simulate the highly constrained crack growth in the laminated composites, the fracture behavior of the ductile matrix material is studied in a deeply cracked double edge notch geometry (DEN) by using the same mesh shown in Fig. 1b. The resulting load and load-line displacements for this case are summarized in Fig. 2. The evolution of the maximum damage in terms of nucleation and growth of voids ahead of the crack tip region is monitored at the third ring of elements as shown in Fig.1 c. The variation of the void volume fraction with axial displacements at this location is given in Fig. 3. As can be seen from Figs. 2 and 3, the failure of the elements in the third row and the crack extension took place in the linear region of the load load-line displacement curve.

The load load-line displacement curve during the fracture process of the ductile layer in the laminates composite is also given in Fig. 2. Although for both cases the uncracked ligament were the same, in the case of the composite, larger load levels were required to achieve the same amount of axial displacement, as seen in Fig. 2. The evolution of maximum damage at the same location, third ring elements, is also given in Fig. 3. Due to the development of large scale triaxial stress state for the crack at the interface region, much faster evolution of the void growth and early crack extension can be seen in the figure.

For these cases, the evolution of the maximum principal stress at the third ring elements up to first crack extension are summarized in Fig. 4. As can be seen in the composite, the maximum principal stress increases at a faster rate, later with the development of the plastic zone and faster damage evolution high stress triaxiallity is reduced and maximum stress values become very similar to that seen during the fracture process of the ductile material in the DEN specimen.

In the following two simulations, the crack tip location is moved from the interface to the brittle layer in the amount of 0.034 times the original crack length. For these simulations it is assumed that the crack extension in the brittle layer occurs with the attainment of a critical stress σ_f; its values were chosen as 5 and 10 times the yield strength of the ductile layer. The resulting load load-displacement curve for the $\sigma_f = 10\sigma_{ys}$ also shown in Fig. 2. After the large drop in the applied load as result of the failure of the brittle ligament, the resulting load load-line displacement for this case is very similar to the previous case. The evolution of the maximum principal stress values for these cases, at the third ring elements, up to formation of the first cracks in the ductile ligament are also shown in Fig. 4. As can be seen, the crack tip is now only 0.034 times the original crack length away from the crack tip, the behavior of the ductile region is completely elastic and the maximum stress levels increase linearly with axial displacement. After the failure of the brittle ligament, the maximum stress levels attained are larger than those seen earlier, and the magnitude increases with the increasing σ_f / σ_{ys} ratio. Due to this large stress elevation, as can be seen from Fig. 3, the void growth occurs very fast causing relaxation of the stresses very quickly, as seen in Fig. 4.

It is now well established that in order for brittle fracture (cleavage or intergranular) to take place it is necessary that not only a critical stress should be attained ahead of the crack tip but this stress level should extend some microstructurally characteristic distance [12]. Considering the stress relaxation resulting from the void formation, from the results seen in Fig. 4, it appears that for nondebonding interfaces the brittle behavior of the ductile layer is controlled by the extent of the cracks in the brittle layers and the fracture strength of the brittle layers.

CONCLUSIONS

In this study the failure of the ductile layers in a laminated composite system was studied by using a constitutive relationship that accounts for strength degradation resulting from the nucleation and growth of voids. The results indicate that similar maximum stress values develop in the ductile layers as in the fracture test of the same ductile material if the crack tip is already at the interface. For nondebonding interfaces, brittle behavior of the ductile layers is dependent upon the extent of the cracks and the fracture characteristic of the brittle layers.

a

σ

2a

2W

σ

b

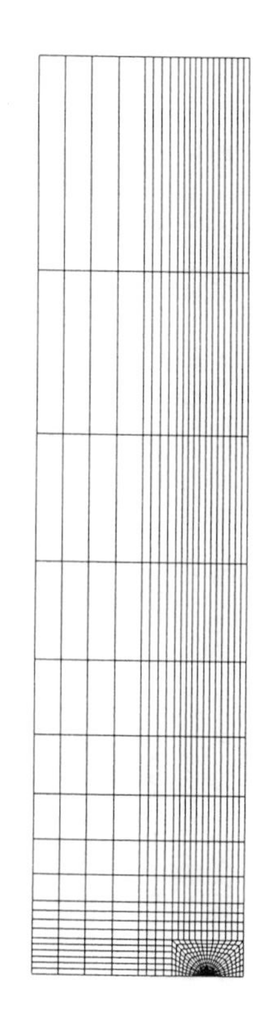

c

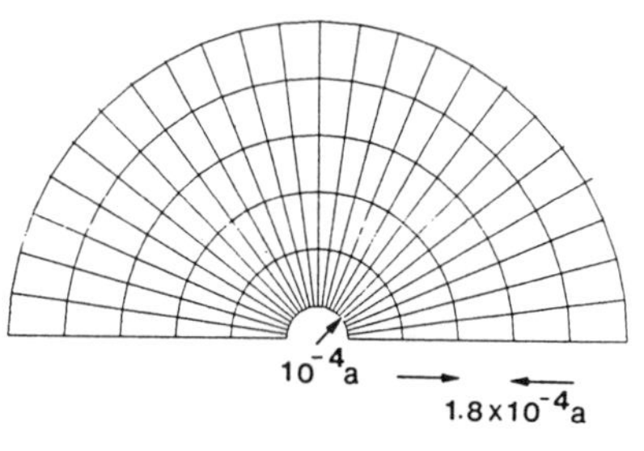

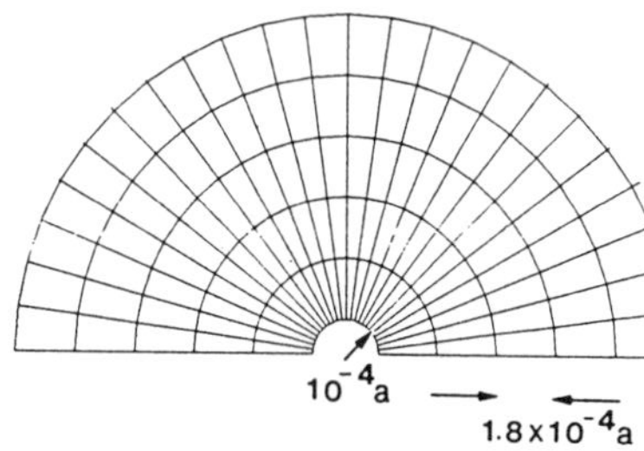

Fig.1 a-) Schematic representation of collinear cracks at the interface of the ductile layer in the laminated composite, b-) FEM mesh used in the analysis and c-) the details of the crack tip region.

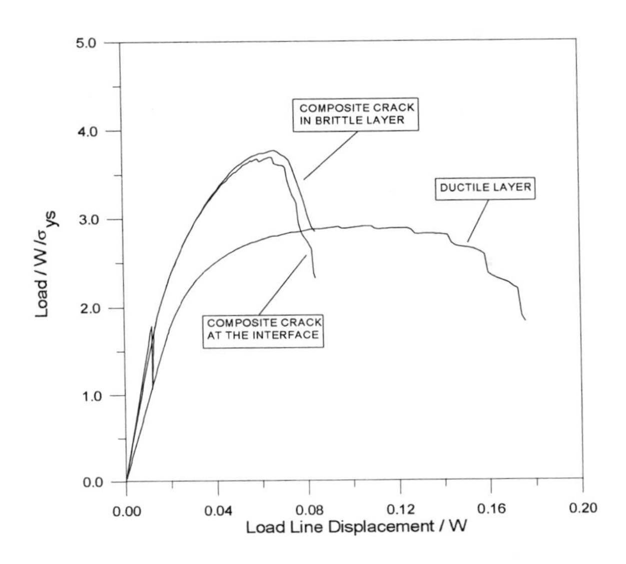

Fig.2 Load and load line displacement curves.

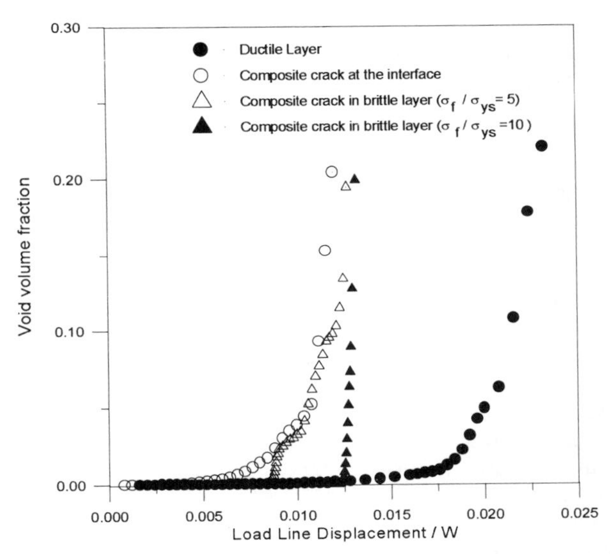

Fig.3 Variation of the void volume fraction ahead of the crack tip at normalized distance (x / W) 2.3×10^{-3} during the fracture process.

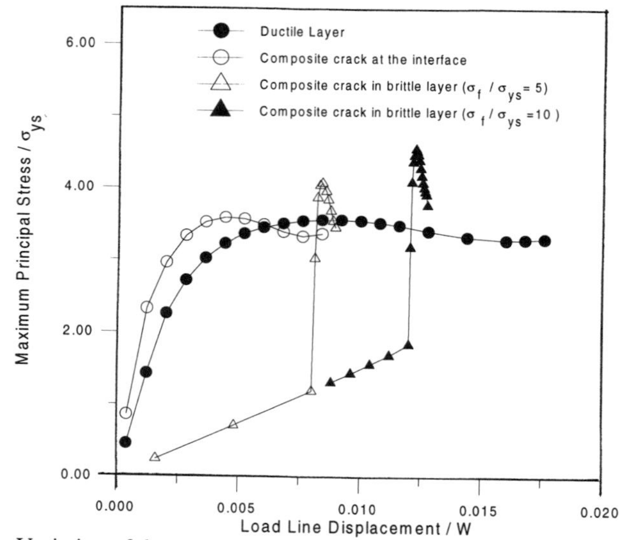

Fig.4. Variation of the maximum principal stress ahead of the crack tip at normalized distance (x / W) 2.3×10^{-3} during the fracture process.

ACKNOWLEDGMENT

This work was performed for the United States Department of Energy under contract W-7405-Eng-82 and supported by the Director of Energy Research, Office of Basic Energy Sciences.

REFERENCES

1. J.T Beals and V.C. Nardone, J. Mat. Sci. **29**, 2526 (1994).
2. L. Shaw and R. Abbaschian, Acta Metall. Mater. **42**, 213 (1994).
3. H.C. Cao, B.J. Dalgleish, H. Deve, C. Elliot, A.G. Evans, R. Mehrabian and G.R. Odette, Acta Metall. Meter. **38**, 2969 (1990).
4. C.K. Syn, S. Stoner, D.R. Lesuer and O.D. Sherby, "Influance of volume fraction and interlayer bond strength on fracture toughness of multilayer Al6090-25%SiC and Al5812 laminates" in High performance metal and ceramic matrix composites ed. K. Upadhya, (TMS Warrandale, PA, 1994) p.125.
5. W.H. Hunt, Jr., T.M. Osman and J.J. Lewandowski, J. Metals, **45**, 30 (1993).
6. J. Kajuch, J. Short and J.J. Lewandowski, Acta. Metall. Meter. **43**, 1955 (1995).
7. A.L. Gurson, PhD thesis, Brown University (1975).
8. A.L. Gurson, J. Eng. Mater. Technol. **99**, 2 (1977).
9. V. Tvergaard, Int. J. Fract. **17**, 389 (1981).
10. V. Tvergaard, Int. J. Fract, **18**, 237 (1982).
11. V. Tvergaard and A. Needleman, Acta Metall. Mater. **32**, 157 (1988).
12. R.O. Ritchie, J.F. Knott and J.R. Rice, J. Phys. Solids. **21**, 395 (1973).

RESIDUAL STRESS DISTRIBUTION IN AN Al_2O_3-Ni JOINT BONDED WITH A COMPOSITE LAYER

X.-L. Wang[1)], B. H. Rabin[2)], R. L. Williamson[2)], H. A. Bruck[2)], and T. R. Watkins[1)]
[1)]Oak Ridge National Laboratory, Oak Ridge, TN 37831-6064
[2)]Idaho National Engineering Laboratory, Idaho Falls, ID 83415-2218

ABSTRACT

Neutron diffraction was used to investigate the residual stress distribution in an axisymmetric Al_2O_3-Ni joint bonded with a 40 vol%Al_2O_3-60 vol%Ni composite layer. A series of measurements was taken along the axis of symmetry through the Al_2O_3 and composite layers. It is shown that after taking into account the finite neutron diffraction sampling volume, both the trends and peak values of the experimentally determined strain distribution were in excellent agreement with calculations of a simple finite element model, where the rule-of-mixtures approach was used to describe the constitutive behavior of the composite interlayer. In particular, the predicted steep strain gradient near the interface was confirmed by the experimental data.

INTRODUCTION

High-strength ceramic-metal joints are being developed for use in a great variety of industrial applications, ranging from structural components in heat engines to coatings in electronic devices. However, upon cooling from the fabrication temperature, residual stresses develop due to the thermal expansion mismatch between the metal and ceramic components. Moreover, although no experimental evidence has been reported, a tensile stress concentration was repeatedly predicted by various finite element models to occur near the edges of the ceramic component close to the interface [1-2]. In some cases, these residual stresses exceed the bond strength and promote mechanical failure along the ceramic-metal interface. In cases when the bond is strong, the residual stresses tend to cause fracture in the ceramic. Earlier experimental studies [3-6], principally by X-ray and neutron diffraction, have shown that the residual stress distribution in directly bonded ceramic-metal joints is reasonably understood within the frame work of elasto-plastic finite element calculations.

One approach to reduce the residual stress in ceramic-metal joints is through the use of functionally graded materials, where materials properties including thermal expansion vary continuously from one end to the other. A variety of analytical and numerical models have been developed to understand and optimize the residual stress state in these materials. Due to the complexity of the microstructures involved and the associated difficulties in describing the constitutive behavior of the composite layers, simplifying approximations, such as rule-of-mixtures, are typically used. Critical needs, therefore, exist for experimental verification of these models before they can be used with confidence for design purposes. In this paper, we report a neutron diffraction determination of the residual stress distribution in a prototype functionally graded material and a comparison of the experimental data with the results of a finite element analysis [2]. The measurements focus on the vicinity of the interface between the ceramic and composite layers where a steep strain gradient has been predicted.

EXPERIMENTAL DETAILS

The specimen, in the shape of a rod, is a three-layer material, consisting of an Al_2O_3 layer, a Ni layer, and a 40 vol% Al_2O_3-60 vol% Ni composite interlayer. It was fabricated using powder processing techniques [7], followed by controlled cooling to room temperature. The specimen

Mat. Res. Soc. Symp. Proc. Vol. 434

thus fabricated exhibits sharp interfaces between layers. Micrographs taken from the composite layer indicate a uniform microstructure with a rather continuous Al_2O_3 phase. A schematic of the specimen, along with the dimensions, is shown in Fig. 1.

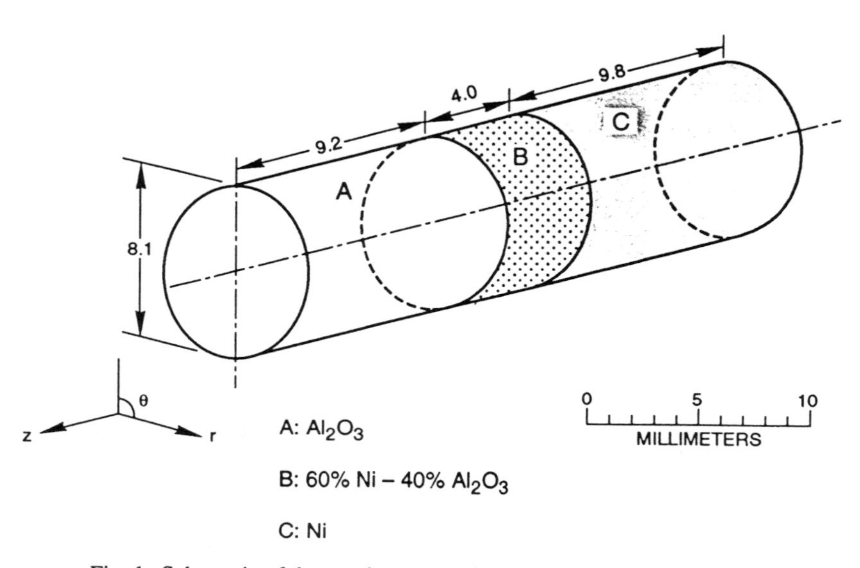

Fig. 1 Schematic of the specimen used for neutron diffraction measurements.

The neutron diffraction measurements were conducted at the High Flux Isotope Reactor of Oak Ridge National Laboratory using a triple-axis spectrometer operated in the diffractometer mode. A Be (1 1 0) reflection was used as the monochromator. The take-off angle for the monochromator was 90° and the incident neutron wavelength was 1.615 Å. To facilitate fast data collection, a position sensitive detector was mounted at the analyzer position of the spectrometer. Slits of dimensions $0.8 \times 4\ mm^2$ and $0.8 \times 30\ mm^2$ were inserted before and after the specimen, which together defined a sampling volume of approximately $2.6\ mm^3$. Radial and axial strain (ε_{rr} and ε_{zz}) distributions were investigated in this study. For each strain component, a series of measurements was taken along the specimen axis of symmetry through the Al_2O_3 and composite layers. For the purpose of illustration, the specimen orientation for the measurement of ε_{zz} is shown in Fig. 2.

In order to observe the predicted steep strain gradient near the interface, high spatial resolution is essential. In principle, this would require a small sampling volume at the expense of reduced scattering intensity. Fortunately, earlier neutron diffraction measurements and finite element modeling studies have demonstrated that in cylindrical ceramic-metal joint systems, strains along major axes (ε_{rr}, $\varepsilon_{\theta\theta}$, and ε_{zz}) are almost constant over half of the radius. This has permitted the use of narrow but tall slits to obtain reasonable scattering intensity without sacrificing the spatial resolution in the desirable directions, as illustrated in Fig. 2. To further improve the ability of detecting the predicted strain gradient, the specimen was over-stepped (sampling volume overlapping), with a step size of 0.25 mm, in the vicinity of the interface (see Fig. 2).

DATA ANALYSIS AND RESULTS

The Al_2O_3 (3 0 0) reflection ($2\theta \approx 72.0°$) was used for strain determination. The recorded diffraction profile was fitted to a Gaussian function to yield the position, width, and intensity of the peak. Strains at a given location were then determined using

$$\varepsilon = \frac{d - d_0}{d_0} = \frac{\sin\theta_0}{\sin\theta} - 1, \qquad (1)$$

where d is the lattice spacing and 2θ the diffraction angle of Al_2O_3 (3 0 0). d_0 and $2\theta_0$ are the corresponding stress-free values.

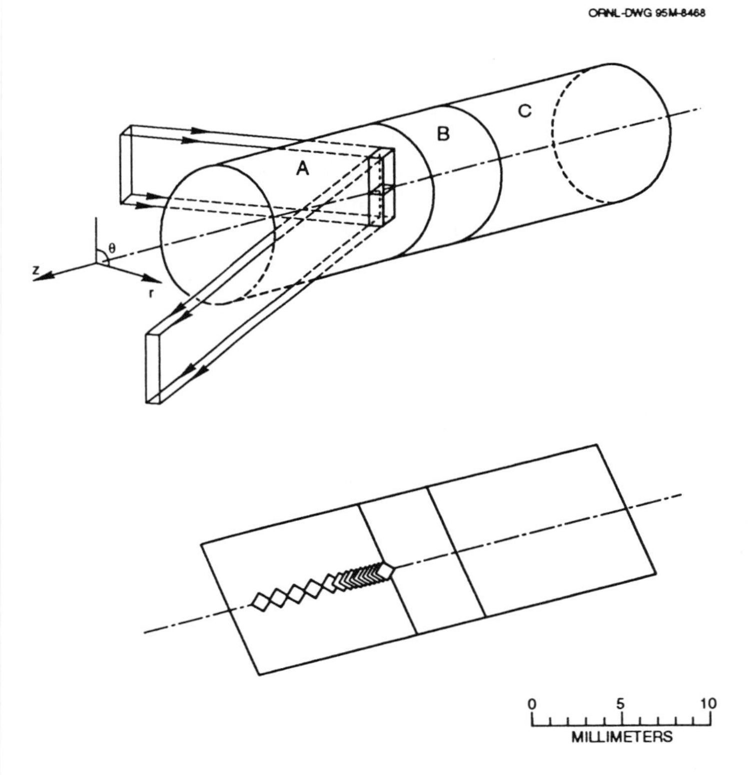

Fig. 2 Schematic illustrating the orientation of the specimen for axial strain (ε_{zz}) measurements and the overlap of sampling volume to improve the ability of detecting the predicted strain gradient.

In general, strains measured with diffraction methods are a superposition of macro- and microstrains. Since in this study we were solely concerned with the macrostrain distribution resulting from the thermal expansion mismatch between bonded dissimilar materials, the data points farthest from the interface were chosen to be zero. In this way, effects of any microstrains due to the thermal expansion anisotropy within the single phase Al_2O_3 were removed. However, data in the composite layer still contain a contribution of microstrains from the thermal expansion mismatch between Al_2O_3 and Ni phases, which is expected to be compressive and of the order of 10^{-4}.

Fig. 3 shows the experimentally determined ε_{rr} and ε_{zz} along the axis of symmetry. The error bars are estimated standard deviations from least-squares fitting of the recorded diffraction profiles, which in this experiment were dominated by the unfavorable scattering intensity due to the small sampling volume used.

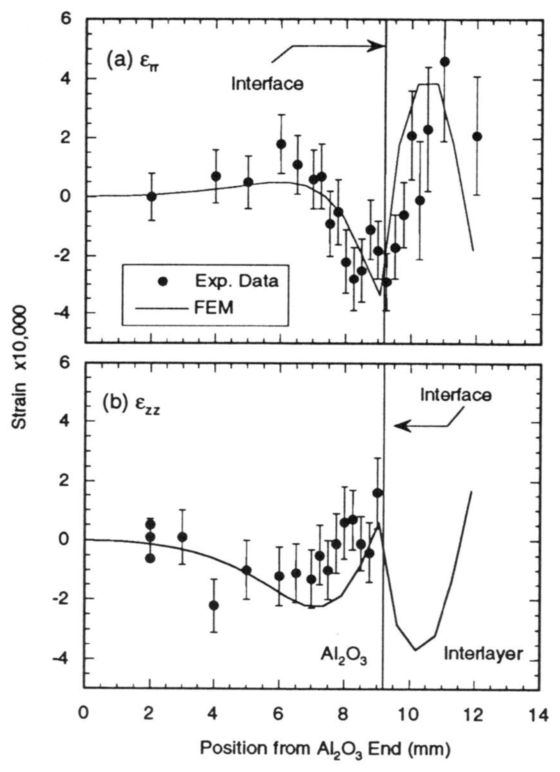

Fig. 3 Experimentally determined strain distribution and the results of finite element modeling along the axis of symmetry for (a) ε_{rr} and (b) ε_{zz}. The symbols are experimental data; the solid curves are finite element results which have been averaged to take into account the finite neutron diffraction sampling volume.

Within the experimental precision, the experimental data provided evidence of a sharp strain gradient through the interface. Overall, the magnitudes of the measured strains are quite small, on the order of 10^{-4}. In the Al_2O_3 layer, ε_{rr} becomes increasingly compressive as the interface is approached. The maximum compressive strain is located on the Al_2O_3 side adjacent to the interface. Upon entering the composite layer, ε_{rr} changes from compressive to tensile at approximately 1 mm from the interface. Given the error bars, the experimental data also suggest that ε_{rr} reaches a tensile maximum in the middle of the composite layer. ε_{zz}, on the other hand, shows a quite different axial dependence. It is mostly compressive in the Al_2O_3 layer, becoming tensile only when the interface is approached. Measurements of ε_{zz} across the interface were not attempted because in this measurement geometry, an artificial peak shift was anticipated when the sampling volume was partially buried in the Al_2O_3 layer [8]. This artifact leads to an apparent strain and adds ambiguity to the determination of ε_{zz}.

COMPARISON WITH FINITE ELEMENT MODELING

The finite element model described in Ref. [2] was utilized to evaluate the residual stress and strain distributions for the specimen measured with neutron diffraction. Spatially uniform cooling and perfect bonding at materials interface were assumed. Because the specimen remains axisymmetric during cooling, the model was reduced to two-dimensional computation. Fine meshing was employed in the vicinity of interfaces and the radial free surface, due to the expected large stress and strain gradients in these regions. All materials were assumed to be isotropic. The Al_2O_3 was required to remain elastic, while plasticity was allowed in the Ni and composite layers. Creep behaviors were not considered in the present model, i.e., materials response was assumed to be independent of time. Mechanical and thermophysical properties for the composite layers were assigned using a modified rule-of-mixture approach [9]. Additional information concerning the general modeling approach, materials properties, and particularly the rule-of-mixture formation, can be found in Ref. [2]. Numerical solutions were obtained using the ABAQUS computer program [10].

For comparison, the calculated strain values along the axis of symmetry are also plotted in Fig. 3 (solid lines). Note that here the numerical results have been averaged to take into account the finite neutron diffraction sampling volume. As can be seen, the calculations and experimental data are in excellent agreement in both the trends and peak values. In particular, the predicted steep strain gradient near the interface was confirmed by the experimental data.

As the finite element results were compared with the experimental data, it became clear that the finite neutron diffraction sampling volume had to be considered in order to seek quantitative agreement. In an initial comparison, the numerical results were simply plotted for the column of elements adjacent to the axis of symmetry. In this case, the magnitudes of calculated strains were significantly higher than the experimental values, particularly in the vicinity of the interface Only when the finite element results were averaged over the neutron diffraction sampling volume, was quantitative agreement realized.

DISCUSSIONS

The present experiment demonstrates that with appropriate experimental arrangement, strain changes over the range of 1-2 mm can be spatially resolved by neutron diffraction. This level of spatial resolution is required for mapping the residual stress distribution near the interface of ceramic-metal joint structures, as was evidenced in this study. Because neutrons are highly penetrating in most materials, the measurements are non-destructive and, in general, no special specimen preparation is required.

As stated earlier, the specimen under study is a prototype functionally graded structure and the experiment was designed to verify the constitutive relationship used for calculating the residual stress distribution. The neutron diffraction data show that the modified rule-of-mixtures approach [9] provides an adequate description of the residual strain distribution along the axis of symmetry. This gives the confidence of using the model to address key issues in the design of ceramic-metal joints. For example, controlling of the tensile stress concentration, located near the edges of the ceramic component close to the interface, is highly desirable. There, the plasticity in the composite layers is expected to play an important role. Also, the creep behavior in the metal as well as the composite layers may have to be considered. An experimental investigation is currently in progress to determine the mechanical and thermophysical properties of the composite materials containing various amount of Al_2O_3 to further verify the accuracy of the rule-of-mixtures approach used in the model.

Using the present model, it can be shown that by inserting a composite layer between the Ni and Al_2O_3 layer, the peak stress values near the interface are reduced. Although the calculations are made for rod-shaped specimen, the model can be readily adapted to calculate the residual stresses in disk-shaped specimens [3] and perhaps in laminated materials as well so long as there is no atomic diffusion between layers. Because of the limited intensity, the present mapping system does not yet have sufficient spatial resolution to determine the distribution of residual stresses in thin laminated materials. However, progress is underway to improve the instrument so that neutron diffraction data can be collected with a much smaller sampling volume (hence higher spatial resolution), which will enable direct determination of the residual stress distribution in laminated materials.

ACKNOWLEDGMENTS

This research was sponsored in part by the U. S. Department of Energy, Assistant Secretary for Energy Efficiency and Renewable Energy, Office of Transportation Technologies, as part of the High Temperature Materials Laboratory User Program. Partial support was also provided by U. S. Department of Energy, Office of Energy Research, Office of Basic Energy Sciences, Division of Materials Sciences. Oak Ridge National Laboratory is managed by Lockheed Martin Energy Research Corp. for the U.S. Department of Energy under contract number DE-AC05-96OR22464.

REFERENCES

1) P. O. Charreyron, N. J. Bylina, and J. G. Hannoosh, in Fracture Mechanics of Ceramics, Vol. **8** (Plenum Press, New York, 1986) p. 225; P. O. Charreyron, D. O. Patten, and B. J. Miller, *Ceram. Eng. Sci. Proc.*, **10**, 1801 (1989).
2) R. L. Williamson, B. H. Rabin, and J. T. Drake, *J. Appl. Phys.*, **74**, 1310-20 (1993).
3) X.-L. Wang, C. R. Hubbard, S. Spooner, S. A. David, B. H. Rabin, and R. L. Williamson, *Mat. Sci. Eng.* (in press).
4) K. Masanori, M. Sato, I. Ihara, and A. Saito, in Advances in X-ray Analysis, Vol. **33**, Edited by C. S. Barret et al. (Plenum Press, New York, 1990) p. 353.
5) O. T. Iancu, D. Munz, B. Eigenmann, B. Scholtes, and E. Macherauch, *J. Am. Ceram. Soc.*, **73**, 1144 (1990).
6) L. Pintschovius, N. Pyka, R. Kubmaul, D. Munz, B. Eigenmann, and B. Scholtes, *Mat. Sci. Eng.*, **A177**, 55 (1994).
7) B. H. Rabin and R. J. Heaps, *Ceram. Trans.*, **34**, 173 (1993).
8) S. Spooner and X.-L. Wang, unpublished.
9) Tamura, Y. Tomota, and H. Ozawa, in Proceedings of the Third International Conference on Strength of Metals and Alloys (Institute of Metal and Iron and Steel Institute, London, 1973) p. 611.
10) ABAQUS, Habbitt, Karlssan, and Sorensen, Inc., Pawtucket, Rhode Island (1993).

A FINITE ELEMENT STUDY ON CONSTRAINED DEFORMATION IN AN INTERMETALLIC / METALLIC MICROLAMINATE COMPOSITE

J. HEATHCOTE,* G. R. ODETTE,** G. E. LUCAS
*Materials Department, University of California, Santa Barbara, CA 93106,
johnh@engineering.ucsb.edu
**Department of Mechanical Engineering, University of California, Santa Barbara, CA 93106

ABSTRACT

The mechanical properties of intermetallic / metallic microlaminates were studied by determining the fundamental composite properties that control the fracture behavior: namely, the stress-displacement functions of the metal layers. Finite element methods were used to model the stress-displacement function of a constrained metal layer and to examine the effect of constituent properties, residual stress, offset cracks in adjacent intermetallic layers, and debonding inclusions in the metal layer. Finally, FEM models representative of four specific microlaminates were developed and the results were compared to experimentally determined $\sigma(u)$'s for those composites. Determining these fundamental composite properties and showing how they control the mechanical behavior gives insight into the optimum design of this composite system.

INTRODUCTION

The incorporation of ductile metal reinforcements in an intermetallic matrix has been shown to increase the fracture resistance from that of the matrix. Ligaments of ductile metal left in the wake of a growing crack plastically deform during crack extension, imparting a closing force on the crack. The bridging stress - crack opening displacement function, $\sigma(u)$, of these ligaments is a fundamental composite property, determining the fracture resistance curve[1-3]. $\sigma(u)$ can differ significantly from uniaxial stress-displacement behavior due to the constraint imposed on the deforming metal by the intermetallic matrix.

Recent work on a set of intermetallic / metallic microlaminate composites based on Nb_3Al/Nb and Cr_2Nb/Nb systems has demonstrated considerable toughening imparted by the metal layers[4,5]. Several methods were used to construct $\sigma(u)$ functions, including constructions based on microhardness and confocal/fracture reconstruction and fits to resistance curves based on a bridging model. The resulting $\sigma(u)$ functions were found to be self consistent, and they could be used to predict resistance curve behavior in other geometries as well as fracture strengths.

The purpose of this study was to further examine the nature of deformation and fracture of the ductile layers and to use finite element methods (FEM) to model the effects of various observed processes on the $\sigma(u)$ behavior.

EXPERIMENT

Fracture surfaces were observed by both scanning electron and confocal microscope. Fracture reconstruction methods based on the work of Kobayashi[6,7] have been combined with confocal microscopy (CM) [8] to determine the nature of the fracture process and quantify displacement to failure in the ductile layers. Quantitative topographical measurements of conjugate surfaces were obtained by confocal microscopy. By overlaying matching height profiles taken from corresponding lines on each surface, the point of separation of those surfaces can be recreated. This recreation provides information about the deformation of the metal layers and about the cracking patterns in the intermetallic layers.

Both CM and SEM revealed a number of characteristic features associated with deformation and fracture of the constrained metal layers. These are illustrated schematically in Figure 1 and included: a) offset cracks, b) slanted cracks, c) splitting cracks in the adjacent intermetallic layers and d) microvoid growth and coalescence in Nb(Cr) alloy (vs. pure Nb) layers. From CM reconstruction profiles, measurements were made of crack offset and crack slanting. These distributions for the four microlaminates are shown in Figure 2. Also shown on these graphs are the levels of offset and slanting used in the FEM models for these microlaminates.

Mat. Res. Soc. Symp. Proc. Vol. 434

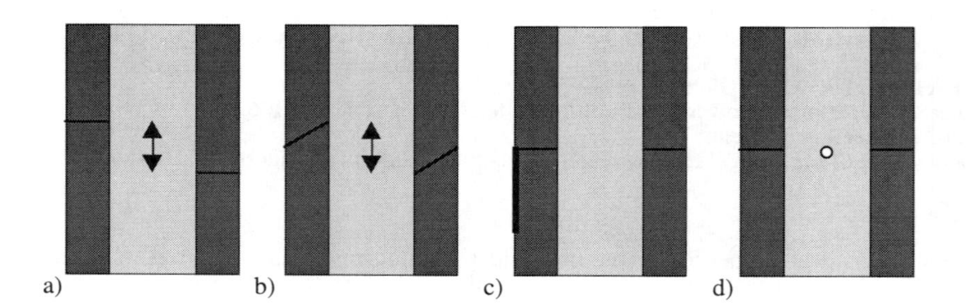

Figure 1. FEM cells used to model observed processes, including a) offset cracks, b) slanted cracks, c) splitting cracks in intermetallic layers, and d) microvoid growth and coalescence (modeled as debonding inclusion) in Nb(Cr) alloy (vs. pure Nb) layers.

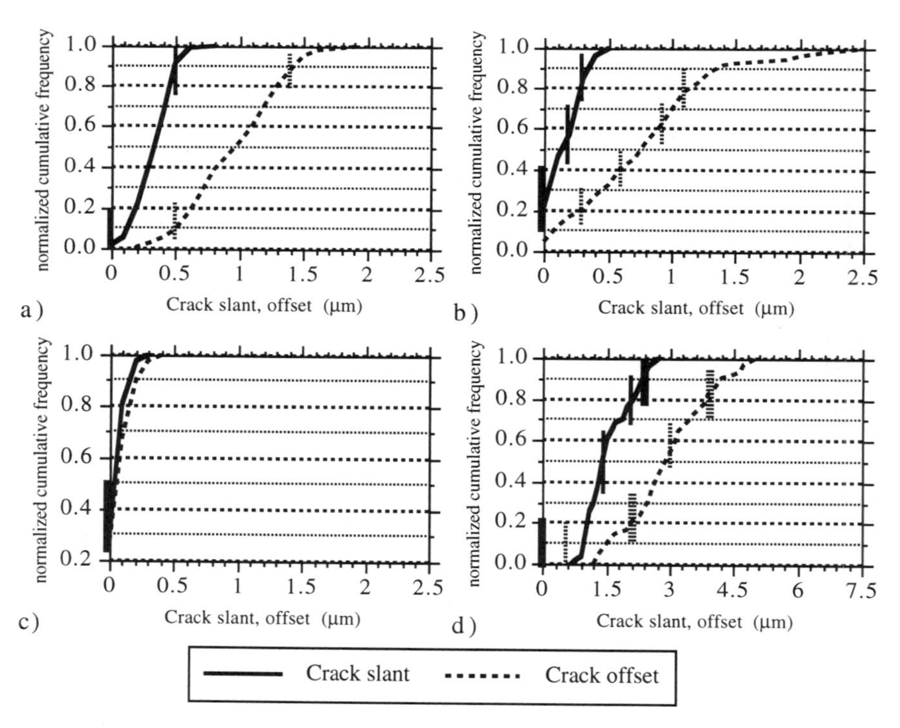

Figure 2. Cumulative probability distributions of the crack slant and crack offset measured by confocal microscopy for a) L9 (Nb_3Al / Nb), b) L20 [$(Nb,Ti)_3Al$ / Nb(Ti,Al)], c) L17 [Cr_2Nb / Nb(Cr) / 2μm], and d) L60 [Cr_2Nb / Nb(Cr) / 6μm].

Using SEM observations, the mode of metal failure and the extent of intermetallic splitting were determined for each of the microlaminates. The Nb and the Nb(Ti,Al) layers of the L9 and L20 microlaminates failed by deforming to a single ridge. In contrast, the Nb(Cr) layers of the L17 and L60 microlaminates failed with internal void growth.

The extent of intermetallic cracking was measured from SEM micrographs. Lengths of each metal layer were measured that corresponded to one of three regions: no intermetallic cracking; cracking in one intermetallic layer on one side of the fracture surface; or, finally, cracking in two layers or in one layer on both sides of the fracture surface. The fractions of each region for all of the microlaminates are listed in Table I.

Table I
Fractions of metal layers with cracking in adjacent intermetallic layers

	No cracking	One layer / side	Two layers/sides
L9	1.00	0.00	0.00
L20	0.211	0.542	0.247
L17	0.763	0.196	0.04
L60	0.691	0.282	0.027

ANALYSIS

The ABAQUS© finite element code was used in the modeling of this system[9]. The mesh used in the FEM model represented a single metal layer surrounded by two intermetallic half layers. The left and right borders had reflective symmetry, so that the mesh was a cell, repeated to represent the multilayer geometry.

The effect of certain variables on the stress-displacement function were analyzed. These included constituent properties, the presence of residual stresses, crack offset in adjacent intermetallic layers, splitting cracks in the intermetallic layers, and the presence of holes or inclusions in the metal layers. In addition, specific models that combined the applicable variables for each microlaminate were studied. A number of two metal layer modeling cells were used for these to approximately represent the distributions of crack slanting and offset.

Some of the constitutive behavior of the Nb layers is not exactly known. It was parameterized. Figure 3a shows three stress-strain curves used in the FEM model to determine the effect of the metal's constitutive behavior on the $\sigma(u)$ function. The stress at 8% strain was kept at a fixed value, and the hardening exponent, n, was varied. This stress corresponds to the estimated ultimate tensile strength from microhardness testing reported in the companion paper. The $\sigma(u)$ predictions are shown in Figure 3b. In each case, $\sigma_{peak}/\sigma_{8\%}$ is approximately 3. For the rest of the modeling, n is set to 0.24, a reported value for niobium[10].

In the microlaminates being modeled, residual stresses are present due to the thermal mismatch between the metal and the intermetallic. These stresses are compressive in the intermetallic layers and tensile in the metal layers. The effect of a 500MPa residual stress was modeled. The results are plotted in Figure 4a. For the rest of the modeling, the residual stresses are present. The effect of an offset between the cracks in the adjacent intermetallic layers was explored. Figure 4b shows the effect of a half- and a full-thickness offset on the $\sigma(u)$ curve. Likewise, model predictions were made for a cell with a splitting crack at the midthickness of one intermetallic layer on one side of the crack and for a cell with splitting cracks on both sides of the major crack. The predicted curves are shown in Figure 5a.

The Nb(Cr) layers in the L17 and L60 microlaminates failed by the growth of internal voids. For this reason, the effect of a debonding inclusion on $\sigma(u)$ was tested. Figure 5b shows the effect of a 0.05xthickness radius bound inclusion and for a 0.05xthickness radius hole. Along with these curves are the functions for inclusions that debond at various interfacial stresses.

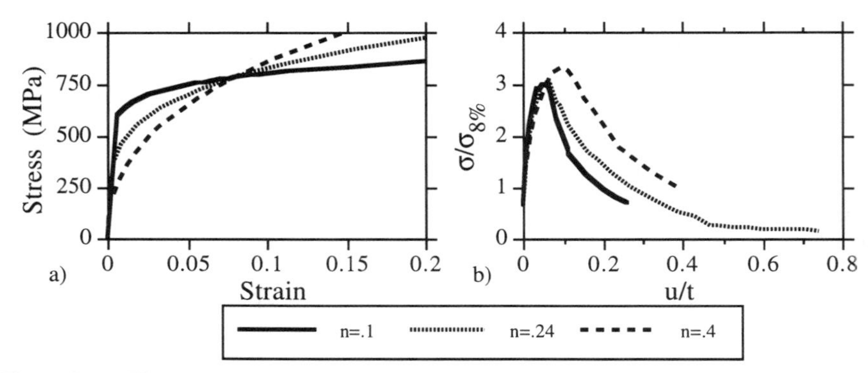

Figure 3. a) Three stress-strain curves used to define the metal layers unconstrained properties in the FEM model. The stress at 8% strain is fixed and the values for n are varied. b) The corresponding FEM predictions for constrained metal layers with those properties.

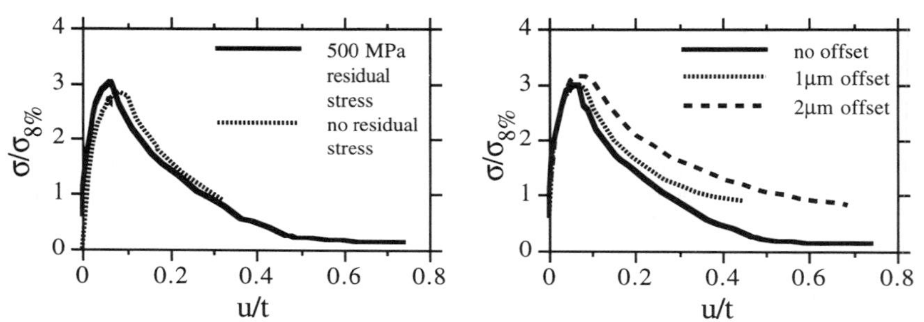

Figure 4. a) FEM σ(u) predictions for a constrained metal layer with and without a residual stress. b) Predictions for metal layers with varying degrees of offset between cracks in adjacent intermetallic layers.

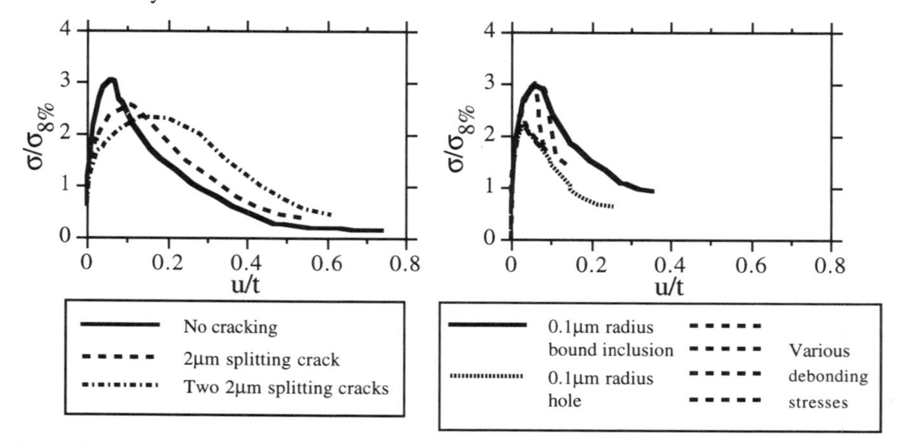

Figure 5. a) FEM σ(u) predictions for constrained metal layers with splitting cracks in the intermetallic layers. b) Predictions for a metal layer with a centered 0.1μm hole or rigid inclusion.

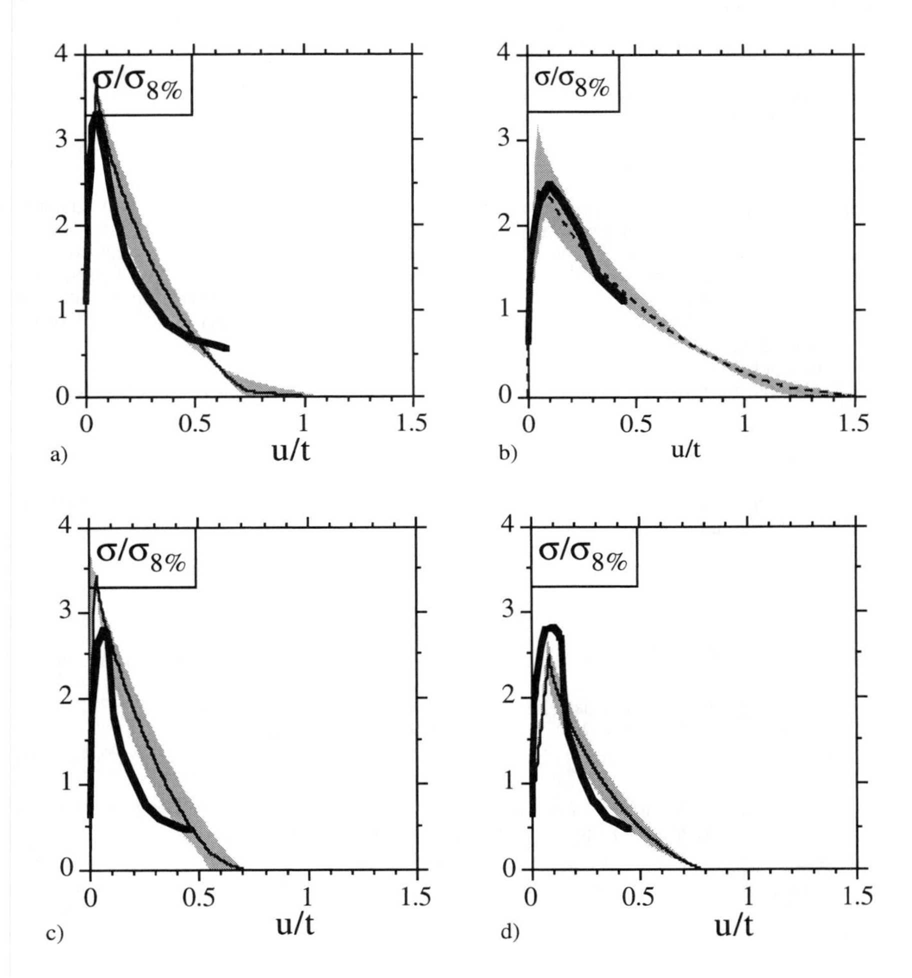

Figure 6. Solid lines are stress-displacement functions [σ(u)] for: a) L9 -- Nb_3Al/Nb; b) L20 -- $(Nb,Ti)_3Al/Nb(Ti,Al)$; c) L17 -- $Cr_2Nb/Nb(Cr)/2\mu m$; and d) L60 -- $Cr_2Nb/Nb(Cr)/6\mu m$ as determined by finite element models representative of each microlaminate. In some cases, two or more model results are averaged together to capture the correct distribution of crack offset and slanting. Also shown on each figure are ranges of σ(u) functions determined by use of a bridging model fit to experimental data, as explained in a companion paper[4].

Each of the microlaminates analyzed has a different combination of variables affecting the σ(u) function of the deforming ligaments. For this reason, models specific to each microlaminate were developed. Models were developed that captured the distribution of crack offset and slanting, the correct percentage of intermetallic cracking, and the correct metal failure mechanism. The predictions from each of these models are shown in Figure 6 along with σ(u) functions fit to experimental results by use of a bridging model as described in another paper[4].

CONCLUSIONS

A combination of fracture reconstruction and scanning electron microscopy was used to develop a finite element model of the constrained deformation of the metal layers of an intermetallic / metallic microlaminate composite. The effect of a number of variables on the σ(u) function of the deforming ligament has been shown. In addition, σ(u) functions determined by a procedure described by a companion paper were compared to FEM models representative of each microlaminate. The fracture processes described in this work have an effect on the shape of the σ(u) function. This is important as the shape of this function controls the resistance curve behavior and fracture strength of these materials. Understanding these relationships provides a basis for understanding the behavior of these composites and designing structures with optimal properties.

ACKNOWLEDGMENTS

This work was funded by Wright-Patterson Air Force Base under subcontract from General Electric. (Contract number FY1457-91-01001)

REFERENCES

1. G.R. Odette, B.L. Chao, J.W. Sheckherd, G.E. Lucas., Acta Metall. Mater., 40 (1992) 2381.
2. F.E. Heredia, M.Y. He., G.E. Lucas, A.G. Evans, D. Konitser, Acta Metall. Mater., 41 (1993) 505.
3. K.T. Venkateswara, G.R. Odette, R.O. Ritchie, Acta Metall. Mater., 40 (1992) 353.
4. J.Heathcote, G.R.Odette, G.E. Lucas, R.G. Rowe, these proceedings.
5. J. Heathcote, G. R. Odette, G. E. Lucas, R. G. Rowe, D. Skelly, Acta Met Metall (in press).
6. K. Ogawa, X. J. Zhang, T. Kobayashi, R. W. Armstrong, G. R. Irwin, ASTM-STP-833, American Society for Testing and Materials, Philadelphia, PA, p. 393 (1984).
7. T. Kobayashi, and D. A Shockey, Met. Trans., **18A**, p. 1941 (1987).
8. Edsinger, K., Odette, G.R., Lucas, G.E., and Wirth, B., Effects of Radiation in Materials: 17th International, Gelles, Nanstad, Kumar, Little, Editors, ASTM-STP, (in press).
9. ABAQUS, © Hibbitt, Karlsson, and Sorenson, Inc., (1994).
10. Metals Handbook, 9th Edition, Vol.3, ASM, (1980) 777-779.

SYNTHESIS OF CARBON NITRIDE COMPOSITE THIN FILMS PREPARED BY PULSED LASER DEPOSITION METHOD

ASHOK KUMAR*, R. B. INTURI**, U. EKANAYAKE*, H. L. CHAN*, Q. You*, G. WATTUHEWA*, and J. A. BARNARD**
*Department of Electrical Engineering, University of South Alabama, Mobile, AL 36688
**Department of Metallurgical and Materials Engineering, University of Alabama, Tuscaloosa, AL 35487-0202

ABSTRACT

Carbon nitride/titanium nitride composite coatings have been deposited on Si (100) and 7059 corning glass by in-situ pulsed laser deposition technique. A pulsed laser (λ = 248 nm) has been used to ablate the both pyrolytic graphite and TiN targets. It has been shown that TiN provides a lattice-matched structural template to seed the growth of carbon nitride crystallites (W. D. Sproul et. al., Appl. Phys. Lett. Vol. 67, 203-205; 1995). This paper describes the same approach to grow carbon nitride composite coatings with varying thicknesses of buffer layer and carbon nitride films at different temperatures and pressures. Our preliminary results show the superior mechanical properties (hardness and modulii). The films have been characterized by X-ray diffractometer, SEM, FTIR and Raman spectroscopic techniques.

INTRODUCTION

Ever since the theoretical study of Liu and Cohen [1] indicated the possible existence of superhard β-phase carbon nitride (β-C_3N_4), many attempts have been made to synthesize this hypothetical material by various techniques [2-5]. These β-C_3N_4 materials have attracted the attention of various investigators worldwide due to their inserting characteristics and wide range of applications, such as cutting tools, wear resistance, and barrier against corrosion. However, no one has successful in preparing crystalline thin films of β-C_3N_4 phase. Recently, W. D. Sproul et. al. [6] has reported the formation of CN_x/TiN coatings using dual-cathode magnetron sputtering. The TiN provides a lattice-matched structural template to seed the growth of carbon nitride crystallites. The pulsed laser deposition (PLD) technique has proven to be one of the best technique to deposit excellent quality of TiN films on Si(100) and metal substrates [7-8]. This paper also describes the same approach to carbon nitride composite coatings with varying thickness of buffer layers of TiN and CN_x films at different processing conditions by PLD method. Mechanical properties of thin films, such as film cohesiveness and bonding to the substrate, are influenced by film stresses which are characteristic to the deposition condition. This paper presents the detailed evaluation of mechanical properties by nanoindentation method of composite CN_x films prepared under a wide range of conditions by PLD technique.

EXPERIMENTAL

The deposition of the films was carried out in a recently dedicated "Laser Processing of Materials" research laboratory at the University of South Alabama. The detailed description

Mat. Res. Soc. Symp. Proc. Vol. 434 © 1996 Materials Research Society

about the deposition system is described somewhere else [9]. The carbon nitride films were deposited on Si (100) substrates at 100^0C with varying partial pressure (50 mTorr to 300 mTorr) in N_2 environments. Composite carbon nitride/titanium nitride coatings were deposited on Si (100) substrates. For depositing composite CN_x coatings, TiN films were first deposited as a buffer layer at 600^0C [8] in high vacuum on Si (100) substrates at 10 J/cm^2. After depositing 3000 Å thick TiN films, the graphite target was ablated at an energy density of 3 J/cm^2. The thickness of CN_x was 6000 Å. In one sample , the thickness of both TiN and CN_x was increased to investigate the effects of mechanical properties of the composite films. The films were characterized by FTIR to analyze the bonding between carbon and nitrogen. The mechanical properties such as hardness and Young's modulus of these films were determined using the Nanoindenter. Microstructural features of these films were investigated to understand the effects of the processing parameters to correlate with the mechanical properties. The carbon-nitrogen bonding characteristics of the films were studied using FTIR spectroscopy.

RESULTS

The IR absorption band of the CN_x band is shown in Figure 1. Their locations were listed and compared with the published data in literature. The IR band was observed in the range between 2300 cm^{-1} and 2400 cm^{-1} , which is attributed to the stretching mode of C≡N bonding. Figure 2 shows the band of the composite CN_x films deposited on Si (100). Here, we also find the IR bands in the range of 2300 cm^{-1} and 2400 cm^{-1} . The IR bands in composite films (with TiN under layer) have a strong peak of C≡N bands, which indicates the formation of dominant phase of CN_x.

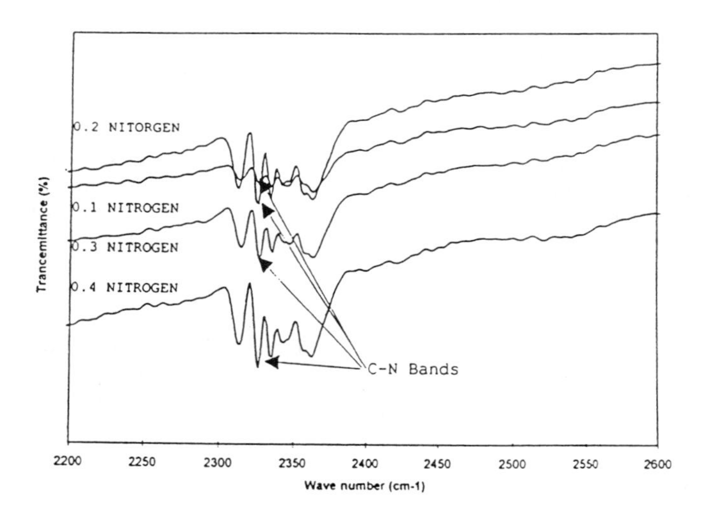

Fig. 1 FTIR spectrum of carbon nitride films on Si (100) substrates at different partial pressure of nitrogen environment

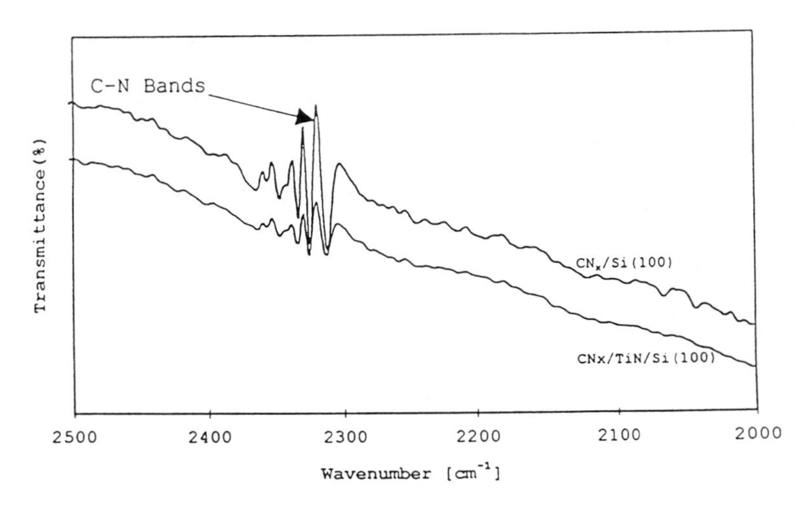

Fig. 2 FTIR spectrum of carbon nitride composite films deposited on Si (100) substrates

The diffraction spectra taken from carbon nitride films deposited under different conditions are given in Figure 3. In all the cases, the peak that appears in each scan can be indexed either to Si substrate or to TiN underlayers. Therefore, it is suggested that the structure of carbon nitride films prepared in this study are amorphous. It has been recently reported that crystalline carbon nitride/TiN composite coatings deposited by dual cathode magnetron sputtering system have remarkably enhanced the hardness value of the films. Carbon and nitrogen species are delivered to substrate surfaces that bear no structural relationship to CN_x.

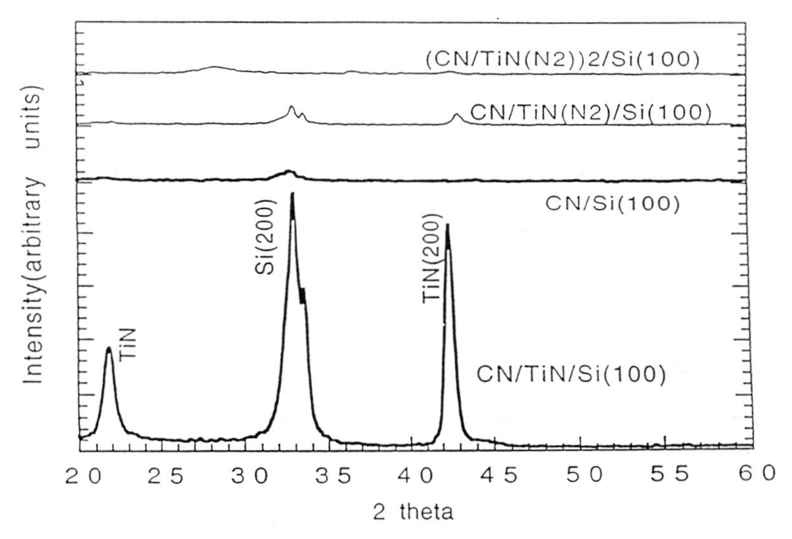

Fig. 3 X-ray diffraction pattern of composite carbon nitride films on Si (100) substrates

Based on existing literature on the growth of metastable phases, it is important to provide a structural template to seed the growth of CN_x. An ideal structural template [10] to seed the growth of CN_x is one with at least one low free energy crystal plane lattice matched to some low free energy plane of CN_x. TiN provides a lattice matched structural template to seed the growth of carbon nitride crystallites.

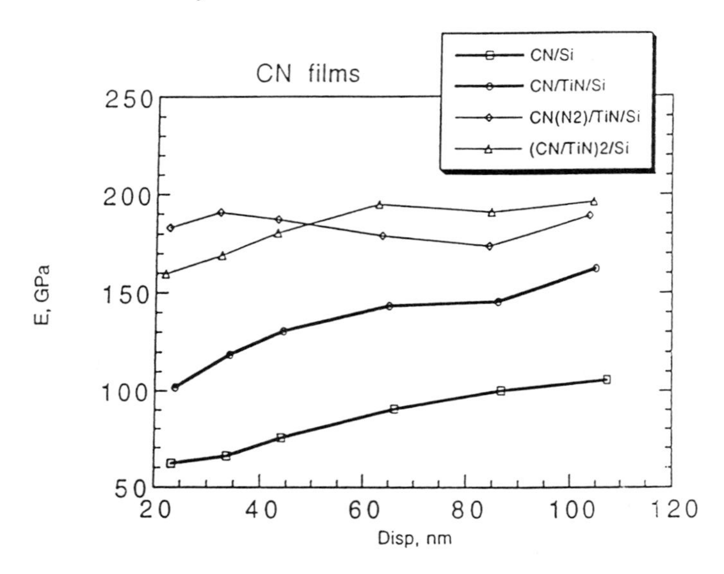

Fig. 4 Young's modulus of composite carbon nitride coatings on Si (100) substrates as a function of penetration depth

The elastic modulus and hardness of carbon nitride films on Si (100) substrates determined as a function of displacement are presented in Figure 4 and Figure 5 respectively. In both figures, the substrate effect is found to be absent. The deposition of a seed TiN layer shows a beneficial effect on hardness and modulus. A three to four fold increase in both hardness and modulus is found, when carbon nitride film is deposited with TiN is underlayer on Si (100). When the double layer of carbon nitride and TiN is deposited on Si (100), the hardness is slightly lower compared to the case of a deposition of single carbon nitride layer on TiN. However, the modulus is unaffected. Load displacement traces determined for CN_x films deposited in this study is plotted in Figure 6. This figure shows that all films have elastic properties. The hardness and modulus of carbon nitride films deposited in this study are higher than those of amorphous carbon nitride films prepared by a dc magnetron sputtering [11] of graphite target in a nitrogen atmosphere

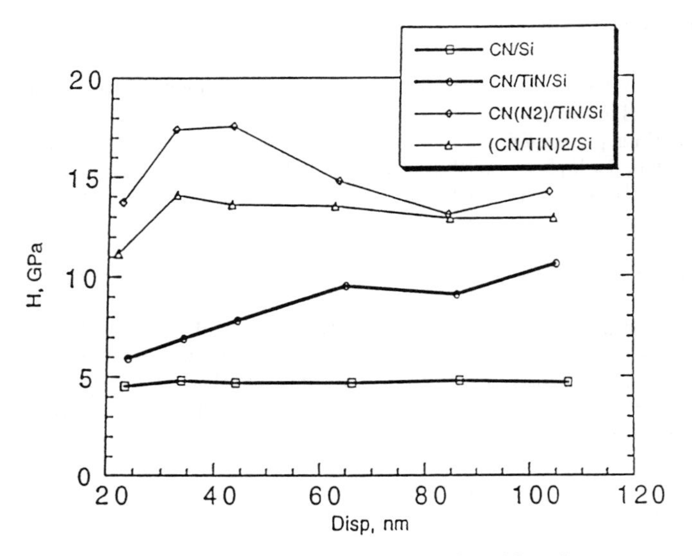

Fig. 5 Hardness of composite carbon nitride coatings on Si (100) substrates as a function of penetration depth

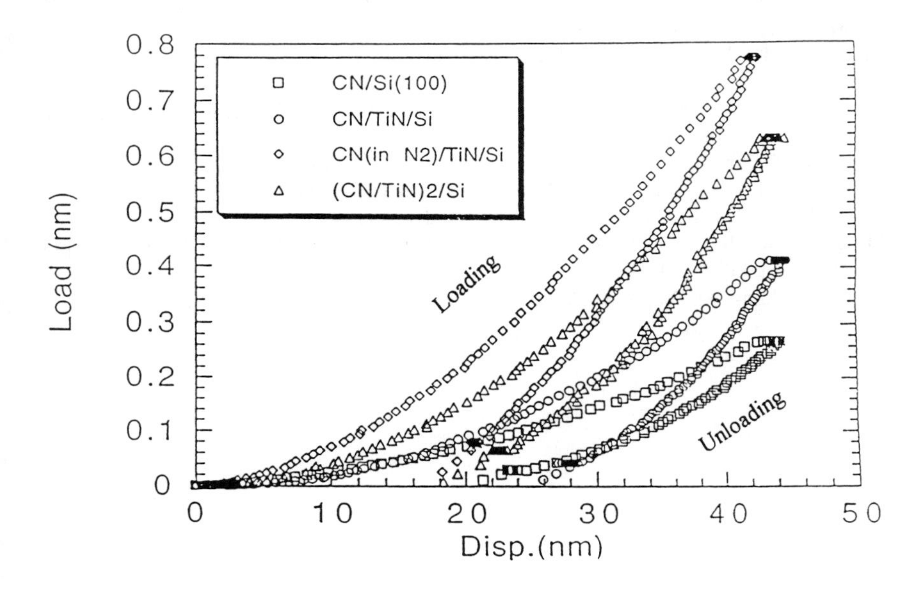

Fig. 6 Load displacement of composite carbon coatings on (100) substrates as a function of penetration depth

CONCLUSIONS

IR bands of the carbon nitride films are observed in the range 2300 cm-1 to 2400 cm-1, which is attributed to the C ≡ N bonding. X-ray diffraction suggested that the structure of the carbon nitride films prepared in this study are amorphous. It has been also demonstrated that using TiN as an under layer on Si(100) substrate increases the hardness and modulus values of carbon nitride composite coatings three to four fold compared to the films of CN_x on Si(100), without any buffer layer.

ACKNOWLEDGEMENTS

This research was supported by Alabama NASA EPSCoR program. Part of this research was supported by Alabama University/TVA Research Consortium and NSF SIP programs. AK also acknowledges the partial support from DOE EPSCoR Young Investigator Award. We are grateful to Mr. L. McCormick (Technician, College of Engineering at the University of South Alabama) for his technical assistance.

REFERENCES

[1] A. Y. Liu and M. L. Cohen, Science 245, (1989) p. 849
[2] F. Xiong, R. P. Chang, and C. W. White, Mater. Res. Soc. Symp. Proc. 280, (1993) p. 587
[3] D. Li, M. S. Wong, Y. W. Chung, S. C. Cheng, X. Chu, X. W. Lin, V. Dravid, and W. D. Sproul, Appl. Phys. Lett. 67, 203 (1995)
[4] S. Kumar and T. L. Tansely, Solid State Commun. 88 (1993) p. 803
[5] C. Niu, Y. Z. Lu, and C. M. Liber, Science 261 (1993) p.334
[6] D. Li, X. Chu, S. Cheng, X. Lin, V. P. Dravid, Y. Chung, M Wong, and W. D. Sproul, Appl. Phys. Lett. 67 (1995) p. 203
[7] J. Narayan, P. Tiwari, X. Chen, et.al. Appl. Phys. Lett. 61, (1992) p. 1290
[8] Ashok Kumar, J. Narayan, and X. Chen, Appl. Phys. Lett. 6, (1992) p. 976
[9] D. Kjendal, M. S. Thesis, University of South Alabama (1995)
[10] D. Li, X. Lin, S. C. Cheng, V. P. Dravid, Y Chung, M. Wong and W. D. Sproul, Appl. Phys. Lett. 68, (1995) p. 1211
[11] D. Li, Y. Chung, M. Wong, and W. D. Sproul, J. Appl. Phys., 74, (1993) p. 219

DESIGN AND PROPERTIES OF MULTILAYERED CERAMIC COMPOSITES

D. B. MARSHALL
Rockwell Science Center, 1049 Camino Dos Rios, Thousand Oaks, CA 91360

ABSTRACT

The design of multilayered ceramic composites is reviewed, with the aim of relating the properties that can be achieved to the microstructure of the composite. Limitations on some properties, such as damage tolerance and strength, are discussed. Failure mechanism maps that define some of these limits are given.

INTRODUCTION

Considerable interest has been shown recently in multilayered ceramic composites with layer dimensions in the range of one to several hundred microns. The purpose of the multilayered microstructure is to make use of interactions between the layers to introduce non-linear behavior and thereby overcome the inherent brittleness of the materials within the individual layers.

Several classes of multilayered ceramic composites may be usefully distinguished on the basis of the nature of the bonding between the layers (Fig. 1):
(1) Weakly bonded systems, in which continuous crack growth normal to the layers is prevented by debonding of the layer interfaces. Such composites, which are common in natural systems (e.g. shells), are capable of exhibiting damage-tolerant, non-catastrophic failure modes, especially in bending. However, their transverse strengths are low. Examples of ceramic systems include SiC/C,[1,2] Si_3N_4/BN,[3] and several oxide systems (e.g. Al_2O_3, ZrO_2) with weak-bond layers of porous oxides[4] β-aluminas[5,6] or rare-earth phosphates.[7,8]
(2) Strongly bonded systems in which the layers interact to increase the fracture toughness, albeit with failure occurring by growth of a single crack. Two toughening mechanisms have been explored. One involves enhancement of the degree of transformation toughening in materials containing ZrO_2 by the interaction of a crack-tip transformation zone with layers of Al_2O_3.[9-11] The other is based on alternating layers of metal and ceramic (e.g. Al/Al_2O_3),[12,13] or intermetallic compound and ceramic, in which the metal phase provides ductile bridging ligaments between the faces of a crack growing in the ceramic layers. Further discussion of strongly bonded systems here will be restricted to transformation toughed composites.
(3) Hybrid composites, consisting of alternating layers of homogeneous ceramic and fiber reinforced ceramic composite (Fig. 1(c)). This system requires strong bonding between the layers (i.e. between the homogeneous ceramic layers and the matrix of the fibrous composite layer) and weak bonding between the fibers and matrix of the composite layer. With proper design, hybrid composites exhibit many of the properties of fiber-reinforced composites (including non-catastrophic failure), but with higher first matrix cracking stress, higher stiffness, and improved wear and abrasion resistance of the surface. They have also been found to exhibit remarkably good impact resistance. Examples include composites with homogeneous layers of SiC, Si_3N_4 or SiAlON alternating with fibrous composite layers of SiC-reinforced glass,[14] as well as Al_2O_3 alternating with graphite epoxy composite.[15]

The following sections provide a brief review of current work on these three multilayered composite systems, along with brief of discussion of the rationale for microstructural design and an estimate of the properties and limitations that may be expected for each system. While we will consider only composites with planar multilayered structures, it should be noted that blocks of such microstructures may be used as sub-units of more complex hierarchical structures. Two examples currently being examined are fiber reinforced composites in which the matrix consists

Mat. Res. Soc. Symp. Proc. Vol. 434

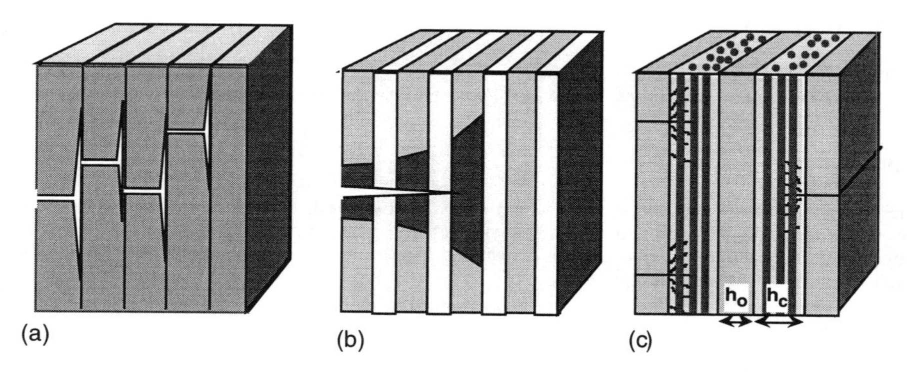

Fig.1 Schematic of three classes of multilayered ceramic composites:(a) weakly bonded layers; (b) strongly bonded transformation toughened systems; (c) hybrid composites.

of a multilayered structure concentric with the fibers,[16,17] and polycrystalline titanium aluminide materials in which the grains consist of colonies of alternating layers of γ-TiAl and α - Ti_3Al.[18]

2. WEAKLY BONDED SYSTEMS

2.1 Graphite/SiC and Si_3N_4/BN

The development of weakly bonded multilayered ceramic composites was stimulated by the work of Clegg et. al.[1,2,19,20] on composites containing layers of SiC separated by thin layers of carbon. They demonstrated a non-catastrophic failure mechanism when such composites were loaded in bending, as shown in Fig. 2. Each layer of SiC fractures independently, with extensive delamination occurring at the interface. Similar behavior has been observed in Si_3N_4/BN composites.[8] However, under tensile loading parallel to the layers in this type of composite, the stress-strain curve is linear to the peak load, and failure is catastrophic with no load bearing capacity beyond the peak. The strength normal to the layers is very small. Because of these characteristics, these composites are best suited to components loaded in bending or thermally loaded components.

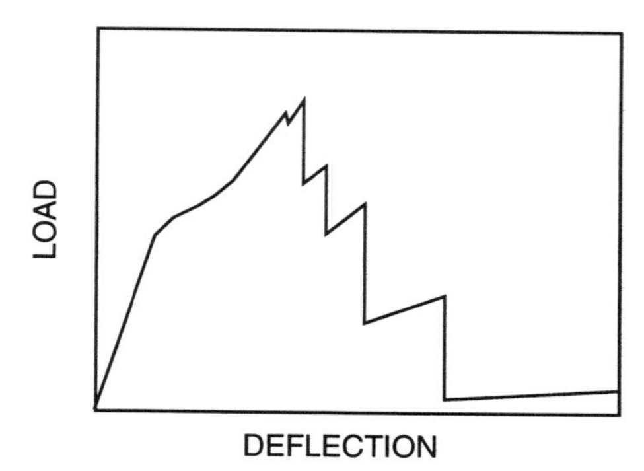

Fig. 2. Response of weakly bonded composite in flexural loading (re-drawn from Ref. 1).

2.2 Microstructural Design

Folsom et. al.[21] have provided analytical solutions for the stress-strain curve for composites such as that depicted in Fig. 2, with interfaces that are sufficiently weak to allow complete debonding as each layer fractures. The results indicate that SiC layers with a single deterministic strength give a stress-strain curve in bending that is linear to the peak load, as determined by fracture of the outermost layer. The load then decreases in a sawtooth manner (as in Fig. 2) as the remaining layers fracture in turn. In tensile loading parallel to the layers, failure is catastrophic in all layers simultaneously. If the layers have a statistical distribution of strengths, some non-linearity before the peak load and load-bearing capacity beyond the peak in tensile loading is possible. However, very low values of Wiebull modulus are required, with a concomitant decrease in the peak stress.

Optimum properties will most likely be found in systems that do not debond completely (Fig. 3). This can be achieved either by having an interface toughness that increases with increasing debond length (Fig. 3(a)) or by inducing other damage mechanisms within an interphase layer (Fig. 3(b) and (c)). The latter mechanism has been found to be beneficial in certain SiC/SiC fiber reinforced composites with carbon interphase layers between the fibers and matrix.[16]

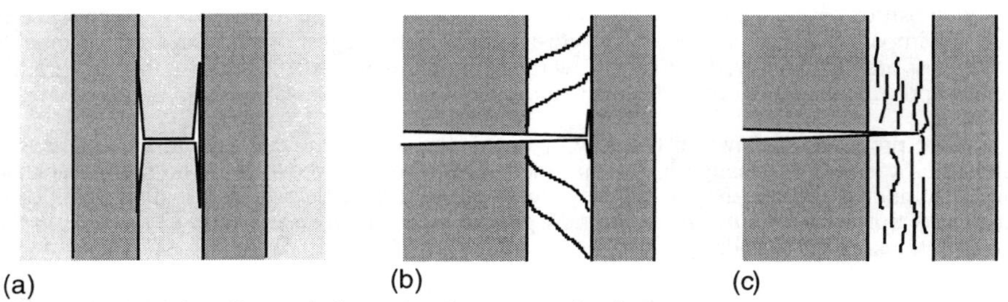

Fig. 3. Partial debonding and alternative damage mechanisms.

2.3 Oxidation-Resistant Systems

The SiC/C and Si_3N_4/BN based systems described in section 2.1 are severely limited by oxidation of the carbon and BN interphases. In an attempt to avoid this limitation, several systems that do not rely on carbon or BN layers are being explored. One makes use of alternating layers of fully dense and porous materials, with debonding occurring within the porous layers. Systems consisting of Si_3N_4/Si_3N_4 and Al_2O_3/Al_2O_3 have shown promising results. Another relies on layers containing materials such as mica and β-aluminas with weak cleavage planes. The third approach makes use of the recently discovered weak bonding between $LaPO_4$ (monazite) and oxides such as Al_2O_3 and ZrO_2.[7]

An example of debonding in a laminar composite consisting of layers of Al_2O_3 and $LaPO_4$ is shown in Fig. 4. The crack growing normal to the layers caused debonding at the $LaPO_4$ - Al_2O_3 interface. After further loading, arrays of sigmoidal shaped microcracks formed within the $LaPO_4$ layer and continued to extend further along the layer with increasing load. Corresponding toughness measurements indicated that the effective debond energy, G_c, increased as the damage propagated along the layer, from $G_c \sim 4$ J/m^2 after initial debonding, to $G_c \sim 12$ J/m^2 after a "well developed" microcrack zone had formed.[7]

3. STRONGLY BONDED LAMINAR CERAMIC COMPOSITES: $Ce\text{-}ZrO_2/Al_2O_3$

3.1 Enhancement of Transformation Toughening

The degree of transformation toughening in $Ce\text{-}ZrO_2$ based materials can be increased by fabricating a laminar microstructure consisting of alternating layers of strongly bonded $Ce\text{-}ZrO_2$ and Al_2O_3 (or a mixture of Al_2O_3 and $Ce\text{-}ZrO_2$) (Fig. 5). The transformation toughening is caused by stress-induced martensitic transformation, which leaves a zone of transformed material (monoclinic phase) around the crack and thereby shields the crack tip stresses. The laminar microstructure modifies the shape of the transformation zone in two ways that improve the effectiveness of this toughening mechanism: the length of the frontal region ahead of the crack tip is reduced and the width of the zone normal to the crack plane is increased (Fig. 6). Although failure always occurs by growth of a single crack (catastrophically), toughnesses as high as 20 $MPa.m^{1/2}$ (with an R-curve extending over several mm of crack growth) have been observed.

3.2 Microstructural Design

The degree of transformation toughening in ZrO_2 containing materials is dependent on the size of the transformation zone and the volume fraction of transformed material within the zone. These are determined by microstructural parameters discussed in detail elsewhere (grain size, composition),[22] as well as the ambient test temperature. The temperature dependence is especially important in the laminar composites: experiments suggest that the large enhancements in toughness are only achieved with a narrow temperature range (~100°C) above the temperature at which spontaneous transformation occurs throughout the material.

The presence of flaws in the Al_2O_3 layers can also limit the amount of toughening achievable by causing a change in fracture mechanism, from one in which the main crack tip grows continuously, to one in which Al_2O_3 layers ahead of the crack fracture and then link back to the crack tip. Once this transition in crack growth mechanism occurs, further increase in the degree of transformation toughening is not possible.

A failure map defining the ranges of parameters for which these two mechanisms occur is shown in Fig. 7. The position of the boundary separating the two mechanisms (and defining the maximum possible toughening for given flaw size in the Al_2O_3 layer) is dependent on the relative residual stress parameter $\alpha = (\sigma_R+\sigma_T)/\sigma_o$, where σ_R and σ_T are residual stresses due to thermal expansion mismatch of the layers and transformation of the ZrO_2 layers, and σ_o is the strength of the Al_2O_3 layer. For flaw sizes that are a large fraction of the layer thickness, we typically find $\alpha \sim 0.5$ and the relative toughness increase achievable (K_a/K_o at the transition) is not much larger than unity. However, for small flaw sizes in the Al_2O_3 layers, large relative increases in toughness are possible.

4. HYBRID COMPOSITES

Hybrid composites combine layers of fiber reinforced ceramic with strongly bonded layers of monolithic ceramic that have significantly higher stiffness than the matrix of the composite. The concept behind this design is that the strain to failure of the monolithic layer can be smaller than the strain for first matrix cracking in the composite layer, while the corresponding applied stress is higher. Then, since tensile loading of the composites before any cracking occurs is in plane strain, cracking occurs first in the monolithic layer at a stress that is higher than the first matrix cracking stress of the composite alone. Moreover, the stiffness of the hybrid composite is larger than that of the fibrous composite alone. With proper design of the composite microstructure, cracks from the monolithic layer penetrate into the adjacent composite layer and arrest because of crack bridging. This requires a strong bond between the matrix of the composite and the monolithic layer: otherwise the interface would debond and the composite would behave as the weakly bonded laminates of Section 2. Further loading causes multiple

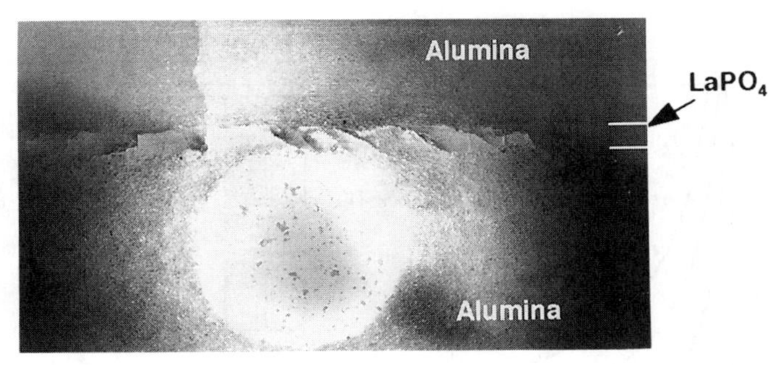

Fig. 4 Partial debonding in $LaPO_4$ - Al_2O_3 layered composite (from Ref 7)

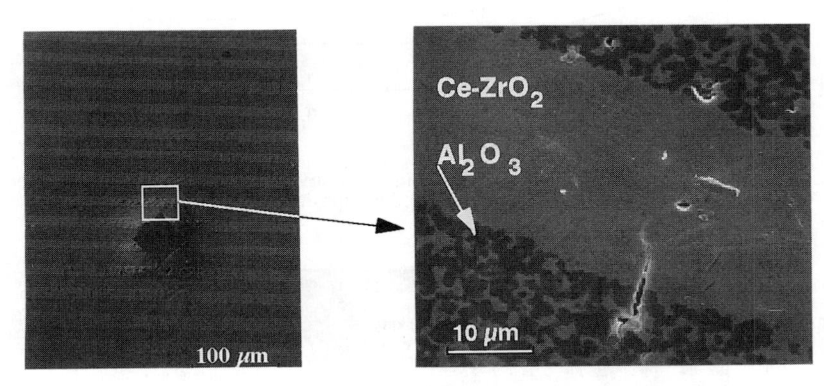

Fig. 5 Multilayered composite of Ce-ZrO_2 and Al_2O_3 /Ce-ZrO_2 Strong bonding indicated by absence of debonding where indentation crack crosses boundary between two layers

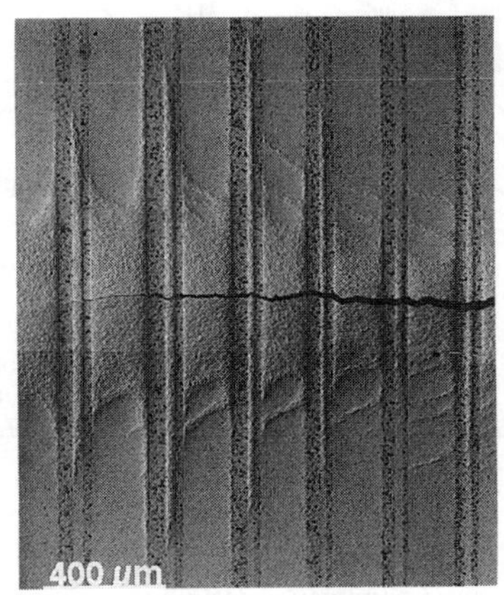

Fig. 6 Transformation zone surrounding crack in multilayered composite of Ce-ZrO_2 and Al_2O_3 /Ce-ZrO_2

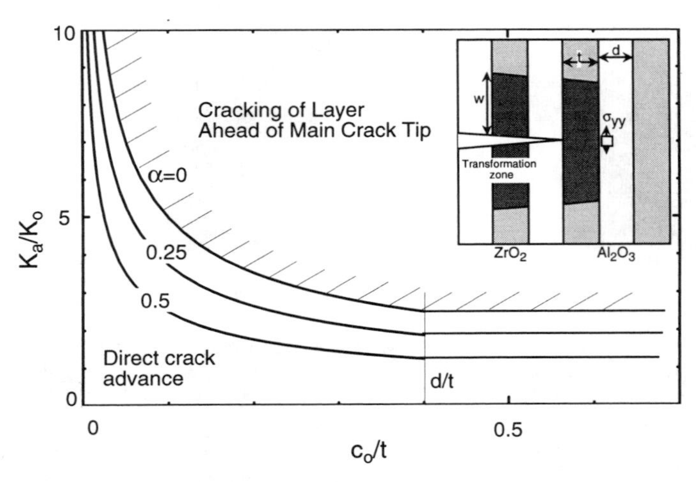

Fig. 7. Failure map defining transition in crack advance mechanism in strongly bonded transformation toughened layered composite.

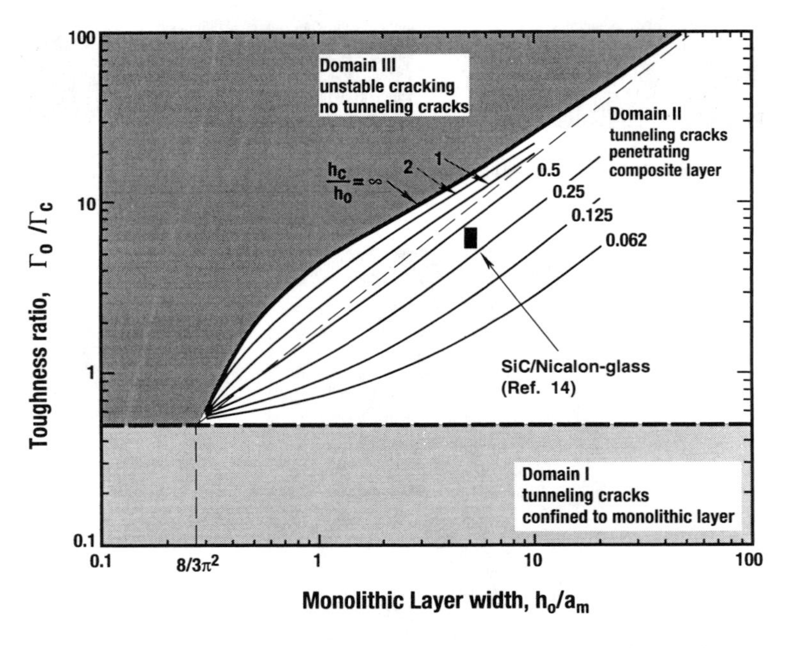

Fig. 8. Failure map for hybrid composites defining conditions for crack arrest within the composite layers after initiating within the monolithic layers.

cracking in both the monolithic layers and in the fibrous composite layers as illustrated in Fig. 1(c) with the peak stress being determined by failure of fibers in the composite layers. This response has been observed in hybrid composites consisting of layers of monolithic Al_2O_3 and fibrous graphite/epoxy composite, and in composites consisting of layers of monolithic SiC and fibrous SiC(Nicalon™)/glass composite.

Two design criteria can be defined for such composites to exhibit non-linear response. The first is that the cracks from the monolithic layers arrest within the adjacent composite layer. The second is that if the cracks do not arrest, the strength of the bridging fibers be sufficiently large to support the applied load. A first approximation for the second condition is simply that the cracking stress for the monolithic layer be smaller than the fiber strength multiplied by the volume fraction of fibers parallel to the loading direction.

The crack arrest condition may be deduced from a recent analysis of tunnel cracking in 0°/90° cross ply composites,[23] with the 90° plies being replaced by the monolithic layers. The results are summarized in Fig. 8, in the form of a failure map with ordinate given by the ratio of the toughnesses of the monolithic layer and the matrix of the composite and abscissa given by the ratio of the thickness of the monolithic layer to a length the scale that characterizes the crack bridging in the fibrous composite. For composites with the bridging constitutive law characteristic of frictional sliding, the bridging length scale is :

$$a_m = \frac{\pi E'}{4}\left(\frac{3\Gamma_c}{2}\right)^{1/3}\left[\frac{4\tau f^2 E_f E^2}{R(1-f)^2 E_m^2}\right]^{2/3} \tag{1}$$

where τ is the interfacial friction stress, f is the volume fraction of fibers in the fibrous composite layers, E_m and E_f are the Young's moduli of the matrix and fibers ($E = fEf + (1-f)Em$), R is the fiber radius, and Γ_c is the toughness of the matrix multiplied by (1-f). Three domains are shown: in region I the cracks from the monolithic layers do not penetrate the composite layers at all; in region II cracks arrest within the composite layers ; and in region III cracks penetrate completely through the composite layer (and hence through the entire specimen, although they may remain bridged by fibers). The boundary between regions II and III is dependent on the relative thicknesses of the fibrous composite and monolithic layers, h_c/h_m, as indicated.

The use of Fig. 8 as a basis for designing hybrid composite microstructures may be illustrated by comparison with some recent data from Cutler et. al.[14] on hybrid composites consisting of monolithic SiC layers and fibrous SiC(Nicalon™)/glass composite layers. The thicknesses of the SiC layers were all 500 µm, and the thicknesses of the fibrous composite layers were chosen to give volume fractions of 0.25, 0.4 and 0.5. Crack arrest was observed for the composites with the second and third compositions, but not for the first. Typical parameters for this system are: $\tau = 2$ MPa, $E_f = 200$ GPa, $E_m = 70$ GPa, R= 8 µm, $\Gamma_o = 40$ J/m^2 and $\Gamma_c = 7$ J/m^2, giving $a_m \sim 100$ µm and $\Gamma_m/\Gamma_o \sim 6$. The location of this system with 500 µm thickness SiC layers ($h_o/a_m = 5$) is shown in Fig. 8: for materials containing 50% fibrous composite ($h_c/h_o=1$) this point falls within domain II, whereas for 25% fibrous composite ($h_c/h_o=1/3$) the material falls within domain III, consistent with observations.

CONCLUSIONS

General properties and microstructural design considerations have been described for three different classes of laminar composites. Each gives a different envelope of properties suited to different applications: (1) weakly bonded systems, suitable for use in flexural or thermal loading such as in combustors; (2) strongly bonded ZrO2 - containing systems with increased toughness, suitable for wear components that would not experience a large range of temperatures during use; and (3) hybrid composites with attractive combinations of properties for wear, armor, and a variety of shell structures.

ACKNOWLEDGMENTS

Funding for this work was supplied by the Air Force Office of Scientific Research, Contract No. F49620-92-C-0028 (Dr. A. Pechenik)

REFERENCES

1. W. J. Clegg, K. Kendall, N. M. Alford, T. W. Button and J. D. Birchall, "A Simple way to Make Tough Ceramics," *Nature,* **347** 455-457 (1990).

2. A. J. Phillipps, W. J. Clegg and T. W. Clyne, "The Correlation of Interfacial and Macroscopic Toughness in SiC Laminates," *Composites,* **24**[2] 166-176 (1993).

3. H. Liu and S. M. Hsu, "Fracture Behavior of Multilayer Silicon Nitride/Boron Nitride Cramics," *J. Amer. Ceram. Soc.,* **in press** (1996).

4. J. B. Davis and W. J. Clegg, "Ceramic Laminates for High Temperature Structural Applications," *Am. Ceram Soc. Annual Meeting,* (1995).

5. P. S. Nicholson, P. Sarkar and X. Haung, "Electrophoretic Deposition and its use to Synthesize ZrO_2/Al_2O_3Micro-Laminate Ceramic/Ceramic Composites," *J. Material Science,* **28** 6274-6278 (1993).

6. P. E. D. Morgan and D. B. Marshall, "Functional Interfaces in Oxide-Oxide Composites," *J. Mat. Sci. Eng.,* **A162**[1-2] 15-25 (1993).

7. P. E. D. Morgan and D. B. Marshall, "Ceramic Composites of Monazite and Alumina," *J. Am. Ceram. Soc,* **78**[6] 1553 - 63 (1995).

8. P. E. D. Morgan, D. B. Marshall and R. M. Housley, "High Temperature Stability of Monazite-Alumina Composites," *J. Mat. Sci. Eng.,* **A195** 215 - 222 (1995).

9. D. B. Marshall, J. J. Ratto and F. F. Lange, "Enhanced Fracture Toughness in Layered Composites of Ce-ZrO_2 and Al_2O_3," *J. Am. Ceram. Soc.,* **74**[12] 2979-2987 (1991).

10. D. B. Marshall and J. J. Ratto, "Crack Resistance Curves in Layered Ce-ZrO_2/Al_2O_3 Ceramics"; pp 517-523 in Science and Technology of Zirconia V. Eds S. P. S. Badwal, M. J. Bannister and R. J. H. Hannink, Technomic Pub. Co., 1993.

11. P. Boch, T. Chartier and M. Huttepain, "Tape Casting of Al_2O_3/ZrO_2 Laminated Composites," *J. Am. Ceram. Soc.,* **69** C-191 (1986).

12. M. C. Shaw, D. B. Marshall, M. S. Dadkhah and A. G. Evans, "Cracking and Damage Mechanisms in Ceramic/Metal Multilayers," *Acta Met.,* **41**[11] 3311-3322 (1993).

13. M. Y. He, F. E. Heredia, D. J. Wissuchek, M. C. Shaw and A. G. Evans, "The Mechanics of Crack Growth in Layered Materials," *Acta. metall. mater.,* (1993).

14. W. A. Cutler, F. W. Zok and F. F. Lange, "Mechanical Behavior of Several Hybrid CMC Laminates," *J. Am. Ceram. Soc.,* (in press).

15. C. A. Folsom, F. W. Zok, F. F. Lange and D. B. Marshall, "Mechanical Behavior of a Laminar Ceramic/Fiber Reinforced Epoxy Composite," *J. Am. Ceram. Soc.,* **75**[11] 2969-2975 (1992).

16. C. Droillard and J. Lamon, "Fracture Toughness of 2-D woven SiC/SiC CVI-Composites with multilayered interphases," *J. Amer. Ceram Soc.,* **79**[4] 849-58 (1996).

17. W. J. Lackey, S. Vaidyaraman and K. L. Moore, "Laminated C-SiC Matrix Composites Produced by CVI," *J. Amer. Ceram. Soc.,* (in press).

18. H. E. Deve, A. G. Evans and D. S. Shih, "A High Toughness γ-Titanium Aluminide," *Acta. metall. mater.,* **40**[6] 1259-1265 (1992).

19. A. J. Phillipps, W. J. Clegg and T. W. Clyne, "Fracture Behavior of Ceramic Laminates in Bending: I. Modelling of Crack Propagation," *Acta Metall. Material,* **41**[3] 805-17 (1993).

20. A. J. Phillipps, W. J. Clegg and T. W. Clyne, "Fracture Behavior of Ceramic Laminates in Bending: II. Comparison of Modelling with Experimental Data,"," *Acta Metall. Material,* **41**[3] 819-27 (1993).

21. C. A. Folsom, F. W. Zok and F. F. Lange, "Flexural Properties of Brittle Multilayer Materials: I. Modeling, "," *J. Americal Ceramic Society,* **77**[3] 689-96 (1994).

22. A. G. Evans and Cannon, "Toughening of Brittle Solids by Martensitic Transformations," *Acta Metall.,* **34**[5] 761-800 (1986).

23. B. N. Cox and D. B. Marshall, "Crack Initiation in Brittle Fiber Reinforced Laminates," *J. Amer. Ceram. Soc.,* in press.

Toughening Mechanisms in Al/Al-SiC Laminated Metal Composites

D.R. Lesuer*, J. Wadsworth*, R.A. Riddle*, C.K. Syn*,
J.J. Lewandowski** and W.H. Hunt Jr.***
* Lawrence Livermore National Laboratory, Livermore, CA 94551
** Case Western Reserve University, Cleveland, Ohio 44106
*** Alcoa Technical Center, Alcoa Center, PA 15069

ABSTRACT

The fracture toughness of laminated metal composites consisting of alternating layers of a metal matrix composite (Al6090/SiC/25p) and a monolithic aluminum alloy (Al5182) has been studied as a function of the volume fraction of the component materials. Finite element simulations of the fracture toughness tests have been used to study the mechanisms of crack growth and extrinsic toughening. The mechanisms responsible for toughening in laminated metal composites are described.

INTRODUCTION

Recent studies have shown that lamination can provide significant improvements to the damage-critical properties (such as fracture toughness, fatigue crack growth behavior and impact response) of metal matrix composites (MMCs) [1, 2, 3]. These laminated metal composites (LMCs) consist of alternating layers of a MMC and a ductile metal layer. Typical combinations have included Al6090/SiC/25p // Al5182 [4], Al6090/SiC/25p // Mg-9Li [5], and X2080/SiC/20p // X2080 [6]. These improvements in properties result from extrinsic toughening mechanisms such as crack blunting, crack deflection and crack bridging. All of these extrinsic mechanisms, which are influenced by local delaminations at interfaces, improve the fracture toughness of LMCs by reducing the local stress intensity at the tip of the crack. However the specific extrinsic toughening mechanisms operating in a given system and their influence on measured toughness depends on the material properties of the component layers, interface properties and laminate architecture (including component material volume fraction and layer thickness).

In this paper we report on studies that have been conducted of toughening mechanisms in an Al / Al-SiC LMC. Specifically, we have studied LMCs consisting of alternating layers of a MMC, Al6090/SiC/25p, and a monolithic aluminum alloy, Al5182. Fracture toughness tests have been performed on LMCs containing different volume fractions of the component materials. Simulations of these experiments using finite element analysis were employed to study the mechanisms of crack growth and the origins of extrinsic toughening in these Al/Al-SiC laminates. The results provide valuable insight into the influence of processing, microstructure and laminate design on the fracture toughness of MMC-based laminates.

MATERIALS, EXPERIMENTS AND RESULTS

The component materials, Al6090/SiC/25p and Al5182, were obtained from commercial sources and chemically cleaned to remove oxide films. Laminates were then made by press bonding alternating layers of these materials at 450°C. After press bonding, the Al/Al-SiC laminates were heat treated by soaking the laminate at 530°C for 75 minutes and then aging at 160°C for 16 hours. This procedure provided a T6 heat treatment to the 6090/SiC/25p layers and had virtually no effect on the microstructure of

Mat. Res. Soc. Symp. Proc. Vol. 434 © 1996 Materials Research Society

the Al5182 layers. For all the laminates in this study, sharp interfaces were maintained between the component layers and no reaction products were formed.

The fracture toughness of these materials was evaluated in both the crack arrester and the crack divider orientations using chevron notch three-point bend bars. Testing and analysis were done according to the procedures specified by Wu [7]. A high speed videocamera was used during fracture toughness testing to monitor the surface component of crack growth. Further details of materials processing and testing procedures have been previously published.[8, 9]

Fracture toughness tests were performed on LMCs containing different volume fractions of the component materials. The volume fraction of the Al6090-T6/SiC/25p component in the laminate was varied from 50% to 97% which produced a variation in the global concentration of SiC in the laminate from 12.5% to 24.3%. The fracture toughness results are plotted in Fig. 1 as a function of global percentage of SiC. The component materials of the laminate differ significantly in fracture toughness (with Al5182 having a higher toughness than Al6090-T6/SiC/25p). However, the LMC shows a modest increase in fracture toughness with increasing global volume percentage of SiC up to a component volume fraction at which the LMC is almost 100% MMC. Clearly in these laminates extrinsic toughening mechanisms are the dominate source of fracture toughness. The fracture toughness at a global volume percentage of 24.3% SiC (97% MMC component) is approximately 35 MPa $m^{.5}$ and particularly noteworthy, since strength and stiffness of the LMC would essentially equal those of the MMC component. The fracture surface and the cross section of the tested sample containing approximately 80 % MMC component are shown in Fig. 2a and 2b respectively.

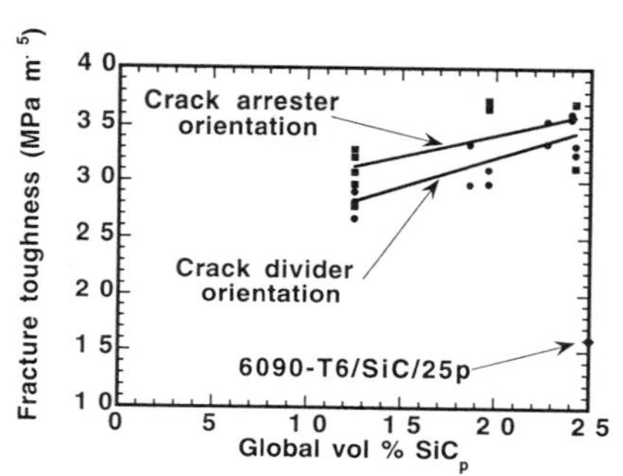

Fig. 1 Fracture toughness versus global volume percent silicon carbide for a laminate consisting of Al6090-T6/SiC/25p and Al5182 in the crack arrester and crack divider orientations. The fracture toughness of Al6090-T6/SiC/25p is shown in the figure.

CRACK GROWTH SIMULATIONS

Crack growth simulations were done using the finite element code NIKE2D, which has the capability to study finite deformation problems and can represent the growth of cracks and the separation of material interfaces (delaminations) along specified nodes in the finite element mesh. The boundary containing these nodes is called a tie-break slideline. Thus the path of crack growth and the interfaces that will be allowed to separate were specified as part of the input to the code. Material separation then occurred along the crack path or material interface, when a critical condition was met. In these problems, separation of the nodes in the finite element mesh occurred when a critical value of effective plastic strain was achieved. As previously discussed by Riddle [10], the effective plastic strain values that were selected for node separation were derived from calculations involving the J-integral. Thus the development of damage and the growth of the crack occurred in a manner consistent with elastic-plastic fracture mechanics.

a b

Fig. 2. Fracture surface of the LMC containing 83% MMC component (a) and cross-section of the LMC containing 79% MMC component (b).

The simulations representing crack growth in the crack arrester orientation were conducted for three layers of the laminate containing 95% of the MMC component. The finite element mesh used in the simulations is shown in Fig. 3. The tie-brake slidelines representing the allowed path of crack growth and the interfaces between layers (that are allowed to separate) are shown in the figure. Contours of effective plastic strain are shown in Figs. 4a-d for four time steps in the problem. The loaded crack produces regions of intense effective plastic strain that are off-axis to the plane of tensile mode opening crack growth. As shown in Fig. 4a, this off-axis deformation results in through-thickness yielding of the ductile layer. This through-thickness yielding produces a delamination at the Al - Al-SiC interface (Fig. 4b) and plastic rupture (Fig. 4c) of the thin Al layer ahead of the advancing crack front. Thus when the advancing crack encounters the ductile layer, the crack is deflected (due to the delamination) and blunted (due to the ruptured layer). The arrested crack is shown in Fig. 4d; further crack growth requires re-nucleation of the crack in the MMC layer. This arresting, re-nucleation process results in significant increase in the amount of energy required for crack growth.

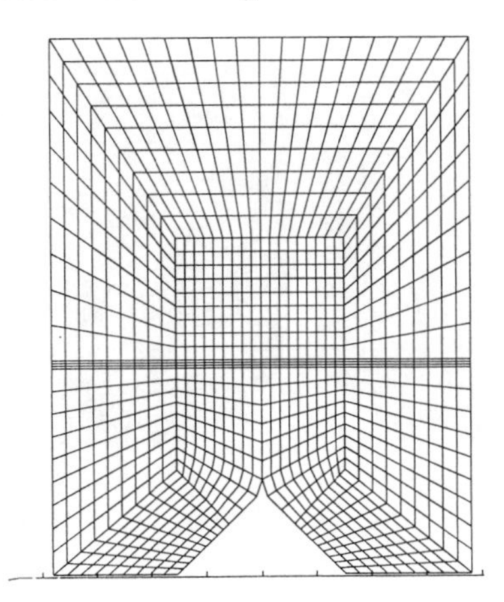

Fig. 3. Finite element mesh used in the simulations of crack growth in an Al / Al-SiC laminate containing 95% MMC component. Tie-break slidelines are shown in the figure.

It is important to note that the crack deflection and blunting mechanisms, as shown in Fig. 4, are independent of volume fraction, which implies, to first order, that the fracture toughness should be independent of volume fraction. This observation is consistent with the data presented in Fig. 1, which shows fracture toughness essentially independent of volume fraction of the component materials. In the simulation, the fracture in the ductile layer (shown in Fig. 4c) occurs at the tie-break slideline. A careful examination of Fig. 4b reveals that deformation in the ductile layer is concentrated in regions significantly off-axis from the plane of mode one crack growth; thus fracture would be expected in off-axis locations. This observation is consistent with the experimental evidence as revealed in Fig. 2, which shows that the origin of fracture in the thin ductile layers was significantly off-axis from the dominant plane of crack growth. The development of damage in regions off-axis to the plane of mode one crack growth is an important source of toughening, since higher loads (and greater energies) will need to be applied to the sample to renucleate the crack in the plane of mode one crack growth in the MMC layer.

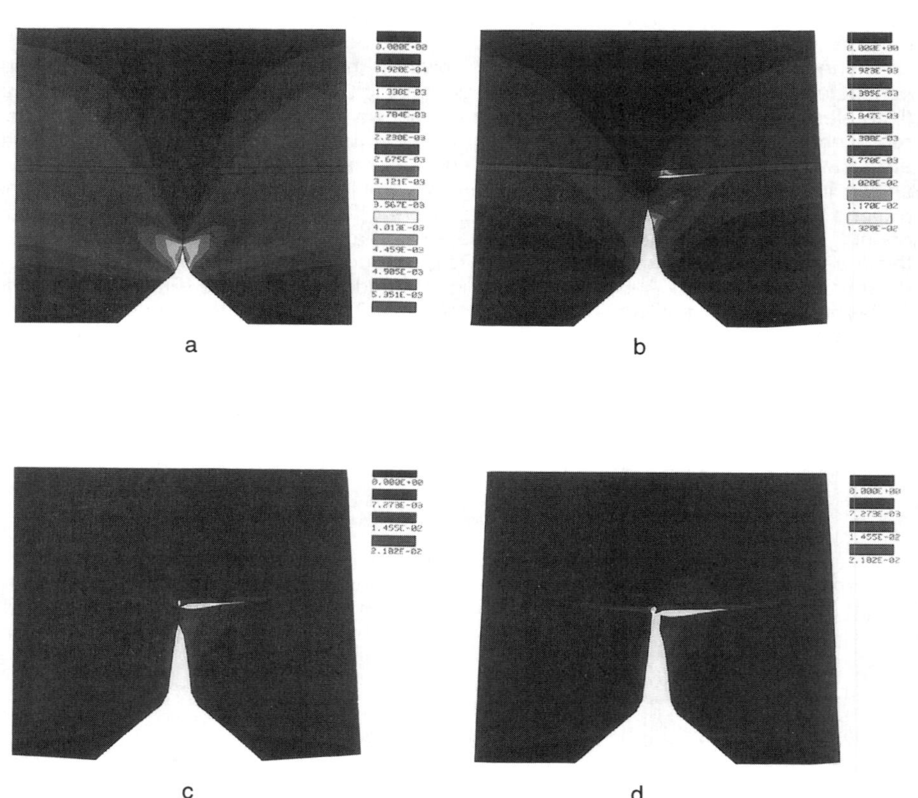

Fig. 4. Contours of effective plastic strain at four time steps in the crack growth simulation.

TOUGHENING MECHANISMS

Figure 5 shows five extrinsic toughening mechanisms that can result from the discrete layers present in LMCs. The laminate orientation that will be influenced by the mechanism and the sensitivity of the mechanism to the volume fraction of the component materials is indicated in the figure. Numerous other extrinsic toughening mechanisms have been identified in composites and monolithic materials and these can provide additional sources of toughening in LMCs [11, 12]. The crack deflection and crack blunting mechanisms have been discussed in the previous section. The crack blunting mechanism resulted in part because of the limited ductility of the Al5182 layer - the layer fractured ahead of the advancing crack tip. For systems in which there is a greater ductility difference between the component materials than the Al6090/SiC/25P // Al5182 LMC discussed here, rupture of the ductile layer can be avoided. In this case the ductile layer can span the wake of the advancing crack and crack bridging will result. Crack bridging in a LMC has been observed in an ultrahigh carbon steel / brass system [13] as shown in Fig. 6. The figure shows three distinct brass layers spanning the wake of the crack in the LMC. The crack bridging mechanism is clearly dependent on the volume fraction of the component materials, because the volume of ductile material spanning the wake of the crack will influence the amount of energy that must be supplied to the material for ligament extension (and eventual rupture) and thus crack growth.

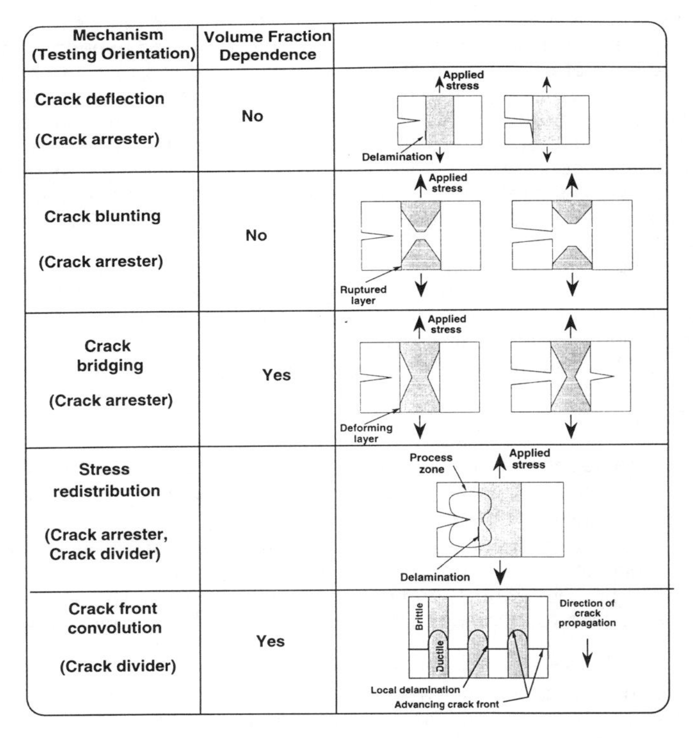

Fig. 5. Extrinsic toughening mechanisms in laminate metal composites.

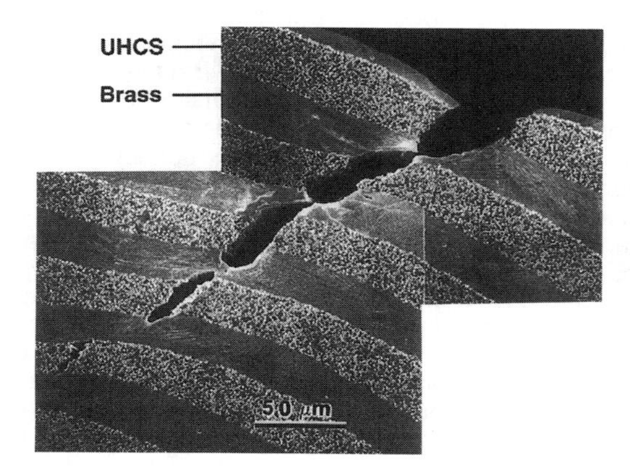

Fig. 6. Crack bridging in a LMC consisting of ultrahigh-carbon steel and brass.

Delamination can also provide extrinsic toughening by reducing the stresses in the layers head of the advancing crack. This mechanism, referred to as stress redistribution in Fig. 5, has been recently studied both theoretically [14)] and experimentally [15)]. for metal/ceramic laminates. In these studies, delamination was found to be significantly more effective than slip in reducing the stress ahead of the crack. Figure 5 also shows a mechanism unique to the crack divider orientation - crack front convolution. In this mechanism, the crack front in the less ductile component leads the crack front in the more ductile component. The shape of the crack front is highly convoluted with the depth of the convolutions related to the extent of delamination at the interfaces. Overall crack front growth is retarded by the plastic tearing required for crack growth in the more ductile layer.

In all the mechanisms shown in Fig. 5, local delamination at layer interfaces is an important prerequisite for extrinsic toughening. As previously described(1), delamination is strongly influenced by residual stresses at the interface as well as the strength and toughness of the interface. These factors are all a function of laminate processing conditions as well as laminate architecture (primarily layer thickness).

CONCLUSIONS

1. Fracture toughness in the Al / Al-Sic laminates studied was found to be essentially independent of the volume fraction of the component materials with K_q values of approximately 35 MPa $m^{.5}$ when the LMC consisted of 97% MMC.

2. As revealed by simulations involving finite element analysis, for the LMCs studied here, delamination at the Al / Al-SiC interface and plastic rupture of the Al layer take place in the process zone ahead of the advancing crack. These mechanisms are expected to introduce toughening that is independent of component material volume fraction.

3. Other toughening mechanisms in LMCs (such as crack bridging and crack front convolution) can produce volume fraction effects. Toughening mechanisms that are active in Al / Al-SiC laminates are strongly dependent on interface delamination.

ACKNOWLEDGMENTS

The authors acknowledge the work of Jim Ferreira for metallography and fractography and Al Shields for mechanical testing. The Al5182 was supplied by Kevin Brown of Kaiser Aluminum, Center for Technology. This work was performed under the auspices of the U. S. Department of Energy by Lawrence Livermore National Laboratory under contract No. W-7405-Eng-48.

REFERENCES

1. D. Lesuer, C. Syn, R. Riddle, O. Sherby, in Intrinsic and Extrinsic Fracture Mechanisms in Organic Composite Systems, edited by J. J. Lewandowski and W. H. Hunt Jr., (The Minerals, Metals & Materials Society, Warrendale, 1995) pp. 93-102.
2. P. B. Hoffman, J. C. Gibeling, *Scripta Metal. et Matl.* **32**, 901 (1995).
3. R. L. Woodward, S. R. Tracey, I. G. Crouch, *J.de Physique IV* **1**, 277 (1991).
4. D. R. Lesuer, C. K. Syn, High Performance Materials in Engine Technology, Advances in Science and Technology 9, edited by P. Vincenzini, (Techna, Florence, Italy, 1995) pp. 27-38.
5. C. K. Syn, D. R. Lesuer, O. D. Sherby, accepted for publication *Matls. Sci. and Eng.*, (1996).
6. T. M. Osman, J. J. Lewandowski, *Scripta Metal. et Mat.l* **31**, 191 (1994).
7. S.-X. Wu, in ASTM STP 855 , edited by J. H. Underwood, S. W. Freiman, F. I. Baratta (American Society for Testing and Materials, Phildelphia, 1984) pp. 176 - 192.
8. C. K. Syn, S. Stoner, D. R. Lesuer, O. D. Sherby, in High Merformance Metal and Ceramic Matrix Composites edited by K. Upadhya, (The Minerals, Metals & Materials Society, Warrendale, 1994) pp. 125-135.
9. C. K. Syn, D. R. Lesuer, O. D. Sherby, in International Conference on Advanced Synthesis of Engineered Structural Materials, edited by J. J. Moore, E. J. Lavernia and F. H. Froes, (ASM International, Materials Park, Ohio, 1992) pp. 149 - 156.
10. R. A. Riddle, Lawrence Livermore National Laboratory Report UCRL-93745, The Effect of the Failure Criterion in the Numerical Modeling of Orthogonal Metal Cutting (1987).
11. R. O. Ritchie, *Matls. Sci. and Eng.* **A103**, 15 (1988).
12. R. O. Ritchie, W. Yu, in *Small Fatigue Cracks* , edited by R. O. Ritchie and J. Langford (The Metallurgical Society, Warrendale, 1986) pp. 167-189.
13. Y. Ohashi, J. Wolfenstine, R. Koch, O. D. Sherby, *Matls. Sci. and Eng.* **A151**, 37 (1992).
14. K. S. Chan, M. Y. He, J. W. Hutchinson, *Matls. Sci. and Eng.* **A167**, 57 (1993).
15. M. C. Shaw, et al., *Acta Metal. et Matl.* **42**, 4091 (1994).

EFFECTS OF DUCTILE PHASE ADDITIONS ON THE FRACTURE BEHAVIOR AND TOUGHNESS OF DRA COMPOSITES

L. YOST ELLIS[1], J.J. LEWANDOWSKI[1], and W.H. HUNT[2]
[1]Case Western Reserve University, The Case School of Engineering, Department of Materials Science and Engineering, Cleveland, OH 44106
[2]ALCOA Technical Center, Alcoa Center, PA 15069

ABSTRACT

Discontinuously reinforced aluminum (DRA) composites have been processed to contain discrete regions of unreinforced aluminum with the objective of enhancing the damage tolerance. The effects of changes in the ductile phase size, shape and strength as well as the SiC_p reinforcement distribution on the toughness were studied. The incorporation of the ductile phase can increase the crack growth resistance of the DRA composite. In such cases, stable crack propagation (i.e. R curve behavior) is observed in contrast to the behavior of the conventional DRA composite which fails catastrophically under the conditions tested. The level of toughening is affected by the size and mechanical properties of the ductile phase as well as the orientation and shape of the ductile regions with respect to the test geometry (i.e. crack arrestor vs crack divider).

INTRODUCTION

Metal matrix composites (MMC's) offer increased specific stiffness and specific strength which can lead to weight savings and improved structural performance. One class of MMC's which has received considerable attention is discontinuously reinforced aluminum (DRA) composites. While desirable specific stiffnesses and strengths may be attained with DRA's, the damage tolerance of DRA's must be increased if these materials are to gain widespread use for structural applications.

While intrinsic toughening approaches have provided many of the improvements in the last decade,[1-4] extrinsic approaches to toughening are currently being explored[2-4] to provide greater improvements in damage tolerance without compromising the stiffness and strength. Material development in this area has focussed on laminate structures consisting of alternating layers of DRA (a "semi-brittle" component) and a monolithic aluminum alloy (a more "ductile" component).[2-9] Such laminates containing discrete interfaces have generally been produced via press bonding or roll bonding of component laminae. Increased fracture toughness and impact energies as compared to the component DRA composites have been reported for these laminates. While the above approaches provide toughening via the introduction of discrete interfaces which may debond during loading, there may be other cost effective toughening approaches where less discrete interfaces are utilized to separate the constituents.

The current paper reports on the preliminary results of a program to improve the damage tolerance of DRA composites by incorporating regions of unreinforced aluminum within a conventional Al/SiC/XXp DRA. The aluminum "toughening" regions are introduced into the material structure via the traditional powder metallurgy processing route which both eliminates the need for additional processing steps as well as producing a strongly bonded interface between the component materials. By altering the unreinforced aluminum/aluminum alloy additions, the influence of several variables can be examined including the size and volume fraction of the "ductile" regions as well as the mechanical properties of the "ductile" phase.

EXPERIMENTAL PROCEDURE

Materials

A DRA composite consisting of a 7093 Al matrix reinforced with 15 volume % SiC particles, henceforth denoted 7093/SiC/15p DRA, (nominal composition of 9 Zn, 2.2 Mg, 1.5 Cu, 0.14 Zr, 0.1 Ni, bal Al reinforced with 10 μm average size SiC particulates)

Mat. Res. Soc. Symp. Proc. Vol. 434

provides the baseline for each material in this work. Production of the DRA followed the traditional powder processing route which involves an initial blending of pre-alloyed 7093 Al powder with SiC particulates. "Toughened" DRA was then produced via additions of aluminum/aluminum alloy particles to the blend, at volume fractions of either 10% or 25%. The final product thus consists of DRA and discrete unreinforced Al regions.

Each material was designed such that the global volume fraction of reinforcement (and hence, modulus) would be the same in all materials and comparable to that of the 7093/SiC/15p DRA. The distribution and strength of the ductile phase was varied independently. The material test matrix is shown in Table I. A material containing "large ductile phase" additions (i.e. LDP) was produced at ALCOA by making additions of commercially pure (c.p.) aluminum particles with a typical size of 10 mm to the initial powder blend. By using a starting ductile particle more than an order of magnitude smaller in size, a material containing "small ductile phase" additions (i.e. SDP) material was also fabricated. A fifth "toughened" material containing 10% by volume of LDP additions of a high strength aluminum alloy possessing a similar composition to that of the 7093 Al matrix of the DRA (i.e. LDP3) was additionally produced.

The behavior and properties of the "toughened" materials are compared to materials with a conventional DRA structure (i.e. C1, C2 and C3). These control materials include the X7093/SiC/15p base material as well as DRA composites containing additions of 10 volume % and 25 volume % c.p. Al powder, respectively. This c.p. Al powder, which is of comparable size to the pre-alloyed X7093 powder used as the baseline composite matrix, contains few alloying elements and therefore produces the same degree of dilution of the solute as the ductile phase additions to the SDP and LDP materials without producing the redistribution of the SiC reinforcement that creates the ductile toughening regions.

Consolidated billets of all materials were extruded at a 22:1 ratio to produce 1" x 3" bars. Figures 1(a-c) show the three major classifications of materials presently studied:

(1) "Control" DRA containing a homogenous distribution of SiC_p (cf Figure 1a).
(2) Materials containing the "small ductile phase", SDP (cf Figure 1b).
The ductile phase thickness in the ST direction ranges from 20-100 μm.
(3) Materials containing the "large ductile phase", LDP (cf Figure 1c).
The ductile phase thickness in the ST direction ranges from 100-500 μm.

All materials, were heat treated prior to testing to produce a slightly overaged condition, T7E92, of the DRA composite. This consisted of a solution heat treatment, 490°C/4 hrs/CWQ, followed by aging, 120°C/24 hrs + 150°C/8 hrs.

Table I : Material Test Matrix (Base material is 7093/SiC/15p).

	Material ID	Type of Ductile Phase Addition	Quantity (vol. %)	Discrete Ductile Phase in Product?
"Control DRA"	C1	None	N/A	No
	C2	commercial purity Al powder	10%	No
	C3	commercial purity Al powder	25%	No
"Small Ductile Phase"	SDP1	small c.p. Al - low strength	10%	Yes
	SDP2	small c.p. Al - low strength	25%	Yes
"Large Ductile Phase"	LDP1	large c.p. Al -low strength	10%	Yes
	LDP2	large c.p. Al - low strength	25%	Yes
	LDP3	large alloyed Al - high strength	10%	Yes

Figure 1: 3-D Microstructures of the three types of materials.

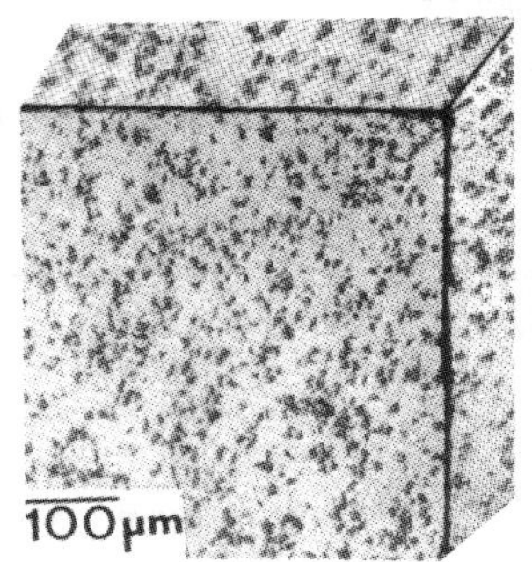

1a) "Control" DRA where dark areas are SiC_p.

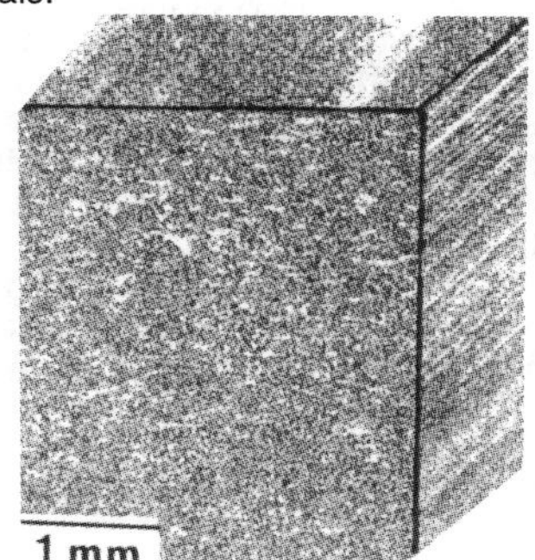

1b) "Small Ductile Phase" toughened DRA (i.e. SDP). The light areas are the ductile phase regions.

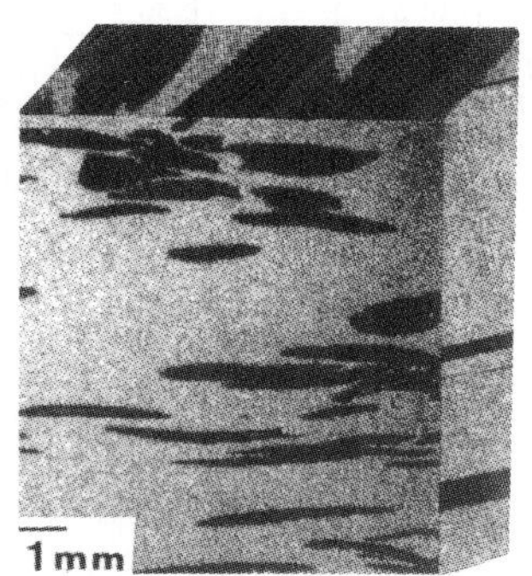

1c) "Large Ductile Phase" toughened DRA (i.e.LDP). Dark areas are the ductile regions.

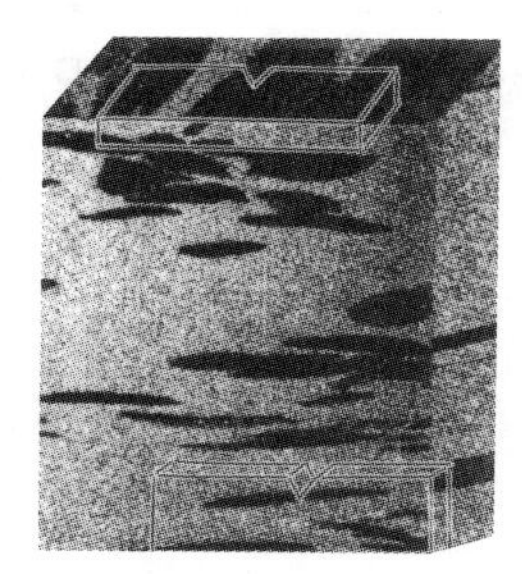

⇐T-L specimen

⇐T-S specimen

1d) Orientations for fracture toughness bend specimens.

Test Methods

Toughness testing was conducted on all materials via a test method determined in conjunction with the ASTM standards for K_{IC} testing (ASTM-E399), J testing (ASTM-E813) and R curve testing (ASTM-E561). Single-edged notched bend specimens (65 mm x 14 mm x 4 mm) containing wire saw notches (root radius 60 μm) were tested in a three point bending configuration (with a 56 mm span width). Bend specimens were removed from the extruded bar in two orientations to the extrusion axis, allowing testing in the T-S and T-L orientations as shown in Figure 1d. These orientations, combined with the locations within the extruded bar from which the specimens were obtained provide specimens which possess a distribution of toughening regions resembling crack arrestor and crack divider laminates, respectively.

During testing, the load (P), load point displacement, crack opening displacement (COD) and crack length on the surface were monitored simultaneously. The crack length was characterized by two methods: (1) monitoring the crack length on the polished specimen surface using optical microscopy and (2) utilizing an unloading/loading sequence to calculate an equivalent crack length based on the compliance technique. Load vs COD traces were subsequently examined for all test records. A toughness, K_{PMAX}, was also calculated using the standard equation for K as specified in ASTM E399, where the maximum load was used for P while the initial crack length was used for a. However, the use of the initial crack length may severely underestimate the actual maximum K, depending upon the instantaneous load and actual crack length. For those specimens which exhibited stable crack growth, resistance curves (i.e. R curves) in terms of crack-growth resistance, K_R, as a function of crack extension, Δa, were prepared. Furthermore, the video recordings of the fracture events were examined to assess the details of the fracture process. Additional details of the fracture process were determined by post-failure examination of the fractured specimens.

RESULTS

Table II summarizes the K_{PMAX} toughness data generated on the various materials, while Figures 2a-2c show three representative P vs COD traces obtained for the various materials tested. Figure 2a illustrates catastrophic crack propagation following a roughly linear P vs COD trace; this type of fracture behavior was obtained in the control DRA's

Table II: Average K_{PMAX} values for each material and orientation tested.

Material ID⇒	C1	C2	C3	SDP1	SDP2	LDP1	LDP2	LDP3
Arrestor (T-S)	19.0	20.8	20.0	19.4	20.8	33.1	30.4	23.3
Divider (T-L)	--	19.3	22.4	18.0	21.7	18.7	18.8	16.8

Figure 2: Representative Load vs COD traces.

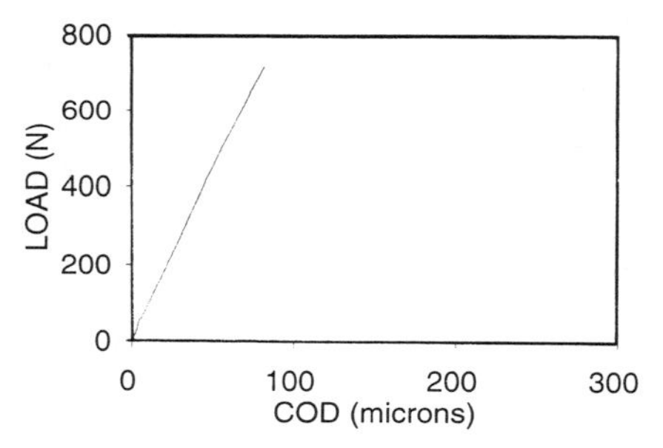

2a) Load vs COD trace demonstrating catastrophic failure as displayed by "Control" DRA's and "Small Ductile Phase" toughened DRA materials. Shown is the curve for a C1 specimen.

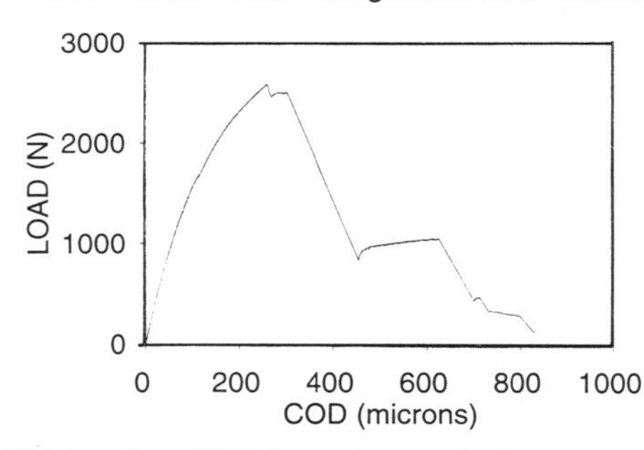

2b) Load vs COD trace demonstrating non-catastrophic fracture with a series of load drops as displayed by "Large Ductile Phase" toughened DRA materials tested in the T-S orientation, specifically, LDP2.

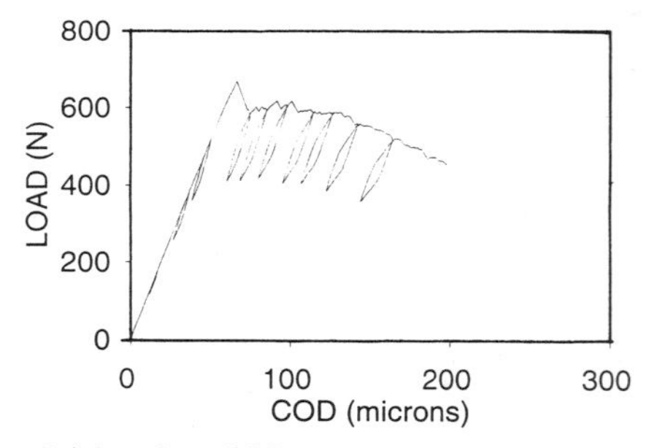

2c) Load vs COD trace demonstrating non-catastrophic fracture with a relatively smooth curve as displayed by "Large Ductile Phase" toughened DRA materials tested in the T-L orientation, specifically, LDP1.

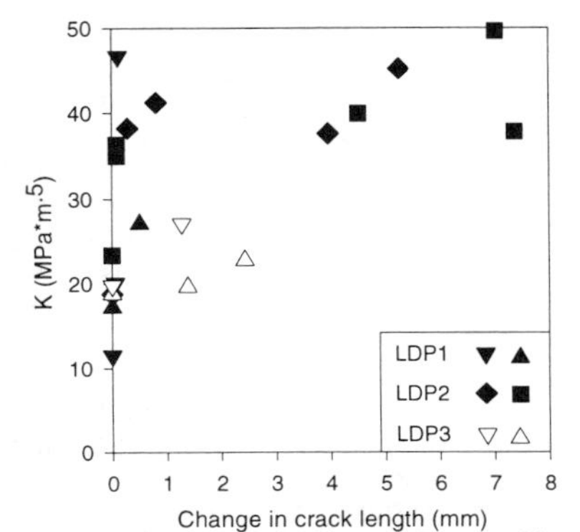

Figure 3: K_R curves for "Large Ductile Phase" toughened DRA materials tested in the T-S orientation.

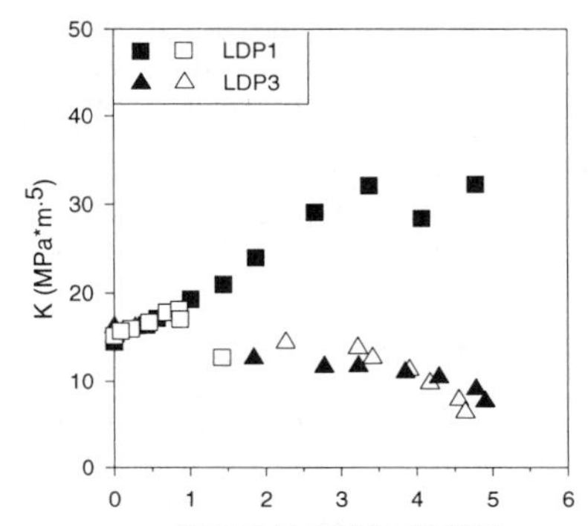

Figure 4: K_R curves for "Large Ductile Phase" toughened DRA materials tested in the T-L orientation. The crack length is estimated via both optical microscopy (■▲) and compliance (□△) methods.

(i.e. materials C1, C2 and C3) as well as the materials containing either 10 volume % or 25 volume % of the "small ductile phase" (i.e. SDP1 and SDP2), regardless of orientation. Furthermore, all the materials which displayed this catastrophic fracture behavior exhibited similar K_{PMAX} values of approximately 19-21 MPa√m. The lack of toughness improvement and catastrophic nature of the crack propagation in such specimens illustrates that the extrinsic toughening afforded by the use of the "small ductile phase" is not effective for the range of small ductile phases and specimen geometries tested presently.

However, the "large ductile phase" materials demonstrated highly nonlinear loading behavior associated with non-catastrophic crack propagation, as shown in Figures 2b and 2c, where the details of the P vs COD trace and the K_{PMAX} values were dependent upon the type and distribution of the ductile phases. Specifically, specimens containing the large ductile phase (i.e. LDP1, LDP2 and LDP3) tested in the T-S orientation (crack arrestor) exhibited crack initiation in the DRA composite at about the same stress intensity as for the control materials, followed by crack arrest when the crack encountered the first ductile phase region. In such specimens, the crack requires additional load and hence energy to further propagate through the specimen (i.e. R curve behavior) as shown by Figure 3. When the ductile phase consisted of unalloyed aluminum (i.e. LDP1 and LDP2) stress intensities of 30 to 50 MPa√m were necessary for continued crack growth in these relatively thin specimens. In-situ observations revealed that each load drop shown in Figure 2b was associated with a rapid advance of the crack through one to three ductile regions and the intervening DRA composite, while post-failure examination showed little to no delamination at the DRA/Al interfaces and highly ductile failure, either by shear or MVC, of the unreinforced aluminum regions. This ductility within the aluminum regions permitted the observed toughening effects to occur despite the presence of strong bonding between the DRA and unreinforced regions. Furthermore, while similar K values were necessary for crack advancement in the LDP1 and LDP2 materials, the higher volume fraction of "ductile phase" in the LDP2 material (and thus the greater number of ductile phase regions) leads to a greater toughening effect as judged by the area under the P vs COD trace, in this material. The LDP3 material which contained highly alloyed unreinforced regions also exhibited step-wise P vs. COD traces, but lower stress intensities (e.g. in the range of 20 to 30 MPa√m) were required for crack advancement. Examination of failed specimens revealed the alloyed unreinforced Al regions failed in a low ductility manner, consistent with the lower level of toughening exhibited.

The "large ductile phase" material tested in the T-L orientation (crack divider) also exhibited stable fracture as revealed in Figure 2c. Serial sectioning of specimens unloaded after some amount of stable crack growth revealed a non-planar crack front, where the crack propagates further in the DRA layers than in the unreinforced Al layers. Rising K vs Δa plots as shown in Figure 4 were obtained for specimens containing unalloyed Al regions (i.e. LDP1 and LDP2) though the details of the curves differ depending which method was used to estimate crack length. As found in the crack arrestor specimens, the unalloyed Al regions failed in a ductile manner while there was little or no delamination at the DRA/Al interfaces. The rising R curves were associated with bridging zones measuring 1 to 5 mm in length as determined via serial sectioning, while the length of the bridging ligament varied depending on the specific size and distribution of the unreinforced Al regions. On the other hand, the LDP3 material, which contained the stronger but less ductile alloyed unreinforced Al regions, displayed flat or non-existent K vs Δa curves regardless of the method used to estimate crack length. Bridge zone lengths of less than 0.5 mm were measured in these specimens.

CONCLUSIONS

Extrinsic toughening approaches have been explored to improve the toughness of DRA materials. It was found that the size, distribution, and properties of the ductile phase additions significantly affected the toughness. Specifically:

1. All "control" DRA composites and all SDP materials failed catastrophically at a K of about 20 $MPa\sqrt{m}$.
2. Materials toughened with LDP particles of unalloyed aluminum exhibited non-catastrophic failure (i.e. R curve behavior) in both the T-S and T-L orientations with evidence of ductile fracture of the LDP particles and a significant amount of crack bridging in the T-L orientation.
3. Materials containing LDP particles of alloyed aluminum exhibited less toughening than those containing pure aluminum. The LDP particles failed in a lower ductility manner and less crack bridging was obtained.

Acknowledgements

The authors wish to thank the Ohio Aerospace Institute and ALCOA for financial support and ALCOA also for supply of material and specimens.

References

1. D.J. Lloyd, *International Metals Review*, **39**, pp. 1-23, 1994.
2. W.H. Hunt, *Intrinsic and Extrinsic Fracture Mechanisms in Inorganic Composites.* (J. J. Lewandowski, and W. Hunt, Jr. eds.), TMS, Warrendale, PA, pp. 31-39, 1995.
3. W. H. Hunt, Jr., T.M. Osman, and J.J. Lewandowski, *JOM*, **45**, pp.30-35, 1993.
4. J. J. Lewandowski and P. M. Singh, *Intrinsic and Extrinsic Fracture Mechanisms in Inorganic Composites.* (J. J. Lewandowski, and W. Hunt, Jr. eds.), TMS, Warrendale, PA, pp. 129-147, 1995.
5. M. Manoharan, L. Yost Ellis, and J.J. Lewandowski, *Scripta Metallurgica*, **24**, pp.1515-1521, 1990.
6. L. Yost Ellis and J.J. Lewandowski, *Materials Science and Engineering*, **A183**, pp. 59-67, 1994.
7. L. Yost Ellis, J.J. Lewandowki, *Journal of Materials Science Letters*, **10**, pp. 461-463, 1991.
8. T.M. Osman and J.J. Lewandowski, *Scripta Metallurgica et Materialia*, **31**, pp. 191-195, 1994.
9. D.Lesuer, C. Syn, R. Riddle, and O. Sherby, *Intrinsic and Extrinsic Fracture Mechanisms in Inorganic Composites.* (J. J. Lewandowski, and W. Hunt, Jr. eds.), TMS, Warrendale, PA, pp. 93, 1995.

IMPACT BEHAVIOR OF EXTRINSICALLY TOUGHENED DISCONTINUOUSLY REINFORCED ALUMINUM COMPOSITES

M.A. IRFAN, N. LIOU, V. PRAKASH
Department of Mechanical and Aerospace Engineering, Case Western Reserve University, Cleveland, OH-44106. email: mai2@po.cwru.edu, nxl11@po.cwru.edu, vxp18@po.cwru.edu

ABSTRACT

Discontinuously reinforced aluminum (DRA) composites with enhanced fracture toughness have recently been developed at ALCOA. The approach consists of producing a composite microstructure in which discrete ductile phases have been incorporated into the DRA through traditional powder processing routes. In the present paper, the high strain rate behavior of these toughened composites is investigated by obtaining (i) the dynamic flow characteristics at various levels of elevated strain rates using a split Hopkinson compression bar, and (ii) energy absorption during dynamic crack initiation and crack propagation using three-point bend specimens loaded on a modified Hopkinson bar configuration.

INTRODUCTION

Discontinuously reinforced aluminum composites are being considered for a range of structural and non-structural applications [1]. Such composites display higher specific stiffness and strength, a lower thermal expansion coefficient, and superior wear resistant properties as compared with the monolithic matrix materials [2]. Despite these positive influences, the brittle reinforcement generally decreases both the initiation and propagation fracture toughness as well as the energy absorbing capabilities during impact [3]. It is important that the damage tolerance of DRA's must be increased if these materials are to gain widespread use for structural applications.

In an attempt to enhance the damage tolerance characteristics of DRA composites, a novel SiC particulate reinforced composite has recently been developed at ALCOA. The material utilizes a microstructural toughening (MT) approach [4] by incorporating discrete regions of unreinforced aluminum within the particulate reinforced DRA composite. The ductile phase reinforcement is introduced into the composite during the powder blending phase of the powder metallurgy process, thus eliminating the need for additional processing steps. The continuous interfaces between the reinforced and unreinforced regions, provide mechanisms for extrinsic toughening of the composite by deflecting propagating cracks [5] and by reducing the driving force at the crack-tip by initiating crack pinning mechanisms such as bridging of crack faces by ductile ligaments.

The current paper provides a brief description of the approach used in the processing and characterization of the *toughened* DRA microstructures. The impact behavior of these novel material systems is investigated by studying the material flow characteristics under dynamic compression using a split Hopkinson compression bar and obtaining the dynamic fracture toughness using three-point bend fracture specimens on a modified compressional Hopkinson bar.

DESCRIPTION OF MATERIALS

Two different microstructurally toughened composite microstructures will be investigated in the present study. The 7093/SiC/15p DRA composite (nominal composition of 9 Zn, 2.2 Mg, 1.5 Cu, 0.14 Zr, 0.1 Ni, balance Al reinforced with 10 μm average size SiC particulates) will provide the

Mat. Res. Soc. Symp. Proc. Vol. 434

base material for each of the two material microstructures. The base material is a powder metallurgy product with sufficient specific stiffness and strength, but with insufficient damage tolerance. The steps involved in powder processing of the base composite involves blending powders of pre-alloyed 7093 aluminum with SiC particulates and then cold isostatically pressing the blend into a solid compact. The *toughened* DRA MMC's are produced by additions of large aluminum particles or aluminum powders to the blend where these additions are of a sufficient size to result in unreinforced ductile regions in the final product. A *large ductile phase toughened* material is produced by making additions of commercial purity aluminum particulates (25 vol. %) with a typical size of 10 mm to the initial powder blend. By using commercial purity aluminum powders (10 vol. %) of more than an order of magnitude smaller in size, a *small ductile phase toughened* material is fabricated. Both composites are extruded and possess uniform microstructures along their entire length. The composites are heat treated to T7E92 prior to mechanical testing. Figure 1 shows the optical micrographs of the two toughened microstructures investigated in the present study.

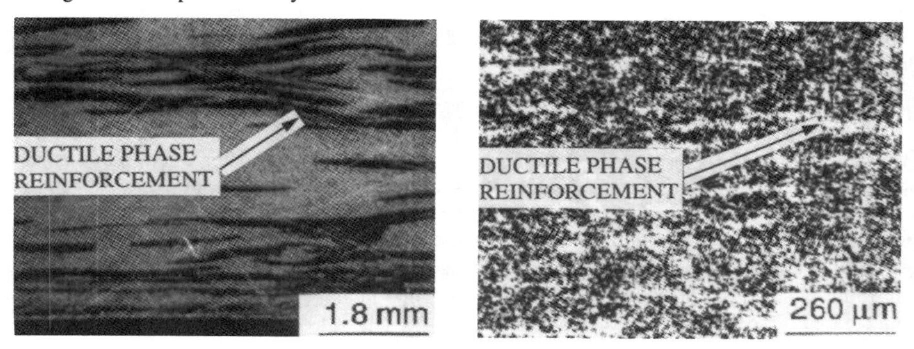

Figure 1: Micrographs of the microstructurally toughened material microstructures.

DYNAMIC STRESS-STRAIN RESPONSE : SPLIT-HOPKINSON PRESSURE BAR (SHPB)

The split Hopkinson pressure bar, for high strain-rate testing of materials, is a well established technique. A schematic of the basic SHPB is shown in Figure 2. Details of this experimental technique can be found elsewhere; see for example, Follansbee [6]. The required impact velocities and stresses are achieved by using a pneumatically-driven projectile guided by a long launch tube. The magnitude of the pulse is directly proportional to the velocity of the striker bar and the duration of the pulse is equal to the round-trip time of an elastic longitudinal wave in the striker bar. Two strain gages are used at each of the strain-gage stations to eliminate bending contributions. Maximum impact velocities are of the order of 100 m/s, limited by the maximum yield stress in the pressure bars. These bars are composed of maraging steel tempered to a yield stress of approximately 2500 MPa. One dimensional calculations by Kolsky [7] show that the strain rate $\dot{\varepsilon}$, strain ε, and the axial stress σ, in the specimen can be estimated using

$$\dot{\varepsilon}(t) = -\frac{2C_o}{L}\varepsilon_R, \quad \varepsilon(t) = -\frac{2C_o}{L}\int_0^t \varepsilon_R \, dt \quad \text{and} \quad \sigma(t) = E_o \frac{A_o}{A}\varepsilon_T(t) \tag{1}$$

where L and A are the original length and the cross-sectional area of the specimen, respectively; A_o, C_o and E_o are the cross-sectional area, the longitudinal bar wave speed and the elastic modulus of the Hopkinson bar, respectively; and ε_R and ε_T are the time dependent reflected and transmitted strains, respectively.

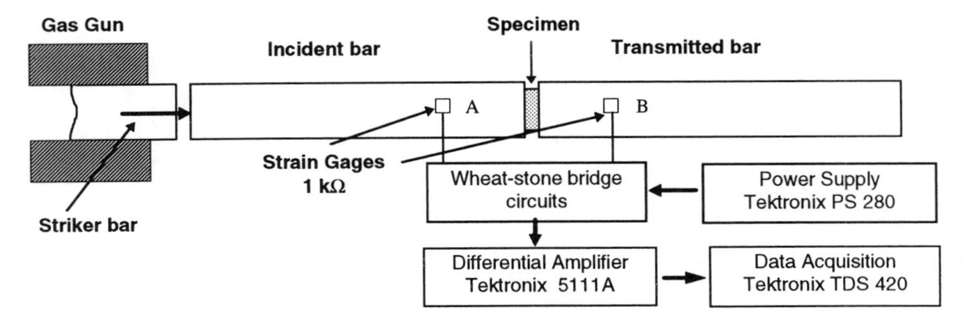

Figure 2: Schematic of the split Hopkinson pressure Bar

Typical experimental results are shown in Figure 3 for materials with large and small ductile phase reinforcements. In both cases, the compressional loading axis of the specimens is maintained along the transverse direction of the extrusion. No elastic constants are deduced from the stress-strain curves, since the calculation of the elastic modulus is extremely sensitive to the arrival times of the reflected and transmitted stress pulses at the strain gage locations.

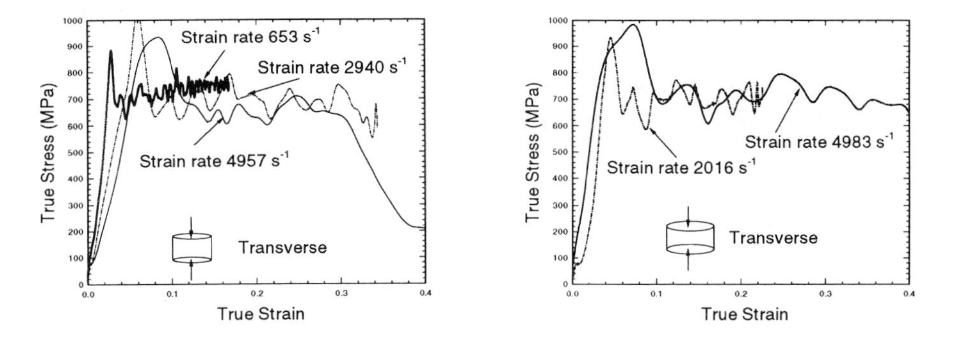

Figure 3: Flow stress curves for (a) 7093/SiC/15p +25% large c.p. Al, and (b) 7093/SiC/15p + 10% small c.p. Al.

The flow stress for 7093/SiC/15p + 25% large commercial purity Al reinforced material shows strain hardening at lower strain rates (653/s), but at higher strain rates, i.e. 2940/s and 4957/s, the evolution of flow stress shows strain softening. This feature of the experimental results can be attributed to the accumulation of damage with strain during dynamic compression. Indeed the specimen subjected to strain rates of 4957/s is observed to fracture at a strain of approximately 0.29. Similar observations are made for 7093/SiC/15p+10% small c.p. Al material where the material flow stress is lower for deformations at higher levels of strain rate. Figure 4 summarizes the dynamic flow stress as a function of the applied strain rate for the 7093 alloy, commercial purity aluminum, and the small and large ductile phase reinforced composites.

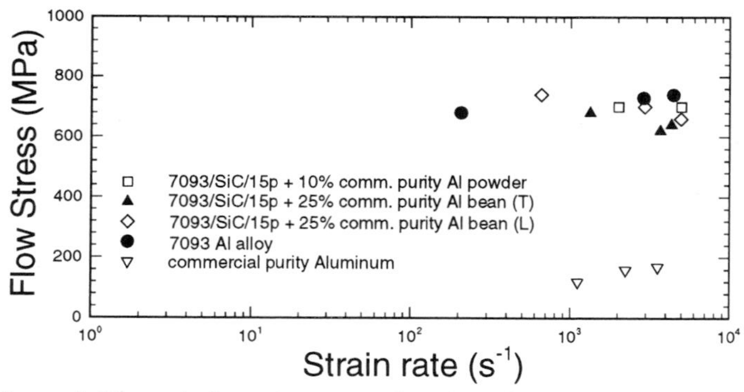

Figure 4: Dynamic flow stress as a function of applied true strain-rate.

DYNAMIC FRACTURE TOUGHNESS: MODIFIED HOPKINSON BAR TECHNIQUE

The experiment involves the dynamic loading of a three point bend specimen by means of a modified Hopkinson bar apparatus [8]; shown schematically in Figure 5. The incident bar is made of 7075 T6 Aluminum with a length of 60.06 in. (152.56 cm) and a diameter of 0.75 in. (1.91 cm). The strain gages are installed on the incident bar at two positions A and B at 13.78 in. (35 cm) and 29.53 in. (75 cm) from the striker bar end respectively. The specimen end of the incident bar is provided with a large radius of curvature of approximately 2 in. (5.08 cm). The striker bar is made of the same material as the incident bar and has a length of 35.50 inches (90.17 cm).

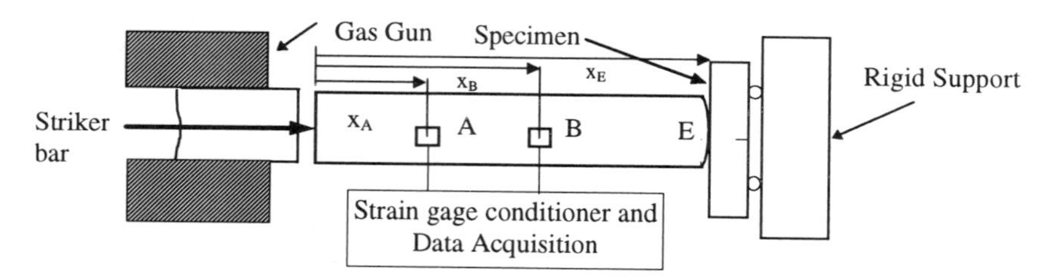

Figure 5: Schematic of the modified split Hopkinson bar configuration.

The anvils supporting the specimen had a length span of 40 mm. The fracture specimens were 9 mm wide and 10 mm deep giving an L/D ratio of 4, in a T-S configuration. Notch depths of 3 mm and 5 mm were machined using a wire saw with a diamond wire of diameter 100 μm. The exact length of the notch was later measured using a traveling microscope.

The strains at the two locations x_A and x_B on the incident bar are converted into normal forces, N_A and N_B, respectively, using

$$N_A(t) = A_o E\varepsilon_A(t) \quad \text{and} \quad N_B(t) = A_o E\varepsilon_B(t) \tag{2}$$

By using the method of characteristics the particle velocity, $v_A(t)$ at x_A can be evaluated to be

$$v_A(t) = v_A(t_P) + \frac{1}{Z}[-N_A(t) - N_A(t_P) + 2N_B(t - T_{BA})] \tag{3}$$

where $t_P = t - 2\,T_{BA}$, $T_{BA} = (x_B - x_A)/C_o$ and Z is the acoustic impedance of the incident bar. Using one dimensional stress wave analysis the particle velocity v_E and the normal force N_E at the loading point x_E can be evaluated as

$$v_E(t) = \frac{1}{2}[v_A(t + T_{EA}) + v_A(t - T_{EA})] + \frac{1}{2Z}[N_A(t + T_{EA}) - N_A(t - T_{EA})] \tag{4)}$$

$$N_E(t) = \frac{1}{2}[N_A(t + T_{EA}) + N_A(t - T_{EA})] + \frac{Z}{2}[v_A(t + T_{EA}) - v_A(t - T_{EA})] \tag{5}$$

where $T_{EA} = (x_E - x_A)/C_o$. The load-point displacement u(t) and the applied force can be expressed as

$$u(t) = \int_0^t v_E(\tau)d\tau \quad \text{and} \quad F(t) = -N_E(t) \tag{6}$$

Figure 6 shows the force displacement curves obtained for the 7093/SiC/15p+10% small c.p. Al toughened composite and the 7093/SiC/15p+25% large c.p. Al toughened composite.

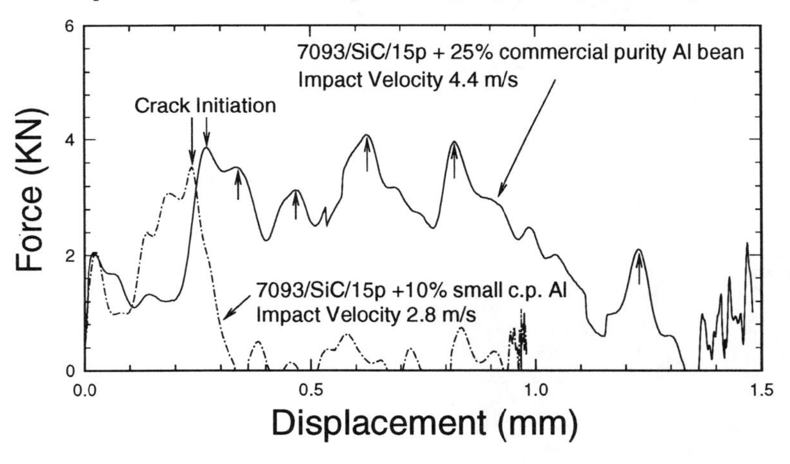

Figure 6: Force-displacement curves for the small and large ductile phase toughened composites

The dynamic crack initiation toughness for the two composites is similar. This is to be expected as the pre-machined notch in the specimens for both experiments resides completely in the base material. On the other hand, the dynamic crack propagation characteristics of the two composites are significantly different. In the small ductile phase reinforced material dynamic crack propagation occurs catastrophically with very little energy being absorbed during crack propagation, whereas the large ductile phase reinforced composite absorbs significant energy before failure.

The enhanced crack propagation toughness of the large ductile phase reinforced composite is attributed to the extrinsic toughening introduced by the large unreinforced ductile phases introduced in the composites. The toughening mechanisms include crack deflection at the brittle/ductile interfaces and crack pinning by the ductile phases in the wake of the crack. Figure 7 shows the material microstructure ahead of the pre-machined notch and the path taken by the crack during the dynamic crack propagation event. Furthermore, each peak in the force-displacement curve (also highlighted by arrows in Figure 6) can be correlated to the extrinsic toughening mechanisms activated every time the crack encounters a ductile ligament in its path.

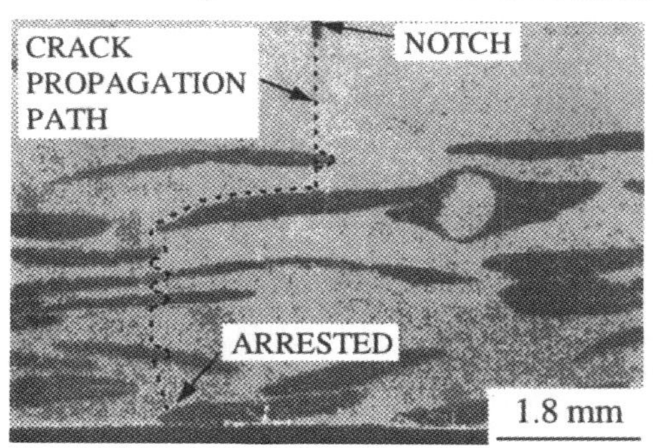

Figure 7: Material microstructure ahead of the pre-machined notch and the path taken by the propagating crack during dynamic failure.

For dynamic fracture experiments conducted at lower impact velocity (2.4 m/s as compared to 4.4 m/s for the experiment discussed in Figure 6) the dynamically propagating crack is observed to be arrested at the very first ductile layer it encounters. Figure 8 shows the force-displacement history for the particular experiment along with the material microstructure ahead of the pre-machined notch. The crack is arrested at approximately 0.5 mm of load point displacement at station E.

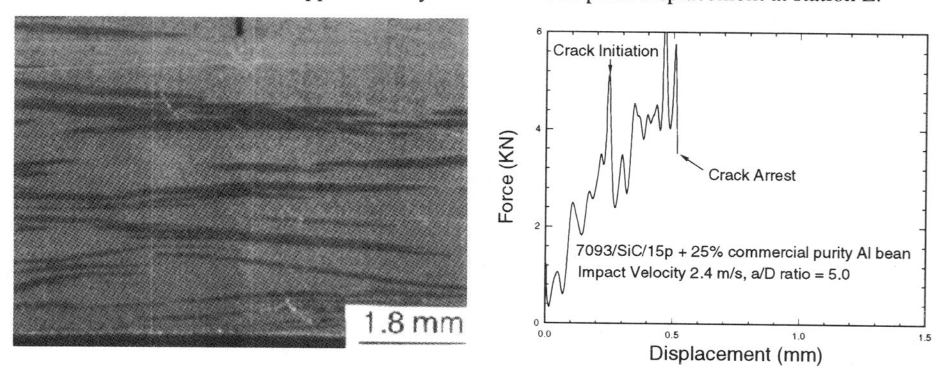

Figure 8: Microstructure ahead of the pre-machined notch and the force-displacement history during crack initiation and propagation.

Figure 9 shows the optical micrograph of the crack propagation path for the experiment described in Figure 8. As the crack approaches the first unreinforced ductile layer it is observed to deflect parallel to the brittle/ductile interface leading to crack arrest.

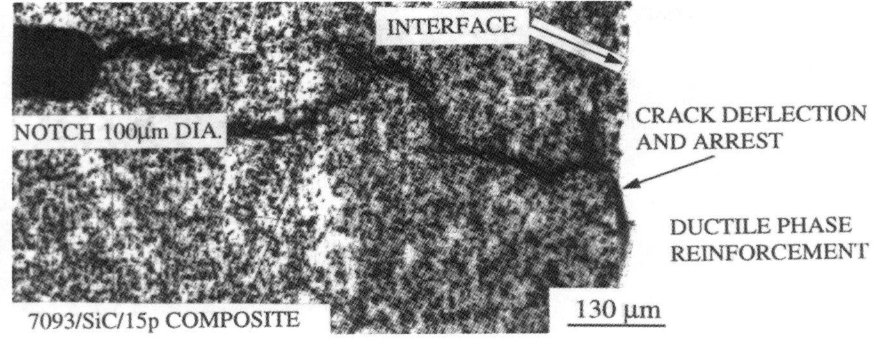

Figure 9: Optical micrograph of the arrested crack at the brittle/ductile interface.

In the present study attempt has also been made to investigate the volume of material participating in the dynamic failure process. Of particular interest is the plastic deformation in the ductile reinforcement layer in the wake of the crack. In view of this, all dynamic fracture experiments were conducted with an orthogonal Moire grid with a pitch equal to 5 μm on the specimen surface. Preliminary results, as shown in Figure 10, indicate that the ductile layers fail predominantly in a shear mode.

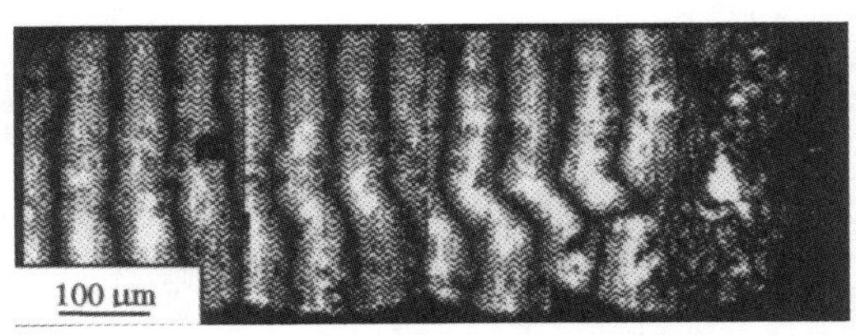

Figure 10: Moire fringes for U displacement in a ductile phase reinforcement layer.

ACKNOWLEDGMENT

The authors would like to thank Prof. J. J. Lewandowski at CWRU and Dr. Warren Hunt at ALCOA for providing the materials used in the present investigation.

REFERENCES

[1] T. W. Clyne and P. J. Withers, '*An Introduction to Metal-Matrix Composites,*' Cambridge University Press, Cambridge, U. K., 1993.
[2] W. H. Hunt, Jr., T. M. Osman, and J. J. Lewandowski, *J. Metals,* **45**, pp. 30-, 1993.
[3] L. Yost Ellis and J. J. Lewandowski, *Mater. Sci. Eng.*, **A183,** pp. 59-, 1994.
[4] V. C. Nardone, J. R. Strife and K. M. Prewo, *Metal. Trans. A.,* **22A**, pp. 171-181, 1991.
[5] R. O. Ritchie, *Mat. Sci. Eng.,* **A103**, pp. 15-28, 1988.
[6] P. S. Follansbee, 'The Hopkinson Bar', *Mechanical Testing, Metals Handbook,* **8**, 9th ed., ASM, Metals Park, Ohio, pp. 198-217, 1985.
[7] H. Kolsky, *Proc. Roy. Soc London,* **B62**, pp. 676-700, 1949.
[8] C. Bacon, J. Farm and J. L. Lataillade, *Experimental Mechanics,* **34**, pp. 217-223, 1994.

FRACTURE OF LAMINATED AND *IN SITU* NIOBIUM SILICIDE-NIOBIUM COMPOSITES

J.D. RIGNEY
The Case School of Engineering, Department of Materials Science and Engineering, Case Western Reserve University, Cleveland, OH 44106 U.S.A.

ABSTRACT

The mechanisms contributing to the fracture resistance of refractory metal intermetallic composites containing a BCC metallic phase (niobium) were investigated using model Nb-Si laminates and *in situ* composites. The controlling influence of ductile phase yield strength and fracture behavior were investigated by varying laminate processing parameters, and/or altering temperatures and applied strain rates during fracture experiments on all materials. The fracture behavior of "ductile" constituents were found to be influenced by phase grain size, solid solution content, constraint (as influenced by interfacial bond strengths), and the testing condition (high strain rates and low temperatures). The measured fracture resistance, when compared to theoretical models, was shown to be controlled by the "toughness" of the "ductile" phase and independent of the fracture behavior promoted (cleavage and ductile). The loss in ductility due to cleavage by high constraint, high strain rates and/or low temperatures was compensated by high yield and cleavage fracture stresses in order to provide a level of toughening similar to that contributed by ligaments which failed with lower yield stresses and greater strains.

INTRODUCTION

Refractory metal intermetallic composites based on the binary Nb-Si system have received considerable attention for high temperature applications (> 1273 K) during the past decade. Primary interest in these systems has been promoted by the high melting temperatures of the 5:3 transition metal silicides, and their superior strength, stiffness and creep resistance.[1-4] Use of these materials in critical structural applications, however, is limited in part by the low ambient fracture resistance of the monolithic materials (i.e. 1 to 3 MPa$\sqrt{m}$ for Nb_5Si_3).[5-7] Among extrinsic techniques, composite incorporation of ductile-phase reinforcement has emerged as a potent method to augment the room temperature fracture resistance of brittle intermetallics and ceramics. The thermodynamic stability[8] and mechanical compatibility of Nb_5Si_3 and terminal Nb(Si) phases imply that a variety of processing routes are available to construct toughened composites. The wide two phase field in the Nb-rich end of the equilibrium Nb-Si phase diagram[8], Figure 1, shows that a range of compositions can be chosen. The stability of the two phase system to 1923 K is also ideal for powder blending or layering constituents prior to consolidation or bonding operations.

The ductile-phase toughening process envisioned for *in situ* composites is illustrated in Figure 2(a) with a non-catastrophic brittle matrix crack propagating into an array of ductile particles. The bridging intact ligaments reduce the stress at the crack tip through elastic loading and plastic deformation. Creation of additional ligaments transfers more of the far field stress to the developing bridged-zone, requiring higher applied stress intensities to propagate the crack tip. The steady-state fracture resistance of materials has been modeled[9] based on microstructural quantities and mechanical properties of the metallic phase. The increase in fracture energy (ΔG) over that of the brittle matrix is controlled by the area fraction of particles on the fracture plane (V_f), their size (a_o), the uniaxial yield strength (σ_y), and the "work of rupture" (χ) of the ductile material, as shown in the following relationship:

$$\Delta G = V_f \cdot a_o \cdot \sigma_y \cdot \chi \qquad \{1\}$$

Previous work has clearly demonstrated that flow and fracture of a ligament can deviate from the uniaxial tensile response due to the constraint of the surrounding elastic matrix.[10,11] The

Mat. Res. Soc. Symp. Proc. Vol. 434

constraint level, in turn, is governed by matrix/particle interfacial debonding as shown in Figure 2(b). Because the area beneath these normalized curves are χ-values, it is apparent that a critical amount of debonding can maximize the "work" performed by the ligaments when comparing experimental conditions where σ_y cannot change. In laminate materials containing a ductile layer, the σ_y and χ parameters will influence the toughness increment in the same way.

In addition to constraint, test conditions (applied strain rate and temperature) and metallurgical state (solid solution content and grain size) can influence deformation and fracture (plastic rupture versus cleavage) of a body-centered cubic metal. The present paper reviews recent works designed to investigate fracture niobium silicide (Nb_5Si_3) - niobium (a body-centered cubic metal) composites produced by lamination or *in situ* casting techniques. Slow strain rate, room temperature fracture experiments were used initially to characterize the development of the fracture processes in all composite types.[3-5,7,12] Modification of laminate fabrication techniques provided a means of altering interfacial strengths, and Nb layer grain sizes and solid solution contents.[6,13] Variation in test temperature (laminates and *in situ* composites) and displacement rate (*in situ* composites only) have been used to change the response of the niobium phases. Composite toughness and fracture behavior (brittle versus ductile) of the Nb interlayers and Nb_p and Nb_s *in situ* phases were quantified extensively.

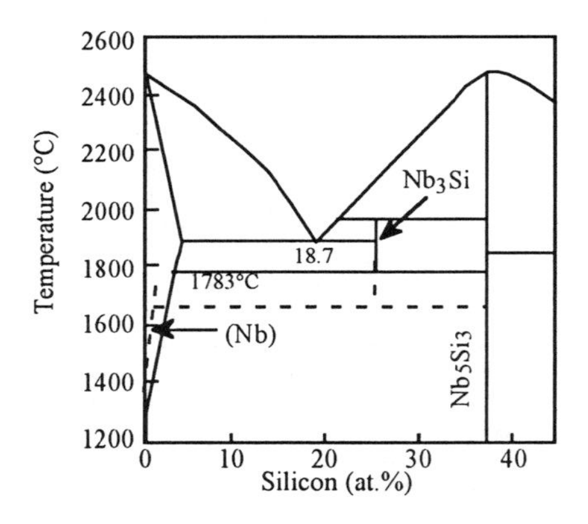

Figure 1: Relevant portion of the modified Nb-Si binary phase diagram.[8]

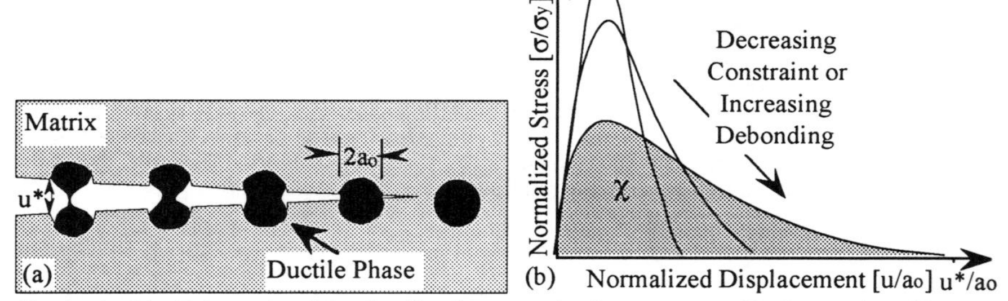

Figure 2: (a) Schematic of the ductile-phase toughening process. (b) Constraint effects on fracture of ductile ligaments.[11]

EXPERIMENTAL METHODS

Nb Foils Heat Treatment and Laminates Production

Recent works[6,13,14] have utilized commercial purity, 250 μm thick Nb foils (Aldrich Chemical Co.) for later lamination with sections of hot pressed Nb_5Si_3 compacts (powder production and consolidation is described elsewhere[6,13]). The tensile deformation and fracture behavior of the foils at 298 and 77 K[13] were evaluated in three conditions to investigate grain size and/or chemistry effects: (a) as received (AR) (10 μm grain size), (b) vacuum heat treated (VHT) (210 μm grain size), and (c) vacuum heat treated with Si in solid solution (Nb(Si)) (210 μm grain size). The heat treatments for (b) and (c) were chosen to mimic microstructural changes (large grains and large grains/ Si solid solution, respectively) that result from the diffusion bonding practice described below for making laminates. The large grained VHT material was produced from AR foils after a 1473 K, 5 h heat treatment. When possible, all heat treatments were conducted with samples wrapped in Ta foil to minimize sample exposure to carbon, oxygen and other impurities.

Tri-layered laminates (i.e. Nb_5Si_3/Nb/Nb_5Si_3) were produced[6,13,14] with Nb foils (a, b or c) between 4 mm slices of Nb_5Si_3 and bonded by one of two methods. Diffusion bonding was accomplished by hot pressing the layered system in vacuum at 1473 K for 5 h under 10 MPa pressure. (High temperature exposure of the layered system without pressure allowed removal of a Nb(Si) layer to be tested as (c) above.) Other laminates with AR or VHT layers were bonded between Nb_5Si_3 layers with a structural adhesive EC1386 (3M Company). The adhesive was cured by heating to 450 K at 5.5 K/ min, holding for 1 h, then slow cooling all under an applied pressure of 0.16 MPa; the result was a 20 μm thick adhesive layer at the interface. Table 1 summarizes the Nb layer and laminate processing parameters. Smooth tensile bars were fabricated from the three Nb layer types. Notched bend bars were fashioned from each laminate such that specimen width (W) by thickness (B) dimensions were 8 x 4 mm^2. Notches were placed in the silicide ≈150 μm from the interface, or to crack depth-to-width ratios(a/W) between 0.45 and 0.55.

Nb-10 at.% Si In Situ *Composite Materials*

Nb-10 at.% Si materials were vacuum arc-cast (Westinghouse, Inc.), extruded in Mo cans at 5.5:1 and 1773 K (Wright-Patterson Air Force Base), and heat treated in vacuum at 1773 K for 100 h.[3,4,12] Notched three point bend bars were machined from the extrusions with the long axis of the bars parallel and the notches perpendicular to the extrusion direction.[12] The smallest specimens, W x B ≈ 6 x 1.5 mm^2 and a/W ≈ 0.5, were fractured in *in situ* monitored tests. These samples were slit-notched by a low speed wire saw providing notch root radii of ≈ 35 μm. Larger specimens were used in remaining toughness tests with typical W x B x (a/W) dimensions as follows: 9 x 6 mm^2 x (0.2 to 0.4) and 6 x 6 mm^2 x (0.45 to 0.55). Mechanical polishing of the samples facilitated fracture monitoring via scanning electron microscopy (SEM) techniques or post-test examination of damage accumulation during crack propagation.

Directionally Solidified (DS) Nb-Si and Nb-Si-Ti In Situ *Composite Materials*

Directionally solidified binary and ternary alloys with compositions listed in Table 2 were produced at General Electric Corporate Research and Development using a Czochralski technique.[15] Materials received were fabricated into single-edge notched bend bars with W x B x a/W dimensions of 2.5 x 1.5 mm^2 x 0.5, conforming to standard sample dimensions.[16] Room temperature toughness values and resistance-curve behavior of these materials were previously quantified by Bewlay et al.,[15] focusing the present work at 77 K and 773 K to investigate the influence of Nb(Si,X) yield strength and fracture behavior on toughness.

Tension Experiments on Nb Foils

Smooth tension experiments were performed on the three Nb layers (AR, VHT or Nb(Si)) at 298 and 77 K (liquid nitrogen).[13] The room temperature samples were machined with gage sections of 6 x 0.25 x 12.7 mm^3, while the lowest temperature samples were prepared with reduced gage sections (3 x 0.25 x 20 mm^3). Samples were loaded on a screw-driven Instron 1125L at an initial strain rate of 10^{-3}/sec.

Table 1: Nb layer heat treatment and laminate production.[6,13,14]

Nb Layer	Layer Processing(*)	Laminate Construction(#)
AR	As Received	Adhesive Bonded *Epoxy: 450K/1h*
VHT	1473 K / 5 h in vacuum	Adhesive Bonded *Epoxy: 450K/1h*
Nb(Si)	1473K, 5h (Si diffusion $Nb_5Si_3 \Rightarrow Nb$)	Diffusion Bonded *Hot Press: 10MPa, 1473K, 5h*

* Processing for tensile testing and laminate fabrication
# Laminates were tri-layered in the form of $Nb_5Si_3/X/Nb_5Si_3$: X = Nb Layer condition to left

Table 2: Compositions of DS alloys manufactured and vol.% of metallic phase.*

Composition at.% Si	Composition at.% Ti	Vol.% Nb(Si) or Nb(Ti,Si)
14	-	57.6
16	-	40.5
18.2	-	37.6
19	-	26.1
22	-	16.4
12	21, 27	54.3, 41.3
16	21, 27, 33	34.5, 27.3, 37.4
15	42.5	36.6
20	33	27.3
22	21	23.3

* Materials produced and microstructures analyzed by Bewlay et al.[15]

Fracture Toughness Experiments on Laminates and In Situ *Composites*

Fracture toughness experiments were performed on the variety of materials/samples described above, in which the single-edge notched specimens were loaded in three point bending using standard guidelines.[16] Initial characterization of fracture processes in the diffusion and adhesively bonded laminates and Nb-10 at.% Si composites involved real time fracture observation within a JEOL 840A SEM equipped with an Oxford Instruments deformation stage. Specimens were loaded with span(S)-to-width(W) ratios of 4:1 and at a load point displacement (LPD) rate of 1 $\mu m \cdot sec^{-1}$. Polished sample surfaces were oriented perpendicular to the electron beam so that surface cracking events extending from the notch root could be monitored. The loads (P_Q) at which initial and subsequent cracking events were observed were used in the following equation[16] to determine the initiation toughness and to construct the resistance(R)-curves (K versus Δa):

$$K_Q = (P_Q S)/(BW^{3/2}) \cdot f(a/W) \qquad \{2\}$$

Loading span (S), thickness (B), width (W), crack length (a) and geometrical factor (f(a/W)) were utilized to determine the stress intensity in ($MPa\sqrt{m}$) units.

Fracture toughness tests on larger samples of laminates or *in situ* composite samples were performed on screw-driven Instron (1125L) or servohydraulic MTS equipment. Although the laminates were tested at a one displacement rate at 298 and 77 K, the Nb-10 at.% Si materials were fractured over a range of LPD rates ($4.2 \cdot 10^{-4}$ to $8 \cdot 10^{1}$ $mm \cdot sec^{-1}$) to induce a variation in initial strain rates (and fracture behavior) at both temperatures. Loading rates ($N \cdot sec^{-1}$) in the linear elastic portion of the test traces were calculated to normalize the testing rate data for specimens of slightly different dimensions. The 77 K tests were conducted by immersing the three point bend samples/fixture in liquid nitrogen. In the 773 K experiments on DS *in situ* composites, the system was sealed by a quartz tube for flow of argon during the test, and temperature controlled with a resistance heated furnace and a thermocouple on the samples. The peak loads (P_Q) recorded and specimen dimensions were used in equation {2} to determine the fracture toughness of each sample.[12]

Post-failure analysis consisted of SEM inspection of the fracture behavior of Nb interlayers and *in situ* formed Nb(Si,X) phases. Montages along the laminated Nb interlayers and from the notch to the back faces along the "plane strain" regions (sample mid-sections) of Nb-10 at.% Si samples were used to quantify the percentage of brittle/ductile fracture of the Nb constituents. Stereo imaging of laminate fracture surfaces provided a means to assess interfacial bonding characteristics on fracture behavior of Nb layers. Quantification of behavior in Nb-10 at.% Si samples was determined in terms of *percent cleaved Nb_p ligaments* by comparing area fraction of cleaved Nb_p to total vol.% Nb_p. The failure of the Nb_s in Nb-10 at.% Si and Nb(Si,X) phases in DS composites was examined qualitatively in a variety of locations on the fracture surfaces.

RESULTS AND DISCUSSION

Microstructural Analyses

Table 3 summarizes the effects of processing condition on the resulting grain size and impurity content of the Nb foil in the laminate materials.[13] The VHT schedule caused recrystallization and grain growth (210 μm) compared to the AR foils (10 μm), and some pickup of C, O and N impurities from the graphite dies and limited vacuum (10^{-4} torr). Preparation of Nb(Si) foils by diffusive reaction between pure Nb and Nb_5Si_3 not only lead to 1050 ppm Si levels in the Nb(Si) but O contamination from the silicide matrix.

The typical microstructure of the Nb-10 at.% Si composite in three orthogonal views is shown in Figure 3[12]. The large Nb_p particles (light phase) and Nb_5Si_3 (dark phase) have been elongated in the direction of extrusion. The Nb_p phases occupy 51.3±4 vol.%, have an average width of 17.0±1.2 μm, and grain size of 12.7±1.8 μm. The aspect ratio of the Nb_p ranged from 5:1 to 10:1 along the extrusion direction. The remaining structure is occupied by eutectic Nb_5Si_3 and secondary niobium (Nb_s). The Nb_s is 24.3±2 vol.% with 2.7±0.3 μm average size. Interstitial contents were relatively low: 230 ppm for oxygen and 50 ppm for nitrogen.

The DS microstructures were analyzed by Bewlay et al.[15] and their observations summarized here; ductile phase vol.% are listed in Table 2. The Nb-Si binary materials varied with Si concentration as dictated by the phase diagram. Hypoeutectic compositions contained Nb_p dendrites with interdendritic eutectic (Nb_s and Nb_3Si); hypereutectic compositions contained primary Nb_3Si with interdendritic eutectic. The presence of Nb_3Si rather than equilibrium Nb_5Si_3 has been shown to result from the sluggish transformation kinetics requiring elevated temperature exposure for long times[8,15] (as was performed at 1773 K for 100 h for the Nb-10 at.% Si composites). The silicide constituent in the present toughness samples was Nb_3Si. Solidification of Nb-Si-Ti ternaries generally resulted in two phase materials consisting of Nb(Si,Ti) and $(Nb,Ti)_3Si$, however the primary solidification phase was dependent on alloy composition. Alloy of Nb-21at.%Ti-22 at.% Si and 32.3 at.% Ti- 19.2 at.% Si also contained low volume fractions of $Nb(Ti)_5Si_3$. Quantitative analysis showed the Nb(Si,Ti) to contain 29-44 at.% Ti and 1.6-2.4 at.% Si.

Mechanical Behavior of Nb Foils and Nb_5Si_3/Nb/ Nb_5Si_3 Laminates

The typical stress-strain curves for the three foil types are shown in Figure 4,[13] while particular stress values and general observations are listed in Table 4.[13] The 298 K behavior was influenced considerably by the heat treatment condition. Exposure of a foil to vacuum heat treatment (VHT with 210 μm grain size) caused a decrease in strengths compared to those of AR foil (10 μm). Because Nb has been shown to exhibit a small Hall-Petch slope,[17,18] the differences in the behavior result primarily from changes in substructure by recovery and recrystallization rather than from grain growth alone. Silicon containing Nb(Si) (210 μm), demonstrated higher strengths than the other two foil types. Silicon is a potent solid solution strengthener of Nb,[12,13,19] increasing the yield stress at 298 K without producing a significant loss in ductility. All specimens at 298 K failed by ductile fracture with high reductions in area. At 77 K, increases in yield stresses were recorded for all specimens consistent with behavior of

Table 3: Microstructural analysis & impurity content of Nb foils & Nb_5Si_3 matrix.[13]

Impurity	AR	VHT	Nb(Si)	Nb_5Si_3
d (μm)(#)	10	210	210	5
C (ppm)	<10	420	210	980
O (ppm)	<125	490	1300	1260
N (ppm)	<50	120	190	*
H (ppm)	<5	*	*	*
Si (ppm)	<100	90	1050	16.9 wt.%

#Grain Size (diameter); *Value not measured

Table 4: Tensile properties of Nb layers.[13]

Nb Layer Condition	Temp [K]	σ_y [MPa]	σ_{UTS} [MPa]	RA* [%]	Fracture Mode
AR	298	268	358	78	Ductile
VHT	298	186	222	79	Ductile
Nb(Si)	298	363	390	76	Ductile
AR	77	730	900	93	Ductile
VHT	77	661	765	6	Cleavage
Nb(Si)	77	844	960	8	Cleavage

*Reduction in area for ductility measurement.

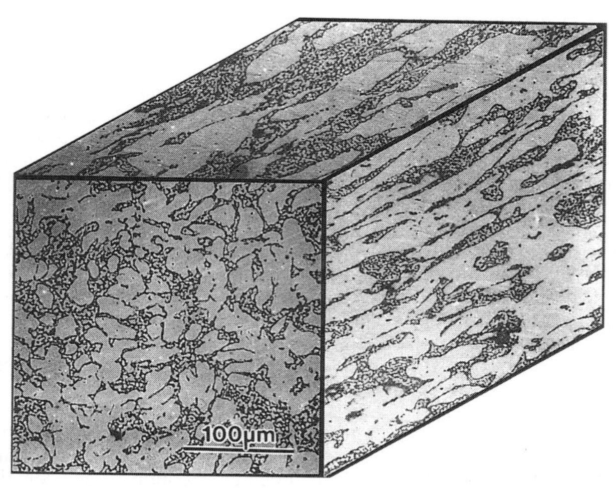

Figure 3: Three dimensional view of arc cast, extruded and heat treated Nb_5Si_3/Nb.[12]

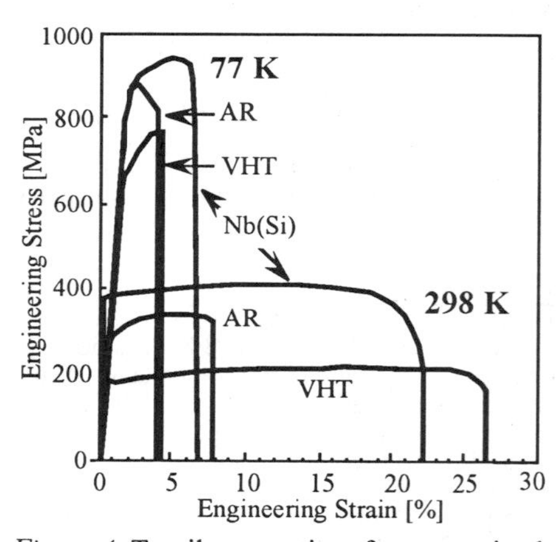

Figure 4: Tensile properties of unconstrained Nb interlayers at 298 and 77 K.[13]

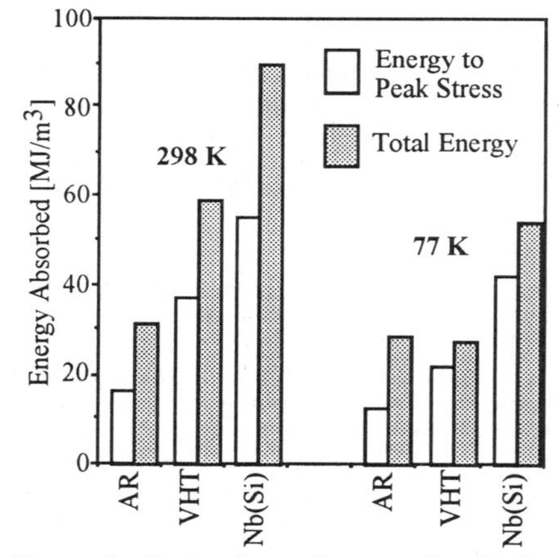

Figure 5: Comparison of energy absorbed during deformation and fracture of Nb layers.[13]

body-centered cubic metals. The failure behaviors were noticeably different, where fine grained AR samples failed by ductile rupture and coarser grained VHT and Nb(Si) samples failed by cleavage. This fracture mode change at lower temperatures is in line with Orowan's[20] proposal of a brittle fracture stress. The grain size and substructure differences on brittle-ductile transition are consistent with Cottrell's model[21] for cleavage fracture strengths, as will be addressed specifically for Nb-alloys in a later section.

Because the mechanical response of ductile ligaments were theoretically shown[9] (Figure 2(b)) to be critical to the toughness increment, the energies required to deform and fracture the Nb foils were calculated by measuring the energy under each stress-strain curve. The values, plotted in Figure 5, summarize the energies for loading specimens to maximum stress and completely to failure at both temperatures. The figure illustrates the beneficial effects of heat treatment on failure energy compared to AR material despite drastic changes in fracture behavior and shape of the stress-strain curve. It was found that Nb(Si) failed with greater energy at peak stress and in total at 77 K compared to the values for AR at 298 K, although Nb(Si) failed by cleavage fracture and the AR by ductile rupture. Figures 4 and 5 and Table 4 show that processing can affect the mechanical properties of the constituents, and that high failure energies are possible even when samples fail by cleavage. Both have important implications on the behavior of these materials systems to predictions of "ductile"-phase toughening models.

The fracture behavior of the laminates were determined from bend experiments at 298 and 77 K.[6,13] In room temperature experiments conducted in the SEM, the typical view of the fracture process in a diffusion bonded laminate at high stress intensity is shown in Figure 6. Cracks have initiated from the notch and in the Nb_5Si_3 layer below the Nb(Si). Little interface debonding is noticeable, however, constraint is relieved by multiple microcracking along several hundred microns of the interface.[13] A significant level of plasticity in the Nb(Si) is observed to the left of the notch where more microcracking is present. Below the notch and to the left are stable cleavage microcracks which have opened due to the high constraint in these locations but have been blunted. Several experiments were conducted at room temperature producing an average toughness of 8.6 ± 1.5 MPa√m.[13] A typical fracture surface, Figure 7, shows a dual

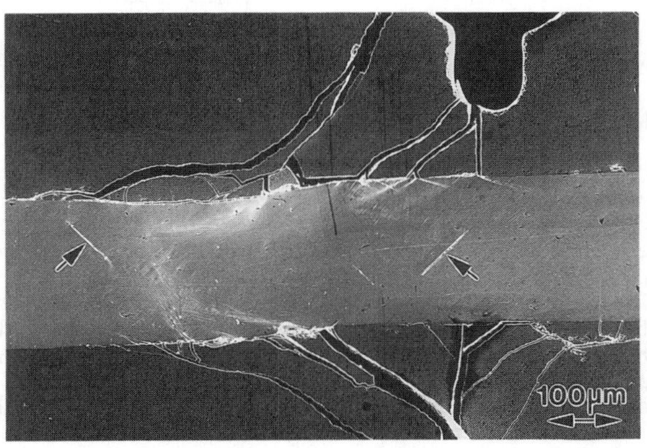

Figure 6: In situ observation of fracture in diffusion bonded laminate. Interfacial microcracking relieves constraint, however, cleavage microcracks (arrowed) were nucleated in some areas.[6,13]

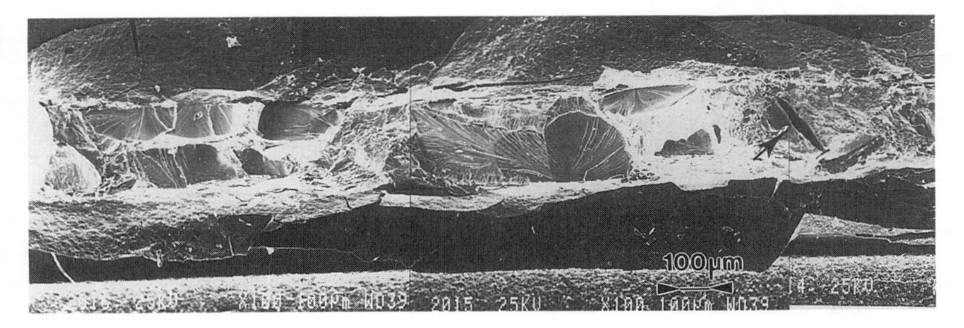

Figure 7: Fracture surface of diffusion bonded Nb(Si) layer tested in bending at 298 K (two-thirds of specimen width is shown) demonstrating mixed-mode failure (cleavage and ductile). As in Fig. 6, arrested cleavage microcracks (arrowed) are also present. [13]

fracture mode existing of 40% cleavage and 60% ductile fracture. The root cause to the dual fracture mode was investigated by examining the interfaces (bonded versus debonded) and local Nb(Si) fracture behavior using stereoimaging. The results of the analysis are presented in Table 5. In general, it was found[13] that at interfacial areas which debonded, whether on the notch- or bottom-side, the Nb(Si) had a greater tendency fail in a ductile manner locally (i.e. 68 and 72% measured), whereas next to well-bonded regions, the Nb(Si) tended to cleave (82 and 96% measured). These results demonstrate the controlling effects of constraint (Figure 2(b)) on the brittle-ductile transition in BCC containing systems. Overall, the lamination process has lead to constraints which have aided a brittle-ductile transition because Nb(Si) at 298 K failed by 100% ductile rupture.[13] Another diffusion bonded laminate tested at 77 K failed entirely by cleavage at 6.2 MPa√m, only a small drop compared to the 298 K value. As demonstrated in Figure 5 for the smooth tension samples, the toughness (or area below the curve) can remain high even at low temperatures where the primary failure mode is cleavage. More recent work to measure fracture toughness has confirmed this finding.[22]

Table 5: Analysis of interface/Nb fracture.[13]

Location/Interfacial Observations		% Debonded
Notch Side , Bottom-Side		63 , 36
		Cleavage [%]
Bonded	Notch , Bottom	82 , 96
Debonded	Notch , Bottom	32 , 28

The adhesively bonded samples constructed with AR and VHT foils fractured at 7.1 and 7.8 MPa√m, respectively, calculated using the peak loads experienced in the 298 K experiments.[13] The fracture process was characterized by the propagation of a single crack from the notch to the interface, interface debonding, fracture reinitiation in the Nb_5Si_3 layer below the Nb layer, and, finally, plastic rupture of the Nb layer.[13] The fine-grained AR layers showed homogeneous deformation, but the coarser-grained VHT foils deformed inhomogeneously.[13,14] Although the fracture behavior of the diffusion bonded and adhesively bonded samples were distinctly different, the toughness in laminates with constrained Nb(Si) were on average higher as expected from the ranking of tensile fracture energies. As suggested earlier, it appears that the "toughness" of the laminate system is controlled by the behavior of the "ductile" Nb layers. The mere appearance of cleavage is not necessarily an indicator of low toughness, provided that large amounts of energy are absorbed during the deformation and fracture process.

Fracture of Nb-10 at.% Si In Situ *Composites: SEM Observation of Crack Growth*

The typical range of R-curve behavior demonstrated by Nb-10 at.% Si samples are shown in Figure 8. Crack initiation in the composites occurred on the bend bar surfaces from 5 to 20 MPa√m depending on the microstructure near the notch tip. Crack initiation was followed by development of a microcracking "damage zone" and ligament bridge formation requiring higher applied stress intensities to propagate cracks. Although some scatter is noted in the R-curves behavior due to slight microstructural variations in the samples tested, an average steady state peak stress intensity value of 28 MPa√m was found at crack extension less than 400 μm. A view of the bridging ligaments and the microcrack damage zone leading to the R-curve behavior is pictured in Figure 9. The microcracking of Nb_5Si_3 (darker contrast phase) and plasticity of the Nb_p (lighter contrast) are clearly visible. The fracture surfaces in the regions of the rising R-curves typically showed extensive plastic stretching and dimpled fracture of Nb_p as well as evidence of interfacial debonding (Figure 10).

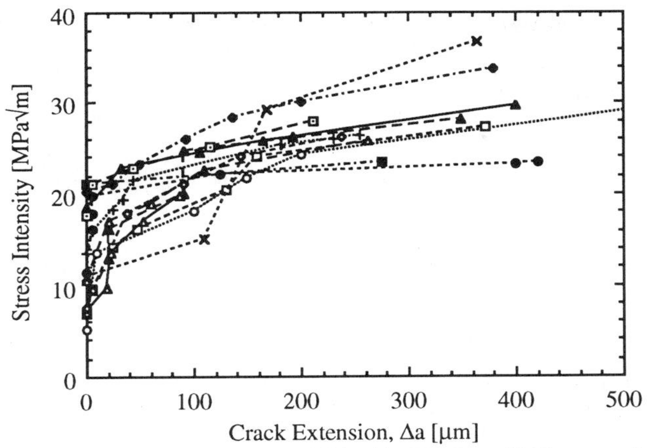

Figure 8: Resistance-curve behavior exhibited by Nb-10 at.% Si composites.[12]

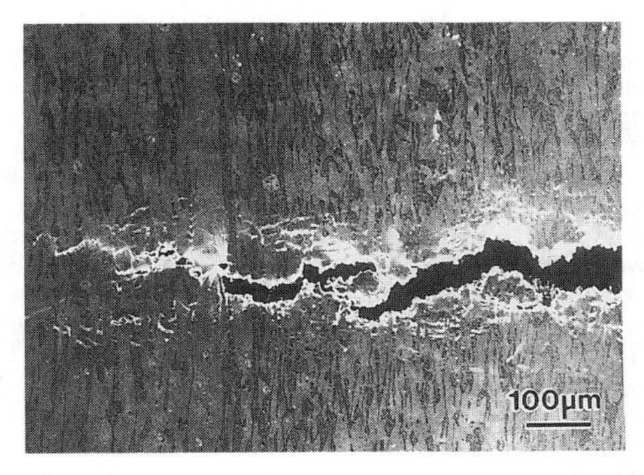

Figure 9: Photomicrograph taken of a well-developed crack in a composite. The large Nb_p phase (light contrast) bridges the crack. A bridged-zone length of 300 to 400 μm is seen.[12]

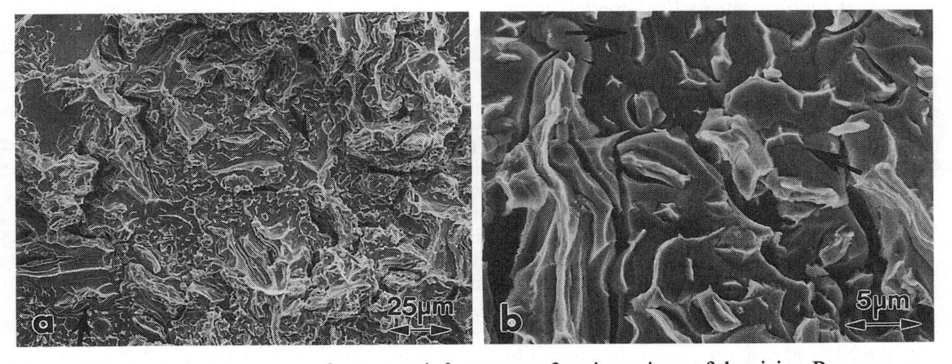

Figure 10: Typical appearance of near notch fracture surface in regions of the rising R-curve (within 300 to 500 μm). (a) Ductile stretching of Nb_p (arrowed) accompanied by some debonding is generally observed. (b) The small Nb_s also fails by ductile rupture (arrowed).[12]

The Nb_p and Nb_s phases provided for R-curve behavior and peak stress intensities exceeding those determined previously for monolithic Nb_5Si_3 (i.e., 1 to 3 MPa$\sqrt{m}$).[4,6,8] The fracture behavior measured was compared to the predictions of toughening models presented earlier. Equation {1} was inverted to estimate the work of rupture (χ) for the Nb phases from bulk properties. After substituting an expression for strain energy release rate, the equation takes the following form:

$$\chi = 1/V_f\sigma_y a_o\left[\left(1-\upsilon^2\right)K^2/E-\left(1-V_f\right)\left(1-\upsilon_m^2\right)K_m^2/E_m\right] \qquad \{3\}$$

The steady-state toughnesses of the composite (K) and matrix (K_m), elastic moduli of the composite (E) and matrix (E_m), and Poisson's ratios for the composite (ν) and matrix (ν_m) were incorporated. Nb_p phases were assumed to provide the majority of fracture resistance, therefore, microstructural quantities, a_o and V_f, were given values of 17 μm and 51 vol.%, respectively. The yield strength (σ_y) for Nb_p phases at low strain rates at 298 K was estimated to be 360 MPa from compression[12] and tension tests[19] with identically processed Nb(Si). The values for K, E, E_m, ν_m and ν_m were 28.4 MPa$\sqrt{m}$, 139 GPa [12], 327 GPa,[4,8,23] 0.2 and 0.35, respectively. Inserting these constants into the expression lead to a calculated χ-value of 1.62; similar values were calculated for other cast and extruded Nb-Si materials.[4,23]

This χ-value suggests that the Nb_p in the bridged-zones deforms under high constraint. Deformation studies of Pb constrained by glass capillaries[11] found χ-values ranging from 1.6 to 6, the former resulting when the Pb was unable to debond from the elastic matrix. As a consequence, flow stresses in the Pb were nearly six times the unconstrained uniaxial yield stresses ($6\cdot\sigma_y$), implying that in the Nb ligaments in the current material could experience (6·360 MPa) or 2160 MPa. Fractography clearly showed that even at these high stress levels, the Nb_p (and Nb_s) were able to remain ductile in the region of the rising R-curve at slow displacement rates. Experiments to be presented in the following section show that higher strain rates and lower temperatures can augment flow stresses to a point that cleavage fracture takes place without necessarily lowering the measured toughnesses.

Fracture of Nb-10 at.% Si In Situ *Composites: Rate and Low Temperature Effects*

The fracture toughnesses determined over six orders of magnitude in loading rate are shown (Figure 11) to be independent of testing rate with an average value of 24.1± 2.2 MPa$\sqrt{m}$. The fracture surfaces created were macroscopically flat. Higher magnification examination, Figure 12, revealed a consistent behavior in the fracture characteristics of the Nb_p: ductile failure in the notched regions and an increasing percentage of cleavage at the sample mid-plane. The fracture behavior of the Nb_p ligaments was quantified from the notch to the back face along the plane strain regions. Figure 13 compares the results from a slow rate test (4 N·sec^{-1}) to that for one tested at a higher rate ($3\cdot10^4$ N·sec^{-1}). In the slower rate sample, the area fraction of cleaved Nb_p increased steadily from 8% near the notch to a peak of 42% half-way along the fracture surface then decreased toward the back face. On the specimen broken at the higher rate, Nb_p had a greater tendency to cleave over the entire surface without an accompanied change in the fracture toughness. It is readily apparent that the testing rate has lead to a substantial difference in the fracture behavior without altering the measured toughness values. On the other hand, examination of Nb_s ligament fracture revealed that they always exhibited ductile rupture in all locations of the fracture surfaces, even to the highest rates tested here. The variation in Nb_p ligament fracture behavior on a single fracture surface is believed to result from a change in the crack tip velocity. In fact, rate of fracture estimates[24] versus a/W and with initial loading rate consistently found this and a similar variation in fracture behavior with the quantified fracture behavior. Higher velocity cracks resulted in a higher percentage of cleavage fracture due to high strain rates and slower cracks similarly allowed plastic rupture of the ligaments.

Toughness tests at 77 K were performed to raise the yield strengths and to induce cleavage of the Nb_p. Measurements made over a similar range of loading rates as those at 298 K are compared in Figure 11. The toughness values again were independent of testing rate and the average value (23.6 MPa$\sqrt{m}$) essentially the same as that measured at 298 K. The load-time

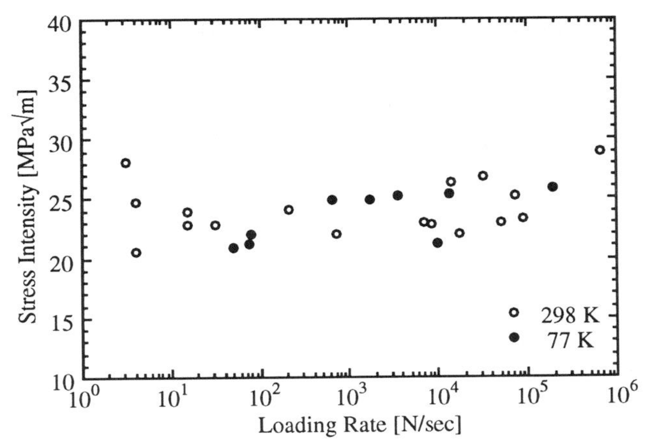

Figure 11: Fracture toughness versus loading rate for Nb-10 at.% Si materials at 298 K and 77 K. The fracture toughness is unaffected by loading rate and temperature.[12]

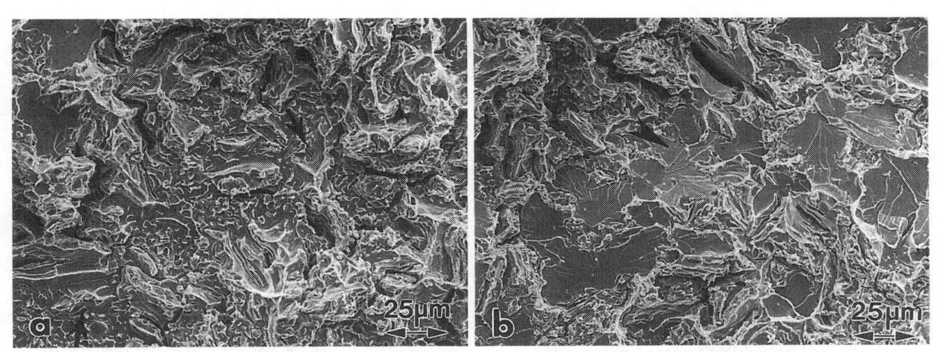

Figure 12: Comparison of (a) near-notch and (b) near-center regions of a sample fractured at 298 K and slow rates. A greater fraction of Nb_p phases are observed to cleave in (b).[12]

test traces indicate that the load drops were severe and unstable fracture took place after the initiation of a crack at the peak load. A higher percentage of cleavage is observed in the notch region of specimen tested at 77 K as compared to the 298 K tests. Unlike at 298 K, the fracture behavior of Nb_p did not change dramatically with position along the fracture surface. Qualitative examination showed that the Nb_s did not appear to cleave at any fracture surface location. Because composition and constraint are expected to be the same for Nb_p and Nb_s phases, the different fracture behavior is possibly due to grain size effect controlling cleavage fracture stresses. This will be addressed in a later section.

Little change in fracture toughness at higher testing rates or lower temperatures was detected although a more "brittle" fracture mode became more dominant in regions near the notch tip. According to expression {1}, the energy absorbed per volume in deforming ligaments is the product of the ligament yield strength (σ_y) and the normalized work of rupture (χ), or ($\sigma_y \cdot \chi$). This ligament "toughness" controls the composite fracture properties. The current results

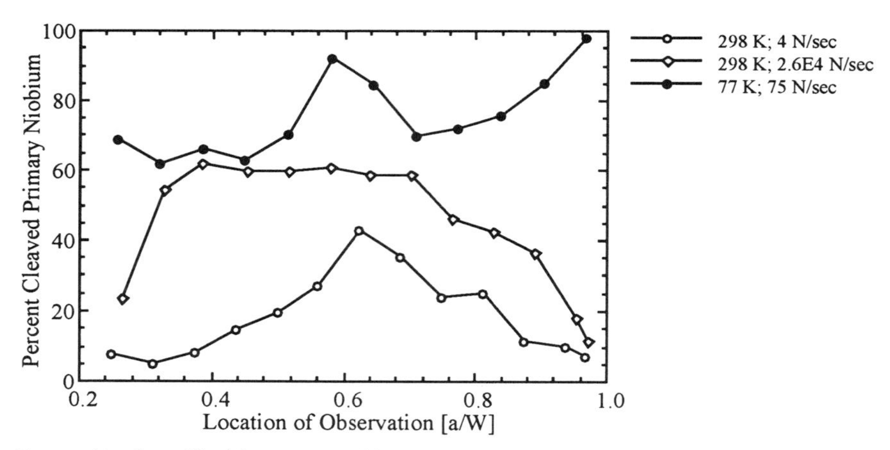

Figure 13: Quantified fracture transition behavior of Nb_p with crack tip position on larger specimens tested at 298 K (two initial loading rates) compared to one tested at 77 K.[12]

suggest that altering experimental conditions to increase yield strength and to promote fracture transitions did not change the "toughness" of the ligaments. The increase in yield strength must be accompanied by a decrease in "work of rupture".[12] Through the strain rate dependence on yielding of pure Nb[25], the strengthening effects provided by Si solid solution,[6,12,13,19] and yield strength measurements for Nb(Si) at 77 K,[12,19] the σ_y values for Nb_p at high strain rates and low temperatures can be estimated to be 560 and 870 MPa,[12] respectively. The flow stresses could increase from 2160 MPa (298 K, low rates) to 3360 MPa (highest rate, 298 K) to 5220 MPa (77 K, low rate) in absence of cleavage fracture. As suggested above, the χ–value at the highest rate at 298 K and the nominal rate at 77 K fell to 0.64 and 0.40 of the slow strain rate, 298 K value, respectively. This decrease is consistent with previous work on pure Nb[26] that demonstrated a decrease in uniform strain at higher rates and lower temperatures (as yield stress increased). The appearance of lower ductility cleavage fracture of Nb_p on the composite fracture surfaces would provide another means for lower χ-values.

The calculations clearly show that χ-values do not need to be large for this composite to remain tough because the yield and cleavage fracture strengths are high values.[12] In reviewing previously published work detailing uniaxial stress-strain curves for Nb and Nb-alloys, and constrained fracture experiments on pure Nb and Nb(Si), the toughness Nb_p is not expected to change dramatically over the range of testing conditions. The measured areas[12] below stress-strain curves from Sargent's work on pure Nb[26] over a range of strain rates are consistent with the temperature effects already presented from Figure 5 for pure Nb and Nb(Si). Other work[17] on pure Nb at 298 K, 196 K and 77 K, showed that fine and coarse pure Nb maintained a high "toughness" value to the lowest temperatures, even when a transition to cleavage fracture occurred. In addition, notched tensile studies[27-29] showed a lack of notch sensitivity to low temperatures. Toughness estimates[12] from notched tensile experiments on Nb(Si) materials indicated toughness values above 20 MPa$\sqrt{m}$ even at 77 K. Recent work[22] has shown this conclusively on precracked fracture toughness samples.

The true cleavage fracture stress (σ_F) for pure and alloyed Nb (Nb(Si) and Nb(Zr)) were previously measured[12] to determine the temperature and solute dependence on σ_F and to predict the stress required to cleave Nb_p and Nb_s phases. The results of this study demonstrated temperature independence from 20 to 148 K for 135 μm grain sized materials and a lack of influence of small additions of Si or Zr on effective surface energies compared to pure Nb.

These results, and estimates from smooth tension[17] and more current notched bend experiments[19,22] provide thirty σ_F measurements to determine the effects of grain sizes (ranging from 48 to 310 μm). The trends compared well with Cottrell's predictions for grain size (d) dependence on tensile fracture stress in body-centered cubic metals[21] where $\sigma_F \propto d^{-0.5}$. A linear regression fit through the data produced the following relationship:

$$\sigma_F = 543 + 7656 \cdot (d[\mu m])^{-0.5} \qquad \{4\}$$

Based on this relationship, the cleavage fracture stress for the Nb_p ligaments (12.7 μm grain size) would be estimated to be 2700 MPa and smaller Nb_s (2.7 μm) would cleave at 5200 MPa.

The fracture experiments produced a significant variation in failure mode of Nb_p on single fracture surface, or when comparing to those produced at other rates or temperatures. The constrained flow stress for Nb_p was discussed earlier to be 2160 MPa at 298 K under slow strain rates, however, fracture was primarily ductile because the cleavage fracture stress was higher, at about 2700 MPa. At higher strain rates or low temperatures, the flow stress is augmented since and the appearance of cleavage becomes more dominant. Four orders magnitude change in strain rate[12] can provide the required increase in yield stress to cause constrained flow stresses to reach σ_F. At 77 K, $6 \cdot \sigma_y$ was 5220 MPa causing an expected transition for 60% to 100% of the Nb_p ligaments in Figure 14. Since the cleavage fracture stress of the smaller Nb_s particles was predicted to be 5200 MPa, it is clear that although they did not exhibit cleavage, they must be on the verge of doing so.

Fracture of DS Nb-Si and Nb-Si-Ti In Situ *Composites*

The fracture toughness of the binary and ternary materials from 77 to 773 K are plotted as separate bar charts in Figure 14 as a function of their metallic phase content. As in the case for Nb-10 at.% Si composites, there was not a significant influence of temperature on the fracture toughnesses of either alloy type. The variation in room temperature fracture toughness of the binary alloys was found to be linear with total vol.% Nb(Si),[15] an unexpected result when considering the that equation {1} suggests $K \propto \sqrt{V_f}$. The deviation from model behavior possibly results from other contributions from microcracking, crack deflections and phase size (a_o in the equation), especially in alloys containing both Nb_s and Nb_p phases. In the alloys with 26.1 and 16.4 vol.% Nb(Si) (hypereutectic compositions) it was proposed that blocky Nb_3Si phases in the microstructures promote crack deflection processes which enhance toughness.

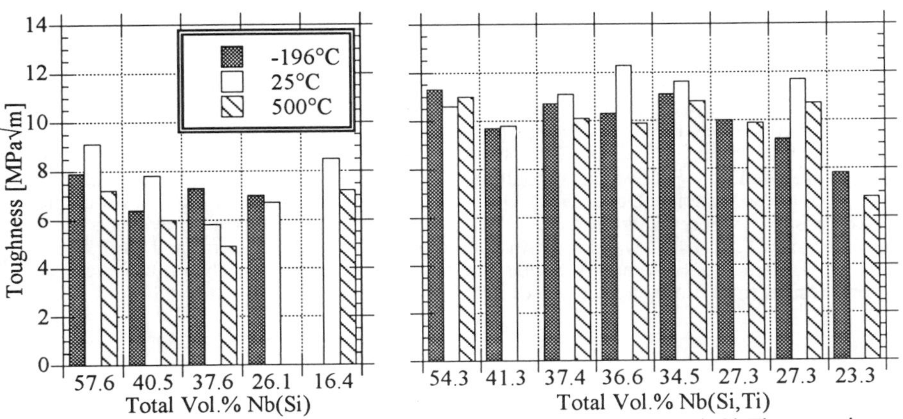

Figure 14: Temperature effects on fracture toughness of DS Nb-Si and Nb-Si-Ti composites plotted relative to the total vol.% of Nb(Si,X) phases within the structure.

Although processes in addition to ductile-phase toughening may be operating, it is clear that changes in temperature should only affect the behavior of the metallic phase or matrix/particle interface as the testing temperatures were well below the brittle-ductile transition for the intermetallic. At the lowest temperatures, the changes in Nb(Si) behavior was characterized by cleavage fracture of larger Nb_p, however, the smaller Nb_s phases exhibited ductile fracture. In the elevated temperature tests, both phases, when present, failed by ductile rupture. The fracture surface and toughness observations are consistent with the conclusions drawn from the work on Nb-10 at.% Si on the phase grain size dependence on cleavage fracture stress and consistency in Nb(Si) toughness over the range of conditions.

The fracture toughness of the ternary Nb-Si-Ti alloys were observed to exceed those values for binary materials of approximately the same vol.% of ductile constituent. This comparison may be somewhat misleading, since the vol.% in Nb(Si) represents phases of two sizes (Nb_p and Nb_s) whereas the Nb(Si,Ti) were approximately the same size in all composites. On all fracture surfaces produced from the various compositions and at 77 and 773 K, the Nb(Si,Ti) phases ($\approx$10 μm in size) were observed to fail in a ductile manner. Based on previous work[30] on Nb-Ti alloys, the 29-44 at.% Ti in the Nb(Si,Ti) phases may have lead to additional solid solution effects (along with that provided by Si) to cause significant increases in yield strengths. In such a case, the higher yield stresses and absence of cleavage fracture have increased fracture toughnesses. These alloys again demonstrate the controlling effects of Nb phase toughness on the resultant composite fracture behavior.

CONCLUSIONS

(1) Processing was observed to affect the mechanical properties of the Nb foils in the laminates through modifications of the interfacial strength (and constraint on the deforming layers), grain size and solid solution content.

(2) Constraint, strain rate and temperature control the fracture behavior of the Nb(Si,X) phases or layers, while grain size and yield strength control the brittle-ductile transition appearance of the phases.

(3) The fracture toughness of the laminates and *in situ* composites is controlled by the toughness of the Nb(Si,X) phases. Ductile phases that exhibit high yield strengths and low failure strains (even as a result of cleavage fracture) whether due to constraint, elevated strain rates or low temperatures, can provide the same level of toughening under conditions where yield strengths are lower and ductilities are greater.

ACKNOWLEDGMENTS

The author would like to thank Dr. Mendiratta (UES Inc.) and Dr. Dimiduk (WPAFB) for the Nb-10 at.% Si materials and Dr. Bewlay (GE, CR&D) for the DS composites examined in this study. The work performed was supported by Air Force Office of Scientific Research under contract numbers F49620-88-0053 (during Graduate Student Research Program, Summer 1990) and 89-0508 (at CWRU, Dr. A. Rosenstein contract monitor). Two years of support was also provided by a Fellowship through MTS Inc. (Minneapolis, MN). The DS materials were tested at the Materials Department, Oxford University (England) during a post-doctoral research position supported by the SERC. Finally, the author gratefully acknowledges the contributions of Prof. J.J. Lewandowski in numerous insightful discussions, and Dr. P.M. Singh, Dr. J. Kajuch, J. Short and A. Samant for allowing presentation of their mechanical property results from niobium foils and tri-layered laminates.

LITERATURE CITED

1. R.L. Fleischer: *J. Mater. Sci.*, 1987, vol. 22, pp. 2281-8.
2. P.R. Subramanian, M.G. Mendiratta, and D.M. Dimiduk: *JOM*, 1996, vol. 48, pp. 33-8.
3. D.M. Dimiduk, P.R. Subramanian, and M.G. Mendiratta: *Acta Metall. Sinica*, 1995, vol. 8, pp. 519-30.
4. M.G. Mendiratta, J.J. Lewandowski, and D.M. Dimiduk: *Metall. Trans. A*, 1991, vol. 22A, pp. 1573-83.
5. R.M. Nekkanti and D.M. Dimiduk: *Intermetallic Matrix Composites, Mater. Res. Soc. Symp. Proc.*, vol. 194, Materials Research Society, Pittsburgh, PA, 1990, pp. 175.
6. J. Kajuch, J.D. Rigney, and J.J. Lewandowski: *Mater. Sci. Eng. A*, 1992, vol. A155, pp. 59-65.
7. J.J. Lewandowski, D.M. Dimiduk, W. Kerr, and M.G. Mendiratta: *High Temperature Composite Materials, Mater. Res. Soc. Symp. Proc*, vol. 120, Materials Research Society, Pittsburgh, PA, 1989, pp. 103.
8. M.G. Mendiratta and D.M. Dimiduk: *Scr. Metall. Mater.*, 1991, vol. 25, pp. 237-42.
9. P.A. Mataga: *Acta Metall.*, 1989, vol. 37, pp. 3349-59.
10. V.D. Krstic, P.S. Nicholson, and R.G. Hoagland: *J. Am. Ceram. Soc.*, 1981, vol. 64, 499.
11. M.F. Ashby, F.J. Blunt, and M. Bannister: *Acta Metall.*, 1989, vol. 37, pp. 1847-57.
12. J.D. Rigney and J.J. Lewandowski: *Metall. Mater. Trans. A*, 1996, in press.
13. J. Kajuch, J. Short, and J.J. Lewandowski: *Acta Metall. Mater.*, 1995, vol. 43, pp. 1955-67.
14. J.A. Short: Master's Thesis, Case Western Reserve University, Cleveland, OH, 1994.
15. B.P. Bewlay, M.R. Jackson, H.A. Lipsitt, W.J. Reeder, and J.A. Sutcliff: 1996 AFOSR Report F49620-93-C-0007. General Electric Corporate R&D, Schnectady, NY, 1996.
16. Standard E-399. *Annual Book of ASTM Standards*. Vol. 03.01. American Society for Testing and Materials, Philadelphia, PA (1988), p. 480.
17. M.A. Adams, A.C. Roberts, and R.E. Smallman: *Acta Metall.*, 1960, vol. 8, pp. 328-37.
18. A.A. Johnson: *Acta Metall.*, 1960, vol. 8, pp. 737-40.
19. M.G. Mendiratta, R. Goetz, D.M. Dimiduk, and J.J. Lewandowski: *Metall. Mater. Trans. A*, 1995, vol. 26A, pp. 1767-77.
20. J.F. Knott: *Fundamentals of Fracture Mechanics*, Butterworths: London, England, 1973.
21. A.H. Cottrell: *Trans. Metall. Soc. AIME*, 1958, vol. 212, pp. 192-202.
22. A.V. Samant and J.J. Lewandowski: *Metall. Mater. Trans. A*, 1996, in press.
23. M.G. Mendiratta and D.M. Dimiduk: *Metall. Trans. A*, 1993, vol. 24A, pp. 501-4.
24. J.D. Rigney: Ph.D. Thesis, Case Western Reserve University, Cleveland, OH, 1994.
25. T.L. Briggs and J.D. Campbell: *Acta Metall.*, 1972, vol. 20, pp. 711-24.
26. G.A. Sargent: *Acta Metall.*, 1965, vol. 13, pp. 663-71.
27. A.G. Imgram, F.C. Holden, H.R. Ogden, and R.I. Jaffee: *Trans. Metall. Soc. AIME*, 1961, vol. 221, pp. 517-26.
28. A.G. Imgram, E.S. Bartlett, and H.R. Ogden: *Trans. Metall. Soc. AIME*, 1963, vol. 227, pp. 131-6.
29. A.L. Mincher and W.F. Sheeley: *Trans. Metall. Soc. AIME*, 1961, vol. 221, pp. 19-25.
30. R.T. Begley and J.H. Bechtold: *J. Less-Common Met.*, 1961, vol. 3, pp. 1-12.

TOUGHNESS AND SUBCRITICAL CRACK GROWTH IN Nb/Nb_3AL LAYERED MATERIALS

D. R. BLOYER, K. T. VENKATESWARA RAO, and R. O. RITCHIE
Department of Materials Science and Mineral Engineering
University of California, Berkeley, CA 94720-1760

ABSTRACT

A brittle intermetallic, Nb_3Al, reinforced with a ductile metal, Nb, has been used to investigate the resistance curve and cyclic fatigue behavior of a relatively coarse laminated composite. With this system, the toughness of Nb_3Al was found to increase from ~1 MPa√m to well over 20 MPa√m after several millimeters of stable crack growth; this was attributed to extensive crack bridging and plastic deformation within the Nb layers in the crack wake. Cyclic fatigue-crack growth resistance was also improved in the laminate microstructures compared to pure Nb_3Al and Nb-particulate reinforced Nb_3Al composites with crack arrester orientations in the laminate providing better fatigue resistance than either the matrix or pure Nb.

INTRODUCTION

Intermetallic materials are currently being developed as potential replacements for superalloys in high performance engines [1-3]. However, due to their brittle nature, extrinsic toughening techniques that invoke crack-tip shielding mechanisms [4,5] have often been used to develop alloys and microstructures with improved fracture resistance, specifically by incorporating fibers, particulates, or laminates as reinforcement phases [1-3,6-25]. In the present study, we examine the fracture and fatigue behavior in a model Nb-reinforced Nb_3Al composite, where toughening is achieved by the addition of ductile Nb layers having layer dimensions in the hundreds of microns. Results are compared with earlier studies which involved Nb/Nb_3Al composites with Nb as ~20 μm particulate reinforcement formed *in situ* by powder metallurgy [6,7], or as ~1-2 μm thick layers magnetron sputtered to form a microlaminate [8,9].

EXPERIMENTAL PROCEDURES

Nb_3Al powder was cold pressed between 125 μm thick Nb foils to yield ~20 vol.% ductile reinforced laminates. These were hot pressed in an argon atmosphere at 1650-1680°C under 37 MPa pressure for 25-40 min to give dense (>98% of theoretical density) composite cylinders. The resultant microstructure consisted of evenly spaced 500 μm thick layers of Nb_3Al separated by 125 μm thick Nb layers (Fig. 1). An interfacial reaction zone, ~20-30 μm in thickness, formed between Nb and Nb_3Al during processing at 1680°C, which resulted in ~30-40% reduction in the Nb layer thickness.

Two orientations were investigated in this study, namely "crack arrester" (0° or C-L), where the crack grows perpendicular to, yet sequentially through, the layers; and the "crack divider" (C-R) where the crack plane is normal to the plane of layers, but the crack advances through all the layers simultaneously. Single-edged notched bend (SEN(B)) beams (with spans ~35-40 mm, thickness B = 3.5 mm and width W = 12.5 mm) were used to study the resistance curve (R-curve) behavior of the arrester orientation; compact tension C(T) specimens (with B = 3.5 mm, W = 25.4 mm) were used for fatigue testing. Similar C(T) specimens were used for both fracture and fatigue testing in the divider orientation.

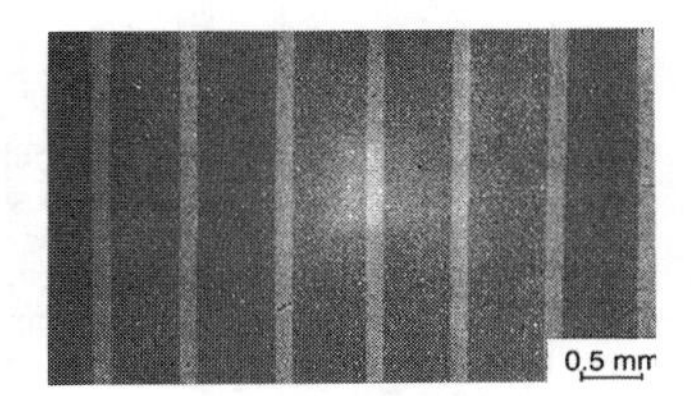

Fig. 1. Coarse-scale Nb/Nb_3Al laminate microstructure after hot pressing.

Fatigue-crack growth tests were performed at 25 Hz using R = 0.1 with crack lengths being monitored using back-face strain measurements supplemented by direct

Mat. Res. Soc. Symp. Proc. Vol. 434 © 1996 Materials Research Society

observation with an optical telescope. Divider specimens were tested under decreasing stress-intensity range (ΔK) loading. Arrester specimens were loaded in a series of constant ΔK experiments to propagate the crack across 2 to 3 layers; the average growth rate across the matrix and Nb is reported. R-curves were subsequently determined by manually loading pre-cracked specimens under displacement control. Scanning electron microscopy (SEM) was used to examine all crack profiles and fracture surfaces.

RESULTS AND DISCUSSION

Fracture Toughness and R-Curve Behavior

Results plotted in Fig. 2 illustrate that the addition of 125 µm thick layers of Nb to Nb_3Al leads to toughnesses exceeding 15-20 MPa√m compared to the intrinsic toughness of only ~1 MPa√m for Nb_3Al (Fig. 2a). Such toughening is achieved in both C-L and C-R orientations. Even though only 20 vol% Nb was employed, these laminates exhibit markedly higher toughness than that reported for *in situ* Nb/Nb_3Al composites [7] that contain 40 vol% of ~20 µm Nb particulates, or microlaminates that contain 50 vol% of ~1-2 µm thick Nb layers (Fig. 2b). For these latter materials, crack growth initiates in microlaminates at about 6 MPa√m and increases to a maximum (steady-state) value of ~10 MPa√m after ~200 µm of crack extension; in the particulate Nb/Nb_3Al composite, the initiation and steady-state toughnesses are respectively ~1 and 6 MPa√m, [7]. In contrast, current results on the coarser Nb/Nb_3Al laminates do not show evidence of crack growth until ~9 MPa√m, and in one case cracks continue to grow stably at stress intensities approaching 70 MPa√m. This implies that oriented, high-aspect ratio reinforcements in the form of fibers and particularly laminates can impart acceptable damage tolerance to brittle materials, as also reported for Nb/$MoSi_2$ composites [12-14]. Furthermore, it is clear that coarse-scale (i.e., hundreds of microns) layer reinforcements are more potent than fine-scale (1-2 µm) layers in enhancing the fracture resistance of Nb_3Al-matrix composites.

As discussed below, mechanistically the elevation in crack-initiation toughness in the coarse-scale laminates (Fig. 2) can be related to crack renucleation across the Nb layer; thereafter, crack growth toughness on the R-curve is associated with crack bridging and plastic deformation in the Nb. The rapid increase in toughness at larger crack extensions, e.g., at stress intensities above 30 MPa√m, is caused by large-scale bridging, where the size of the crack-wake shielding zone becomes comparable to the crack size and specimen dimensions. In this regime, the toughness becomes highly sensitive to specimen geometry.

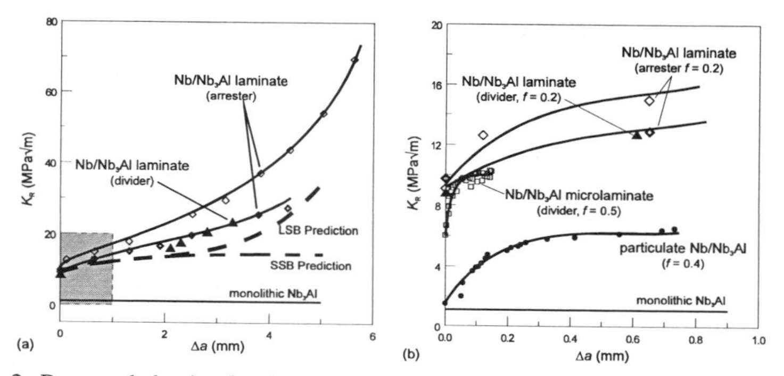

Fig. 2. R-curve behavior for the coarse-scale Nb/Nb_3Al laminates compared to (a) monolithic Nb_3Al, (b) *in situ* Nb_3Al composites reinforced with Nb particulate [7] and Nb/Nb_3Al microlaminates [9]. Predictions for the R-curve behavior of the Nb/Nb_3Al laminates, based on small-scale bridging (SSB) and large-scale bridging (LSB) models, are also shown in (a). Shaded region in (a) is magnified in (b).

Crack/Layer Interactions

Extensive metallography showed that the development of the R-curve is associated with significant crack-reinforcement interactions, involving crack arrest at the Nb layer and crack renucleation ahead of the layer [10,11]. Typically the crack impinges on a reinforcing layer, blunts, and renucleates across the layer with increasing applied stress intensity. This process recurs as the crack advances across several Nb layers leaving them intact, which in turn leads to the formation of large (~3-6 mm) bridging zones in the crack wake. Crack branching on one or both sides of the Nb layer commonly occurred in the matrix near the Nb/Nb_3Al interface, but not at the interface. These branches then linked up to form a single dominant crack as the crack progressed across the specimen. Such near interfacial cracking resulted in "effective debonding" at the Nb/Nb_3Al interface; thus, relieving constraint and promoting ductile failure in the Nb layer [15,16]. Where near interfacial cracking occurred and constraint was relieved, the Nb layer failed by microvoid coalescence. Conversely, when near interfacial cracking was not observed, the constraint imposed on the Nb layer appeared to promote cleavage fracture.

Models for Toughening

Most models for ductile-phase toughening pertain to small-scale bridging conditions, where the bridging zone is small relative to the crack length and remaining uncracked ligament in the sample. However, as large-scale bridging conditions prevailed for these laminates, both small-scale and large-scale bridging models are used below to evaluate the role of layered Nb reinforcements on the fracture toughness of Nb/Nb_3Al laminates.

Small-scale bridging: Bridging models typically require two experimental parameters, the stress-displacement function, $\sigma_c(u)$, for the constrained reinforcement in the matrix and the critical displacement, u_c, at the failure of this reinforcement to estimate the nondimensional work of rupture, χ. The reinforcement toughening contribution, $\Delta G_c = f\sigma_o t\chi$ [16], can then be superposed with the matrix fracture energy to give an expression for the composite toughness [17]:

$$K_{ss} = \sqrt{K_o^2 + E' f \sigma_o t \chi} \quad (1)$$

where K_{ss} is the steady-state (or plateau) toughness, and K_o is the crack-initiation toughness.

To evaluate Eq. 1, we note that values of χ between 1.3-2.5 [14,18] have been reported for Nb/TiAl and Nb/$MoSi_2$ laminates of similar microstructural scale; as these values were influenced by small amounts of debonding, a conservative value of $\chi \sim 1.3$ is assumed here. Taking $K_o \sim 9$ MPa$\sqrt{m}$ (Fig. 2), $f \sim 0.2$, plane strain composite modulus, $E' = 142$ GPa (assuming $E_{composite} = 129$ GPa and $\nu \sim 0.3$), flow stress of Nb, $\sigma_o \sim 245$ MPa [19], and the half thickness of the layer, $t \sim 62.5$ μm, Eq. 1 predicts a steady-state toughness of ~25 MPa$\sqrt{m}$, in reasonable agreement with the measured plateau toughness (Fig. 2a). If allowance is made for the reduction in t and f due to the presence of the reaction layer, the predicted toughness is somewhat lower (~18 MPa$\sqrt{m}$).

Large-scale bridging: An alternative model, developed for particle-reinforced composites [20,21], treats the bridges as rigid plastic springs that provide uniform tractions in the wake. The far-field stress intensity at steady state, K_{ss}, is computed by superposition of the crack-tip stress intensity, K_o, and the shielding stress intensity imparted by the tractions, ΔK_c, given as:

$$\Delta K_c = \frac{2}{\sqrt{\pi a}} \int_0^L \sigma(x)\, F\left(\frac{x}{a}, \frac{a}{W}\right) dx \quad (2)$$

where a is the crack length, x is the distance behind the crack tip, W is the specimen width, L is the bridging zone length, $\sigma(x)$ is a stress function describing the tractions in the wake, and the geometric weight function F is given in refs. [20-22].

For large-scale bridging conditions, Eq. 2 can be used directly, although predictions will be complicated by the fact that the R-curve is also dependent on specimen geometry. A rigorous modeling approach requires the use of the exact stress-displacement function ($\sigma_c(u)$) and crack-opening profile ($u(x)$) for the specific geometry [23]. However, assuming uniform tractions ($\sigma(x) = f\,\sigma_o$) and weight function, F, developed for single-edge notched samples [21,22], a simple estimate for the R-curve under large-scale bridging conditions can be obtained by integrating Eq. 2 at various crack growth increments and superimposing the toughening increment, ΔK_c, over the matrix crack-tip stress intensity K_o. Predictions for the coarse Nb/Nb_3Al laminates are shown in Fig. 2a up to a final bridging zone length of 5 mm. Although the simple model underpredicts the toughness, the calculations are consistent with the observed trend for crack-growth toughness (increasing slope and positive curvature) under large-scale bridging conditions.

Fatigue-Crack Propagation Behavior

Fatigue-crack propagation rates for the coarse Nb/Nb_3Al laminates in the divider and arrester orientations are compared to particulate-reinforced Nb_3Al in Fig. 3. The cyclic crack-growth resistance of Nb_3Al is clearly improved by ductile reinforcement, but to a lesser degree than that observed for the toughness. The incorporation of Nb particulates by *in situ* processing [6] raises the fatigue threshold from ~1 to ~2 MPa√m; this is further increased to above 3 MPa√m with Nb layers in the divider orientation. As the crack is continuously exposed to the intermetallic and ductile phase in both these composites, their crack-growth curves would be expected to lie between those of the unreinforced matrix and of the pure reinforcement. The higher fatigue threshold of the divider laminate reveals the additional benefit of the high aspect ratio of the reinforcement. Despite the increase in fatigue threshold, the improvement in fatigue resistance is not as large compared to the toughening increment of a factor of 6 to 20 or more seen under monotonic loading. This is consistent with behavior in other ductile-particle reinforced composite systems [14,24], and is generally ascribed to a fatigue mechanism that degrades the shielding ability of the reinforcement. Similarly, it has been observed that shielding from Nb-layer bridging in the crack wake is reduced under cyclic loads due to fatigue failure of the Nb bridging ligaments [6,14,17,24]; Fig. 4a clearly depicts the failure of the reinforcing layers under fatigue loading with little deformation compared to the gross plasticity in the layer during quasi-static fracture [6,24].

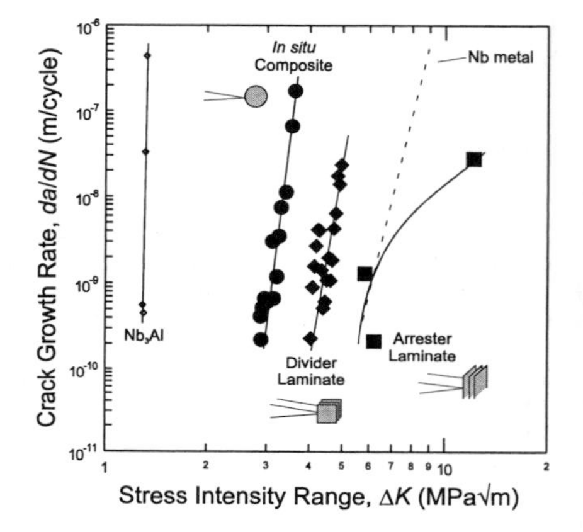

Fig. 3. Fatigue-crack growth behavior of the coarse-scale Nb/Nb_3Al laminates compared to unreinforced Nb_3Al and *in situ* Nb-particulate reinforced Nb_3Al composites [6,12].

Fig. 3 also demonstrates the effect of layer orientation on the fatigue properties of the laminate. The arrester orientation, where the crack tip resides entirely in either the matrix or the reinforcement during growth, shows improved fatigue resistance relative to the particulate and divider composites which have a more continuous matrix. More importantly, growth rates are significantly slower compared to both the Nb and Nb_3Al constituents, particularly at higher applied ΔK levels. At lower ΔK values, growth rates are comparable to rates in monolithic Nb since the maximum stress intensity, K_{max}, is not large enough to reinitiate the crack across the reinforcing layer (Fig. 4a); the crack thus remains trapped in the layer where it can induce fatigue damage and reinforcement failure before appreciable crack bridging can develop. However, at higher applied ΔK values where K_{max} approaches the crack-initiation toughness of the composite, the crack now reinitiates across the layer (Fig. 4b) leaving an intact bridge that contributes to crack-tip shielding. Thus, the arrester orientation is able to utilize similar crack-tip shielding mechanisms as those observed under monotonic loading, although to a lesser degree, leading to significant improvements in the fatigue-crack growth properties of Nb/Nb_3Al composites.

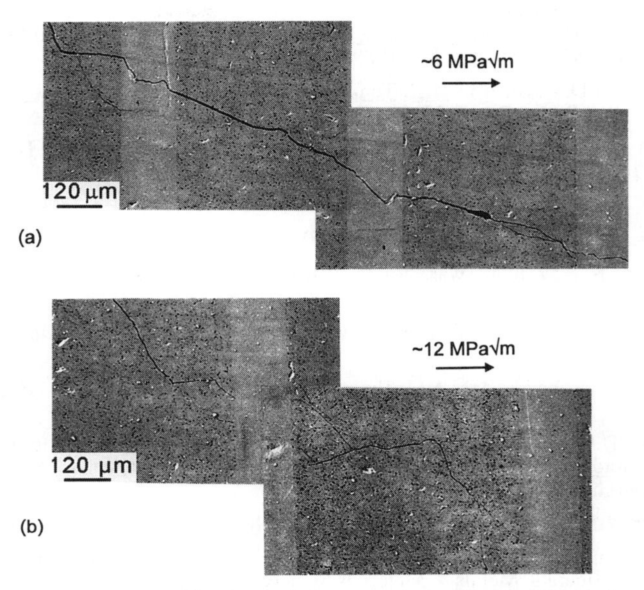

Fig. 4. Crack-path profile for a fatigue crack in the Nb/Nb_3Al (crack-arrester) laminate at (a) $\Delta K = 6$ and (b) $\Delta K = 12$ MPa√m. Arrow indicates direction of crack propagation.

CONCLUSIONS

Based on a study of the fracture toughness and resistance-curve behavior of ductile Nb-reinforced Nb_3Al intermetallic-matrix laminates, the following conclusions can be made:

1. The incorporation of high aspect ratio, coarse-scale, ductile reinforcements in the form of laminates provides significant toughening in Nb_3Al. With 125 µm thick Nb layers, crack-initiation toughnesses of ~9 MPa√m with steady-state toughnesses in excess of 20 MPa√m have been achieved, independent of orientation; this is to be compared to an intrinsic toughness of ~1 MPa√m in Nb_3Al. In the arrester orientation such toughening is associated with crack arrest, crack blunting and crack renucleation across the Nb layers; subsequently, resulting in crack bridging by intact layers.

2. Under cyclic fatigue loading the enhanced toughening in such laminated composites can be retained, although to a lesser degree; properties, however, become markedly dependent upon orientation. Any degradation in toughening under cyclic loads results from the premature fatigue failure of the Nb phase.

ACKNOWLEDGMENTS

This work was funded by the Air Force Office of Scientific Research under the AASERT Program (Grant # F49620-93-1-0441) as a supplement to Grant # F49620-93-1-0107. We thank Dr. C.H.Ward for his support and Drs. B.J.Dalgleish and L.C.DeJonghe for processing help.

REFERENCES

1. R.L. Fleischer in High Temperature Ordered Intermetallic Alloys III, edited by C.T. Liu *et al.* (MRS Symp. Proc. **133**, Pittsburgh, PA, 1989), p. 305.

2. D.L. Anton and D.M. Shah in High Temperature Ordered Intermetallic Alloys III, edited by C.T. Liu *et al.* (MRS Symp. Proc. **133**, Pittsburgh, PA, 1989), p. 361.

3. D.L. Anton and D.M. Shah in Intermetallic Matrix Composites, edited by D.L. Anton, *et al.* (MRS Symp. Proc. **194**, Pittsburgh, PA, 1990), p. 45.

4. A.G. Evans, *J. Am. Ceram. Soc.* **73**, 187 (1990).

5. R.O. Ritchie, *Mater. Sci. Eng.* **A103**, 15 (1988).

6. L. Murugesh, K.T. Venkateswara Rao and R.O. Ritchie, *Scripta Metall. Mater.* **41**, 1107 (1993).

7. C.D. Bencher, A. Sakaida, K.T. Venkateswara Rao and R.O. Ritchie, *Metall. Mater. Trans. A* **26**, 2027 (1995).

8. H.C. Cao, J.P. A. Löfvander, A.G. Evans and R.G. Rowe, *Mater. Sci. Eng. A* **A185**, 87 (1994).

9. R.G. Rowe, D.W. Skelly, M. Larsen, J. Heathcote, G. Lucas and G.R. Odette in High Temperature Silicides and Refractory Metals, edited by C.L. Briant, *et al.* (MRS Symp. Proc., Vol. **322**, Pittsburgh, PA, 1994) p.461.

10. D.R. Bloyer, K.T. Venkateswara Rao and R.O. Ritchie, *Mater. Sci. Eng. A* (1996) in press.

11. D.R. Bloyer, K.T. Venkateswara Rao and R.O. Ritchie in Proc. Johannes Weertman Symposium (The Minerals, Metals & Materials Society, Warrendale, PA, 1996) in press.

12. K.T. Venkateswara Rao and R.O. Ritchie in Fatigue and Fracture of Ordered Intermetallic Materials I, edited by W. O. Soboyejo, T. S. Srivatsan and D. L. Davidson (The Minerals, Metals & Materials Society, Warrendale, PA, 1994), p. 3

13. K.T. Venkateswara Rao, W.O. Soboyejo and R.O. Ritchie, *Metall. Trans. A* **23A**, 2249 (1992).

14. K. Badrinarayanan, A. L. McKelvey, K.T. Venkateswara Rao and R.O. Ritchie, *Metall. Mater.Trans. A* **27A** (1996) in press.

15. M. Bannister and M. F. Ashby, *Acta Metall. Mater.* **39**, 2575 (1991).

16. M.F. Ashby, F.J. Blunt and M. Bannister, *Acta Metall. Mater.* **37**, 1847 (1989).

17. K.T. Venkateswara Rao, G.R. Odette and R.O. Ritchie, *Acta Metall. Mater.* **40**, 353 (1992).

18. H.E. Dève, A.G. Evans, G.R. Odette, R. Mehrabian, M.L. Emiliani and R.J. Hecht, *Acta Metall. Mater.* **38**, 1491 (1990).

19. ASM Metals Handbook, 10th ed., Vol. 2, (ASM Intl., Materials Park, OH, 1994) p. 559.

20. B. Budiansky, J. C. Amazigo and A. G. Evans, *J. Mech. Phys. Solids* **36,** 167 (1988).

21. F. Zok and C. L. Hom, *Acta Metall. Mater.* **38**, 1895 (1990).

22. H. Tada, P. C. Paris and G. R. Irwin, in Stress Analysis of Cracks Handbook, Del Research Corp./Paris Publ., St. Louis, MO, 1985.

23. G.R. Odette, B.L. Chao, J.W. Sheckhard and G.E. Lucas, *Acta Metall. Mater.* **40**, 2381 (1992).

24. K.T. Venkateswara Rao, G.R. Odette and R.O. Ritchie, *Acta Metall. Mater.* **42**, 893 (1994).

The Fracture Behavior of SiCp/Aluminum Alloy Composites With And Without Large Al-Particles

A.B. Pandey[1], B.S. Majumdar[2], and D.B. Miracle[3]

[1]Systran Corporation, 4126 Linden Avenue, Dayton, OH 45432, and on leave from Defense Metallurgical Research Laboratory, Hyderabad-500058, India
[2]UES, Inc., 4401 Dayton-Xenia Road, Dayton, OH 45432
[3]Wright Laboratory, Materials Directorate, Wright-Patterson AFB, OH 45433

ABSTRACT

J_{Ic} measurements were performed on a SiCp/Al-7093 MMC with controlled heat treatments, and the damage mechanisms were evaluated to understand the influence of microstructural parameters on the fracture toughness and crack resistance behavior. The deformed materials showed widely different damage and flow localization for different matrix microstructures. In an effort to improve fracture toughness, large Al particles were incorporated into the powder-metallurgy based MMC, and extruded to obtain pancake shaped Al phases. In the extruded condition, the effect of Al particles on the crack initiation toughness was negligible. However, significant improvement in the toughness was observed when the extruded material was further rolled. These issues are discussed in the context of observed deformation and damage mechanisms.

INTRODUCTION

Discontinuously reinforced aluminum (DRA) composites possess superior stiffness and strength compared to unreinforced aluminum alloys [1], and can be processed via conventional routes such as forging, rolling and extrusion. However, DRAs typically have lower fracture toughness than the unreinforced alloys, and this poses a constraint in the application of DRAs in aerospace structures.

Past studies suggest that the microstructure can significantly influence the fracture toughness of DRA materials [2-7]. However, the exact fracture mechanisms, and in particular, the role of the matrix alloy microstructure on toughness do not appear to be well understood. The presence of large Al particles in the control DRA is expected to show improvement in the toughness without significant loss of strength of composite. The work described herein is part of a larger study to obtain improved understanding of how matrix microstructure influences damage evolution and instability in the highly constrained region ahead of a crack tip. The effect of pancake shaped large aluminum particles on the fracture toughness and damage mechanisms of the DRA composite was also evaluated.

EXPERIMENTAL PROCEDURE

Two DRA materials were considered: (i) a base 15 vol.% SiCp (600 grit)/7093 Al material , designated by the suffix 64 in the tables, and (ii) a 15 vol.% SiCp (600 grit)/7093 Al MMC that contained 10 vol.% pure Al particles, and designated by the suffix 60. In the latter, the incorporation of Al particles necessitated greater SiCp content in the non-Al regions compared with the control (base) material. The MMCs were produced by a powder metallurgy route involving blending of the aluminum alloy powder and SiC particulates, hot consolidation, followed by extrusion at 22:1 ratio resulting in 1"x 3" rectangular bars. The series-60 powder blend contained approximately 10 mm sized pure Al particles. The DRAs were then subjected to different aging treatments to obtain different microstructural conditions. The following heat treatments were considered: solution treated (ST)-490^{0}C/4h/WQ, underaged (UA)-490^{0}C/4h/WQ followed by 120^{0}C/25 min. aging treatment, peak aged (PA)-490^{0}C/4h/WQ followed by 120^0 C/24h, and overaged (OA)-490^{0}C/4h/WQ followed by 120^{0}C/24h+150^{0}C/8h. The solution treated specimens were kept in a refrigerator to minimize any natural aging. The extruded DRA material with the Al particles was further rolled at 450^{0}C for an additional reduction of 70 % to improve the distribution and morphology of the aluminum particles.

Tensile tests of the control DRA material were conducted on flat specimens of gauge length 25.4 mm, width 8.66 mm, and thickness 2.5 mm. The specimens were metallographically polished prior to testing to allow monitoring of damage and surface slip bands at different levels of applied strains. All the tensile specimens were machined parallel to the plane of the plate, and the fracture toughness specimens had both LT and TL orientations. The fracture toughness and J-R curve of each material was evaluated using the single specimen J-integral test method, ASTM E-813 [8]. The compact tension (CT) specimen geometry was employed, with a thickness (B) of

Mat. Res. Soc. Symp. Proc. Vol. 434

7.6 mm , a width (W) of 20.3 mm, and a notch length of 7.6 mm. The CT specimens were metallographically polished on both faces, and replicas were taken at different stages of crack growth to monitor the evolution and sequence of damage. All the specimens were fatigue precracked according to the standard ASTM procedures to an a/W ratio of approximately 0.5. The crack opening displacements (COD) were monitored across the crack mouth using a clip gage mounted between knife edges affixed to the front of the CT specimen. Crack lengths were determined using the unloading compliance technique. For J-Integral determination, the COD measurements were converted to load-line displacements using established formulas. Following fracture, the specimens were examined optically and using scanning electron microscopy. It was found that the final crack lengths based on the fracture surfaces agreed within 10 percent with crack lengths assessed from compliance measurements; this provided confidence in the crack lengths interpreted from compliance data.

RESULTS AND DISCUSSION

(i) Tensile tests

Figs. 1(a) and 1(b) show microstructures of the control DRA and DRA with Al materials, respectively. While the distribution of SiC particulate in the matrix was uniform in both materials, the aluminum particles in the series-60 material were inhomogeneouly distributed. The tensile data for the control DRA are summarized in Table I, and they are in agreement with previous investigations [2-3] on a similar material. The yield strength and tensile strength were minimum in the solution treated condition and were maximum in the peak aged condition.

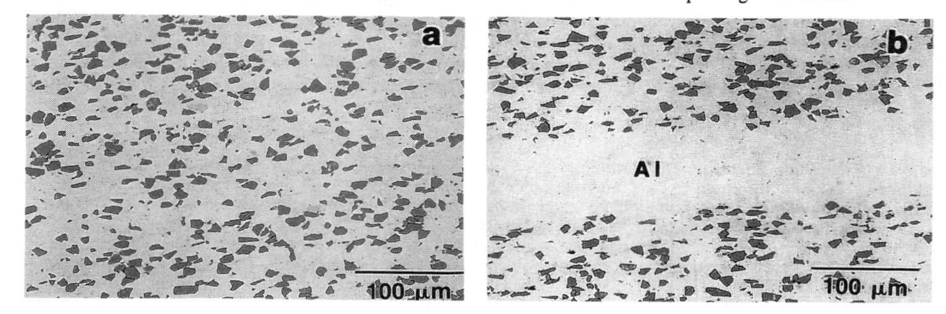

Fig. 1: Microstructures of the (a) control DRA and (b) DRA with Al particles

The ductility of the material showed the anticipated trend of decreased elongation in proceeding from the solution treated to the peak aged condition. The work hardening rates were similar for ST and UA conditions, which decreased significantly for PA and OA conditions. The damaged microstructures suggested that the work hardening rate may have had an important influence on the fracture process, since damage localization was observed to be higher for the microstructures with lower work hardening rates.

Table I : Tensile properties of control DRA composite

Heat Treatment	Young's Modulus, GPa	Yield Strength, MPa	Tensile Strength, MPa	Elongation, %	Work Hardening Exponent
Solution treated	91.1	430	577	8.0	0.084
Under aged	89.9	503	629	5.9	0.089
Peak aged	95.6	642	694	1.8	0.037
Over aged	91.5	591	642	2.4	0.031

(ii) Fracture Toughness

Fig. 2(a) shows the load versus displacement curve for the control DRA material in the solution treated condition. Several unloading lines for crack length estimation are also shown in the figure. Fig. 2(a) also shows steps in the load-displacement curve. The steps were observed in all the samples and suggest that crack growth may occur by a process of repeated material instability ahead of the crack tip, rather than in a continuous manner. The J versus Δ a curve for the solution treated condition shown in Fig. 2(b) indicates a reasonable degree of toughness increase with crack growth. The Keq values are also plotted in the figure, and were obtained from the

J-Integral data using Keq= $\sqrt{(J*E_C/(1-\nu^2))}$, where E_C is the modulus of the composite. The toughness at crack initiation was approximately 25 MPa $\sqrt{m}$, and it increased to approximately 40 MPa $\sqrt{m}$ at 3 mm of crack extension. These values are quite good and may meet toughness requirements in a number of aerospace applications, provided strength values are met.

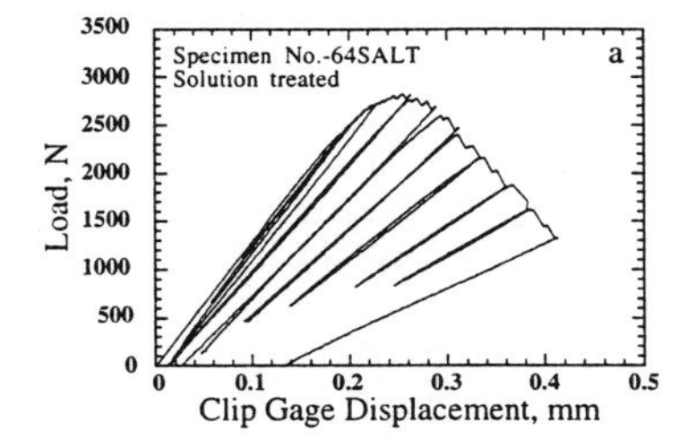

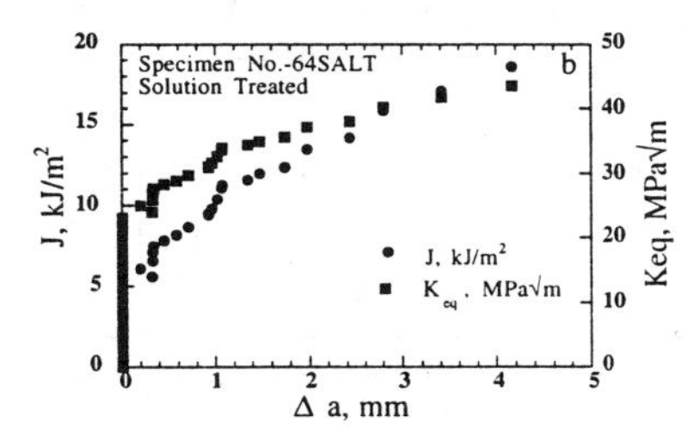

Fig. 2: (a) Load versus displacement and (b) J and Keq versus Δ a curves for the control DRA material in the solution treated condition

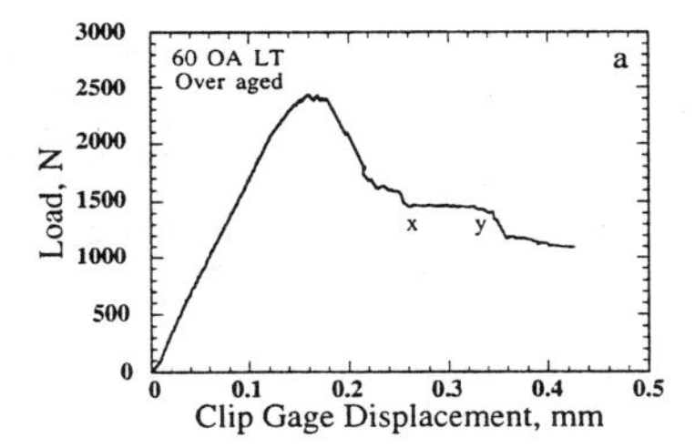

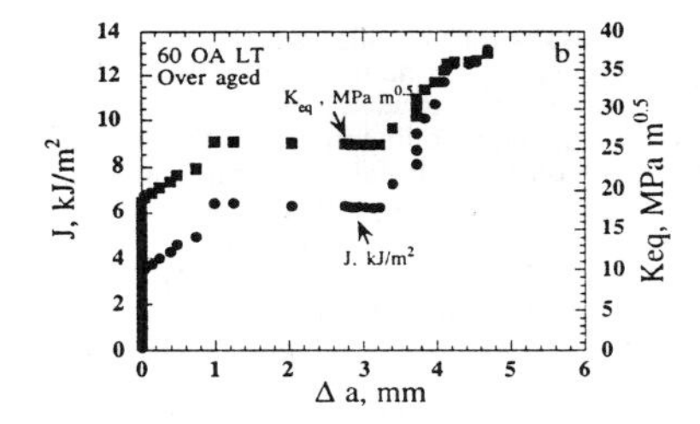

Fig.3(a) Load versus displacement and (b) J and Keq versus Δ a curves for the DRA with Al particles in the overaged condition

Figs. 3(a) and 3(b) show the load-displacement and crack resistance curves, respectively, for the series 60 (DRA with Al) material in the overaged condition. Fig. 3(a) illustrates a region (from x to y) in the crack growth domain where the load remained approximately constant, and the unloading lines showed no crack growth during this period; the net effect was the sharp rise in toughness by approximately 6 MPa $\sqrt{m}$ (see Fig. 3(b)). Visual observation of the side of the sample during crack growth indicated that this region was associated with the crack arriving before a large Al pancake, and being temporarily halted. Further crack growth occurred when this Al-particle ruptured. The mechanism of toughening thus appears to be the plasticity ahead of the crack tip, rather than crack bridging by the Al-particle after the crack had passed the particle. However, the Al particles were inhomogeneously distributed, so that such crack resistance behavior was erratic.

The fracture toughness data are summarized in Table II, where the crack growth resistance is quantitatively represented by the tearing modulus, T ($T=(E/\sigma_y^2)(dJ/da)$). The table shows that for both series of extruded materials, the initiation fracture toughness (Keq or J-initiation) and tearing modulus were the highest in the solution treated condition, and in agreement with the tensile tests which exhibited maximum ductility for this heat treatment condition. It is important to note that precipitate particles in the matrix are essentially absent for the solution treated condition. Thus, the results suggest that precipitate particles are deleterious to fracture toughness, and should be avoided from a toughness perspective. An additional point to note is that the number of cycles needed to grow the precrack over approximately 2.5 mm distance was largest for the solution treated condition. Thus fatigue crack growth resistance may also benefit from elimination of precipitate particles.

Returning to Table II, it may be noted that the peak aged specimens had the lowest initiation toughness and poorest crack resistance behavior, and echoed similar characteristics of unreinforced Al alloys [9-10]. The table shows that in the underaged condition, the toughness behavior of the 64 UA LT and 60 UA LT specimens were similar. With regard to specimen orientation effects, the 64 UA LT specimen had a higher toughness than specimen 64 UA TL, and this behavior is similar to those observed for unreinforced Al alloys [11]. It should also be noted that the crack resistance behavior of 60 UA TL was about 50 percent higher than the 64 UA TL control material, although the initiation toughness was negligibly affected.

Table II : Fracture Toughness of DRA composites in different heat treatment conditions

Materials	J, k J/m^2	K_{eq}, MPa$\sqrt{m}$	Tearing Modulus
64 SA LT	6.4	25.4	2.35
64 UA LT	3.4	19.0	0.78
60 UA LT	3.3	19.5	0.90
60 PA LT	2.2	16.5	0.37
60 OA LT	3.9	20.5	2.57
64 UA TL	2.8	17.4	0.95
60 UA TL	2.9	17.5	1.94
60 UA LT R	6.3	26.0	2.11
60 OA LT R	3.7	20.0	0.47

The results for the rolled specimens for the Al containing material are also provided in Table II, and are designated by specimen numbers 60 UA LT R and 60 OA LT R, where R signifies the rolled condition. As mentioned earlier, the extruded material was rolled to improve the distribution and morphology of the Al particles. Table II illustrates that rolling increased the initiation fracture toughness by approximately 25-percent compared to the extruded control material in the same heat treatment condition. The effect of rolling on the crack growth behavior was even more significant, providing approximately 50 percent improvement. However, no similar improvement was observed for the rolled material in the overaged condition compared to the control extruded overaged material. The reasons are currently under investigation.

It is important to mention that all the fracture toughness data in Table II satisfy plane-strain requirements of ASTM E-813 [8], and therefore represent J_{IC} and equivalent K_{IC} values. However, the present J_{IC} values for the control material are significantly lower than numbers reported in [2] for a similar DRA material in the underaged and overaged conditions. The difference may be ascribed to the two possible reasons. First, the specimens in [2] were tested in the notched condition, rather than with a fatigue precrack. The latter has the effect of significantly raising the local triaxial stresses ahead of the crack tip during the fracture toughness testing, thus initiating crack growth at lower loads and displacements compared with the notched specimen. Second, the CT specimens used in [2] were of 2.5 mm thickness, compared with 7.6 mm employed here. The toughness and strength values indicate that plane strain conditions are satisfied even for the 2.5 mm thick specimens. However, it is important to note that the thickness requirements prescribed in the ASTM procedures were arrived at on the basis of elastic-plastic deformation of a homogeneous material. On the other hand, local plastic strain elevation and hydrostatic stress (see, for example, the theoretical treatments in [12]) and resultant damage (observed) are very large for the composite material, and such damage may have the effect of relaxing stress triaxiality and promoting shear-type fracture even for thicknesses larger than those given by plane-strain requirements. The specimens had a flat fracture profile in the present study. The agreement between the present values and those reported in [13] on fatigue precracked thicker (19 mm thick) bend specimens also suggests the significance of fatigue precracking and specimen thickness on the fracture toughness of DRA materials.

(iii) Micromechanisms

As indicated earlier, the micromechanisms of fracture were evaluated by monitoring damage in tensile and fracture toughness tests. Fig. 4 shows a typical micrograph of damage and plasticity in a tensile tested specimen. It illustrates the two dominant modes of damage, i.e., cracking of a SiC particle, and debonding at a particle/matrix interface. Sharp slip lines are associated with both these types of damage. Additionally, matrix failure was often observed near the end of particles, but not at the interface itself, and may be a result of intense slip activity in a region with the largest von-Mises stress. *The key finding was that all these modes of damage (i.e., particle cracking, and debonding) and accompanying slip behavior were present for all the heat-treatment conditions.* Thus, the different toughness behavior for the different microstructures cannot simply be ascribed to one dominant microscopic damage mode. However, there was one notable difference for the different matrix microstructures. In the solution treated specimen, damage and localized plastic deformation (in the form of slip lines) was uniformly distributed throughout the entire gage length of the tensile sample. Conversely, for the peak-aged condition, damage and localized slip were only observed in a narrow region near the fracture plane. The distribution of damage and slip for the underaged and overaged condition lay intermediate between the solution-treated and peak aged conditions. These results are in agreement with microstructural observations of Lloyd [14] in tensile samples of a SiCp/6061 Al-alloy composite. In reference [15], ductility exhaustion and thermally induced work hardening have been suggested as possible mechanisms controlling failure of DRAs. It is not yet clear whether such mechanisms can explain the large differences in the flow localization behavior of the different microstructures.

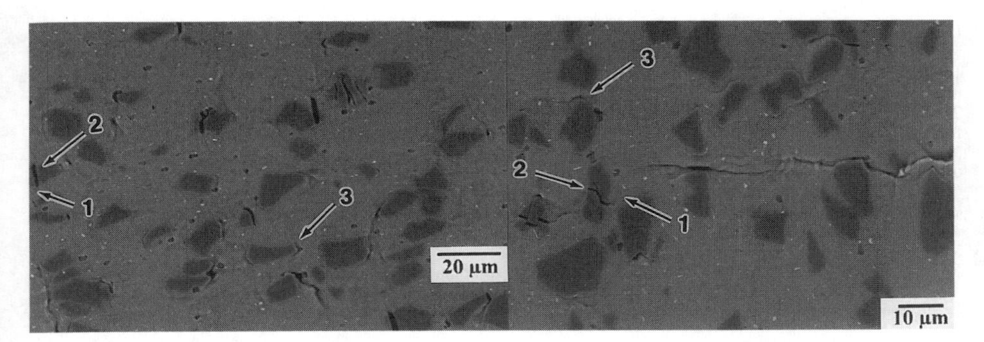

Fig. 4: Damage and plasticity in the form of slip lines marked by 1, particle cracking by 2, and near interface debonding by 3 in a tensile specimen

Fig. 5: Damage ahead of the crack tip showing slip lines by 1, particle cracking by 2, and near interface debonding in a CT specimen

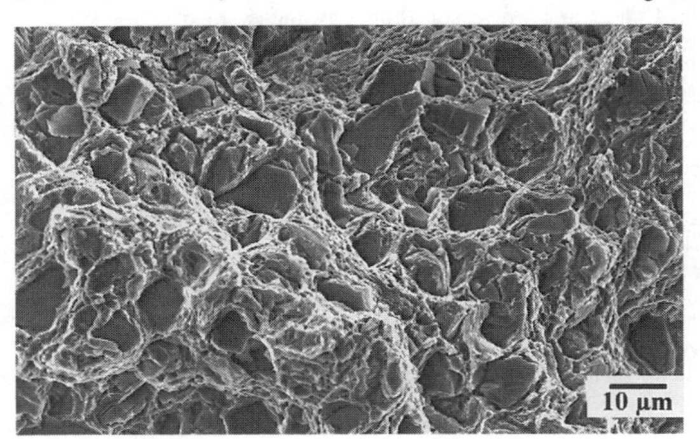

Fig. 6: Fracture surface of the control DRA material

The deformation behavior of the tensile samples also were echoed in the highly stressed region ahead of the crack tip. This is illustrated in Fig. 5, where again both major damage modes are shown to be present, and these damage modes were independent of the heat treatment conditions. These observations are in slight variance with some past studies on DRA materials [2,16]. While particle cracking was dominant in 2xxx based composites [16] in different heat treatment conditions, particle cracking and interface debonding were observed in 7xxx composite in underaged and overaged conditions, respectively [2]. Although we have not yet quantified the frequency of damage modes in different heat treatment conditions, the general appearance showed almost equal proportions of particle cracking and interface decohesion. More significant was the different amounts of flow localization in the different microstructures, and suggest that this aspect of deformation and damage need further consideration. The fractographic features of CT specimens were similar to the tensile samples, and once again the features were independent of the heat treatment condition. Fig. 6 illustrates a typical region of the fracture surface, and its prime features are the higher density of particles on the fracture surface (compared to the microstructure) and ductile dimple rupture of the matrix ligaments. The SiC particles were either cracked or gave the appearance of debonding from the matrix, and Fig. 6 reveals that whatever the damage mode, final fracture localized around particles leading to a high particle density on the fracture surface.

Thus, the overall toughness behavior of the control samples are dominated by localized flow behavior, likely enhanced by the high stress concentration ahead of the crack tip. The sequence of events appear to be the formation of damage sites, be it particle cracks or interface debonds, and the joining up of neighboring damage sites through matrix plasticity, void nucleation and growth, and finally microvoid coalescence. For toughness prediction purposes, a methodology has to be established that incorporates aspects of damage and flow localization under constrained flow conditions. Experiments with geometries that contain different levels of strain concentrations may provide a means of validating any model that is developed. Additional data is required to understand the influence of Al particles on the toughness, because of the non-homogeneous distribution of these particles.

CONCLUSIONS

1. The control DRA showed the maximum ductility and fracture toughness in the solution treated condition and minimum in the peak aged condition. The yield and tensile strengths showed the anticipated opposite trends.
2. Higher work hardening rate and absence of precipitate particles in the matrix appeared to be beneficial for the fracture toughness of DRA material.
3. Damage in the form of particle fracture and matrix/particle interface decohesion were observed for all the heat treatment conditions. Thus, these damage modes cannot explain the different toughness behavior of the materials.
4. The different matrix microstructures showed a distinct difference in flow localization. Thus, for the solution treated condition, both the tensile and CT samples exhibited a rather uniform damage state, connected by matrix slip lines. Conversely, although the peak aged condition exhibited similar damage modes, there was intense flow localization and faster rate of damage coalescence.
5. There was no significant improvement in the initiation toughness of the extruded DRA containing Al particles.
6. The rolled DRA containing Al particles showed significant improvements in both crack initiation and growth toughness. However, this improvement was observed only for the UA condition, and not for the OA condition.

ACKNOWLEDGMENTS

This work was supported by the Materials Directorate, Wright Laboratory, Wright-Patterson AFB, OH. We thank Dr. Mary Lee Gambone of the Wright Laboratory for many helpful discussions, and Dr. Warren Hunt, Jr. of Alcoa Technical Center and Professor John Lewandowski of Case Western Reserve University for supplying the materials and useful discussions.

REFERENCES

1. D.L. McDanels, Metall. Trans. **A16**, 1105 (1985).
2. M. Manoharan and J.J. Lewandowski, Acta Metall. Mater. **38**, 489 (1990).
3. J.J. Lewandowski, C. Liu and W.H. Hunt, Jr., Mater. Sci. Eng., **A107**, 241 (1989).
4. Y. Flom and R. J. Arsenault, Acta Metall., **37**, 2413 (1989).
5. S.V. Kamat, J.P. Hirth and R. Meharabian, Acta Metall., **37**, 2395 (1989).
6. D.L. Davidson, Metall. Trans., **A22**, 113 (1991).
7. N.C. Beck Tan, R. M. Aikin, Jr., and R.M. Briber, Metall. Mater. Trans. **A25**, 2461 (1994).
8. ASTM Annual Book of Standards, American Society for Testing and materials, Philadelphia, PA, 1987, p. 713.
9. G.G. Garret and J.F. Knott, Metall. Trans., **A9**, 1197 (1978).
10. G.T. Hahn and A.R. Rosenfield, Metall. Trans., **A6**, 653 (1975).
11. N.E. Dowling, in Mechanical Behavior of Materials, Engineering Methods for Deformation, Fracture, and Fatigue, Prentice hall, New Jersey, 1993, p. 313.
12. D.B. Zahl, S. Schmauder, and R.M. McMeeking, Acta Metall. Mater., **42**, 2983 (1994).
13. M. Manoharan and J.J. Lewandowski, Scripta Metall., **23**, 301 (1989).
14. D.J. Lloyd, Acta Metall. Mater., **39**, 59 (1991).
15. N. Shi and R.J. Arsenault, in J.J. Lewandowski and W.H. Hunt, Jr. (eds.), Intrinsic and Extrinsic Fracture Mechanisms in Inorganic Composite Systems, The Minerals, Metals & Materials Society, 1995, p. 69.
16. C.P. You, A.W. Thompson and I.M. Bernstein, Scripta Metall., **21**, 181 (1987).

FABRICATION, STRUCTURE AND PROPERTIES OF ALUMINUM-ALUMINIDE LAYERED COMPOSITES

D.E. Alman, U.S. Department of Energy, Albany Research Center, 1450 Queen Ave., S.W., Albany, OR, 97321-2198, alman@alrc.usbm.gov

ABSTRACT

The fabrication of aluminum-aluminide layered composites by reactive bonding of elemental Al and Ni foils was investigated. It was observed that after hot-pressing, thin Ni foils were converted to NiAl. The as-processed Al-NiAl layered structure could be heat-treated to produce an equilibrium Al-Al_3Ni layered composite. Tensile tests revealed that composites could be produced that failed in a "tough" manner and were stronger and stiffer than aluminum.

INTRODUCTION

Composites consisting of both intermetallic and metallic layers, offer an attractive combination of properties from both constituent phases, that is the high strength and stiffness of the intermetallic phase with the high fracture resistance of the metallic phase. Methods to produce such a composite structure include: powder techniques, such as plasma spraying or diffusion bonding, and deposition methods, such as chemical vapor deposition and magnetron sputtering [1-4]. These techniques require sophisticated equipment and/or are limited in the component geometry that can be produced. A potential cost-effective production method for producing metal-intermetallic layered composites is by reactive diffusion bonding of metallic foils or sheets [5-9]. In this process, one of the foils is consumed during bonding, *insitu* synthesizing the intermetallic layer. Three major advantages of this technique are: (1) the metallic foils are readily available, (2) the elemental foils easily can be formed prior to synthesizing the brittle intermetallic phase, and (3) the process utilizes manufacturing equipment that today is used in commercial operations for the production of aerospace components. All of which enhance the potential viability of this process.

This paper emphasizes the fabrication, structure and properties of aluminum-nickel aluminide layered composites produced by reactive bonding Al and Ni foils. This system was selected as a model system, partly because there exists a large body of literature dealing with reactive diffusion of elemental Ni and Al layers for thin-film formation [10]. It should be emphasized that the diffusion distances (i.e., initial Ni and Al layer thicknesses) are significantly greater for the process described in this paper than are utilized in thin-film formation. Also, this system was selected for study because the properties of an aluminum-aluminide layered composite might be attractive for actual engineering purposes. Below the maximum use temperature of Al alloys, nickel aluminide compounds are characterized as possessing high stiffness, high strengths, and low ductility. The latter of which has hindered the utilization of these compounds in bulk form for engineering applications. However, these characteristics are ideal for a reinforcement phase for aluminum matrix composites. In fact, Al-Al_3Ni composites, produced via solidification methods, have been the subject of numerous studies [11-15].

EXPERIMENTAL PROCEDURE

Elemental Ni and Al foils were diffusion bonded in a vacuum, induction heated hot-press.

Mat. Res. Soc. Symp. Proc. Vol. 434

Typically, five Al foils and four Ni foils were stacked in an alternating sequence. The Ni foils utilized in this study were 0.025 mm thick. Samples were produced from Al foils of either 0.500, 0.250 or 0.125 mm thickness. The hot-press parameters consisted of heating the stacked foils to 600°C under an applied pressure of 10 MPa and holding at these conditions for 2 hours. To determine the microstructural stability of the resultant micrsotructures, portions of the laminates were vacuum encapsulated in quartz and heat treated at 450°C for either 1, 10 or 100 hours. Scanning Electron Microscopy (SEM) coupled with energy dispersive X-ray (EDX) was used for microstructural analysis.

Tensile tests were performed at room temperature. Flat dog-bone shaped specimens, with a 12.5 mm gage length and 6.25 mm gage width, were machined from the composite panels. The interface between the layers was oriented parallel to the tensile axis. For comparative purposes, tensile specimens were also prepared from the starting 0.500 mm thick Al foils. Prior to testing, the specimens were mechanically polished to a 1 μm diamond finish. A screw driven tensile testing machine was utilized for testing at a constant cross head speed of 0.5 mm/min, which corresponded to an initial strain rate of 3.3×10^{-4}/s. A clip gauge extensometer was used to monitor strain.

RESULTS AND DISCUSSION

After reactively bonding the elemental foils, the composite consisted of alternating metal and aluminide layers, as illustrated in Figure 1. The Ni foil was almost completely converted to NiAl, leaving only a thin unreacted Ni core in the center, and a relatively thick Al_3Ni interphase region at the boundary with the Al layer (Figure 2). The different starting foils only affected the volume fraction of the phases present (Table I) and not the composition of the phases that formed during hot-pressing. This provides another advantage of this technique, in that the volume fraction of the layers formed can easily be manipulated by varying the thicknesses of the starting foils.

TABLE I
MICROSTRUCTURES OF ALUMINUM-ALUMINIDE LAYERED COMPOSITES

Starting Foil Thickness (mm)		Phase Present [A] After Hot-Pressing	Volume Percent Aluminide Layers	Composite Density(g/cm^3) [B]	Equilibrium Phases
Al	Ni				
0.500	0.025	NiAl*, Al_3Ni, Al**, Ni	18	3.25	Al, Al_3Ni
0.250	0.025	NiAl*, Al_3Ni, Al**, Ni	27	3.53	Al, Al_3Ni
0.125	0.025	NiAl*, Al_3Ni, Al**, Ni	40	3.95	Al, Al_3Ni

[A] Hot-Pressing conditions: 600°C-10MPa-2hrs; *predominate aluminide phase, ** predominate metallic phase

[B] Composite density calculated from the volumetric rule of mixtures (ROM), neglecting minor phases (Ni and Al_3Ni).

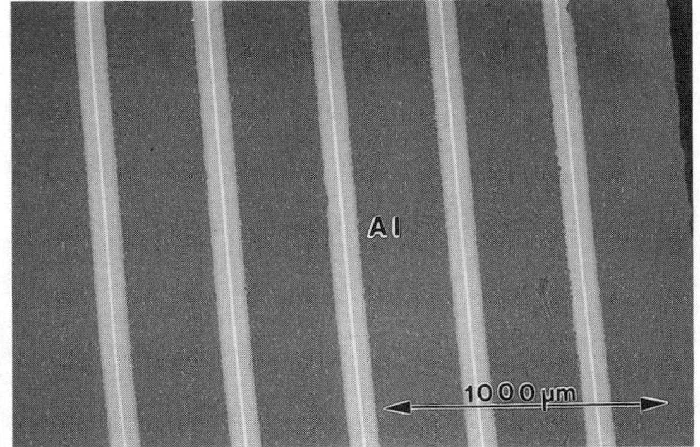

Figure 1. Low magnification view of the microstructure of the Al-aluminide layered composite produced by reactive bonding (at 600°C-10MPa-2hrs).

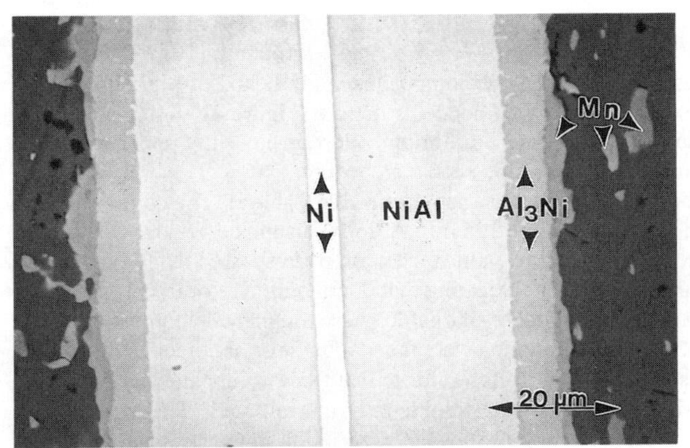

Figure 2. High magnification view of the microstructure of the Al-aluminide layered composite produced by reactive bonding (at 600°C-10MPa-2hrs).

Figure 3. Microstructure of the composite after heat-treating (at 450°C for 100 hrs).

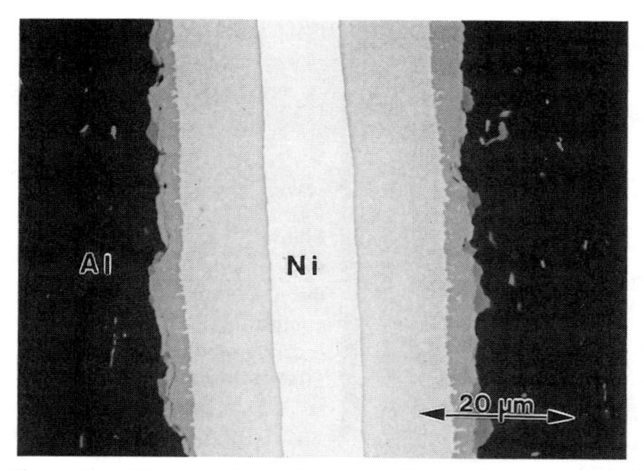

Figure 4. Microstructure of a composite after hot-pressing Al and Ni foils at 600°C and 10 MPa for 15 min. Notice, the formation of aluminide layers.

The nickel core was no longer present in the microstructure of the composite after heat treating at 450°C for 1 hour. Both the NiAl and Al_3Ni layers coarsened. After 10 hours, both aluminide layers continued to coarsen. However, after 100 hours the composite consisted only Al and Al_3Ni layers, the equilibrium phase (Figure 3). It is not too surprising that the high melting phases (such as Ni and NiAl) can be consumed by low temperature anneals, as this occurs during thin-film formation [10]. These results suggest that an optimum processing route for these composites might be to hot-press for a short period of time (such as 15 min) to produce a well bonded structure (Figure 4), followed by heat-treating to drive the structure to thermodynamic equilibrium. Work is presently underway to determine the optimal processing conditions using such a procedure.

The tensile properties of the composites are summarized in Table II. The composite with 40 vol% aluminide layers (i.e., produced from 0.125 mm Al foils) displayed very little strain to failure (<<1 pct). Interestingly, the fracture behavior of the composite showed that the Al layers did in fact fail in a ductile manner (Figure 5) indicating that ductile fracture of the metallic layer may not be sufficient to ensure fracture resistance of ductile phase toughened composites. The composites containing 18 and 27 vol% aluminide layers failed after several percent strain to failure. The stress-strain curves of these composites indicate that these composite can be loaded to a certain stress level, after which slow, non-catastrophic failure is initiated. This failure behavior was evidenced by the sharp yield point followed by serration in the stress strain curves (associated with progressive, sequential cracking of the aluminide layers) with limited or no work-hardening. The yield points of these composites were in essence associated with the failure of the aluminide layers. Initially the load was supported by the brittle, high strength, high modulus aluminide layers. As the aluminide layers began to fail via transverse fracture, the applied tensile load was transferred to the lower modulus, lower strength Al layers. If the Al adjacent to the cracked region in the aluminide layer can absorb the energy released by the propagating crack, then catastrophic fracture of the composite is avoided. If this failure mechanism occurs as envisioned, multiple cracks will form in the brittle aluminide layer, and the composite will fracture in a "tough" manner. This behavior was indeed observed for the 18 and 27 vol% composites (Figure 6). If the Al layer can not absorb the energy released by the propagating crack in the aluminide layer, catastrophic failure will occur, as evidenced by a single crack propagating through the composite structure, as was seen in the 40 vol% composite.

The measured strengths and moduli of the composites were higher (by a factor about 1.5 to 3) than that of Al both in real and density compensated terms (Table II). The exception is the measured strength value of the 40 vol% aluminide composite which fractured in a brittle manner. These results reveal that composites can be produced that fail in a fracture resistant manner and

TABLE II
TENSILE PROPERTIES OF ALUMINUM-ALUMINIDE LAYERED COMPOSITES

Initial Foil Thickness (mm)		Spec. #	Vol% Aluminide	σ_{YS} (MPa)	σ_{UTS} (MPa)	E (GPa)	e_f (%)	σ_{YS}/ρ	E/ρ
Al	Ni								
0.500	0.025	1	18	63	77	100	6	23.7	30.8
		2	18	65	77	96	10	23.7	29.5
		3	18	61	78	107	12	24.0	32.9
0.250	0.025	1	27	93	94	127	5	26.3	36.0
		2	27	88	92	124	8	26.0	35.1
0.125	0.025	1	40	31	31	138	<<1	7.8	34.9
Al foil*		1	0	29	76	48	30	10.8	17.9
Al**			0	30	70	70	43	11.2	26.1

* measured from 0.500 mm thick Al foil; ** data for 1060 Al in an annealed condition [16]

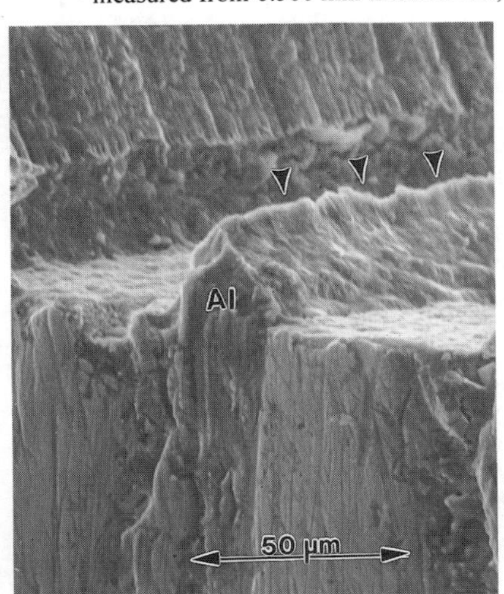

Figure 5. Fracture surface of the Al-40vol% NiAl composite (Al=0.125 mm). Even though this composite displayed little fracture resistance, the Al layers failed in a ductile manner.

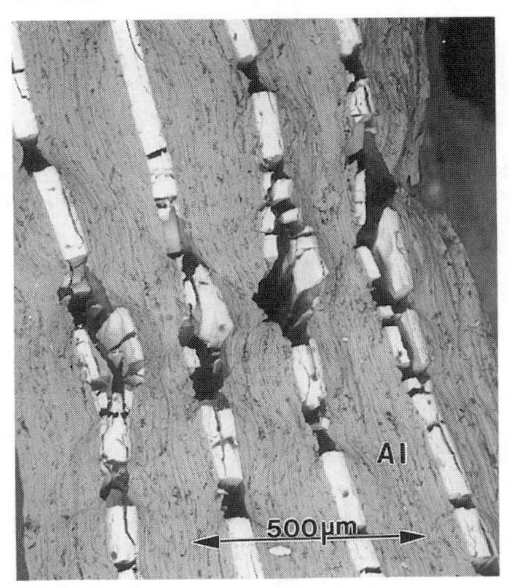

Figure 6. Side view of the fractured tensile specimen of the Al-27vol% NiAl composite (Al=0.250 mm) showing multiple cracks in the aluminide layer. This composite was fracture resistant.

are stronger and stiffer than the base metal.

As the volume fraction of the aluminide layer increased within the composite microstructure, the strength and moduli increased while the strain the failure decreased. These results indicate that for a composite produced from both ductile and brittle phases, there will exist a optimal volume fraction (which for a lamellar composite is equivalent to an optimal layer thickness ratio) that will provide maximum strength and fracture resistance. At low vol% aluminide layers, the soft metal layer will carry most of the load and strengthening due to the addition of the aluminide layer will be minimal. At high aluminide layer vol%, the constraint acting on the metal layer, due to the chemical bond with the aluminide layer, will be sufficiently large to cause the fracture resistance of the composite to be poor.

SUMMARY

The fabrication of aluminum-aluminide layered composites was accomplished by reactive bonding elemental Ni and Al foils. After hot-pressing thin Ni foils were converted primarily to NiAl. The as-processed structure could be heat-treated to produce an equilibrium Al-Al_3Ni layered composite. Composites were produced that were stronger and stiffer than the base Al, and these composites failed in a "tough" manner. Preliminary tensile tests revealed that the aluminide layers strengthen (by a factor of 2 to 3) and stiffen (by a factor of 1.5 to 2) the composite with respect to Al, in both real and density compensated terms.

This method of composite fabrication is attractive, as the continuous reinforcement phase is synthesized *insitu* during processing from inexpensive and easily shaped elemental foils. While this paper emphasized the fabrication and properties of aluminum-aluminide layered composites, this elemental foil, reactive bonding process has been used to produce intermetallic alloyed sheets and other metal-intermetallic (such as Ni-Ni_3Al) layered composites [5-9].

REFERENCES

1 D.E. Alman, K.G. Shaw, N.S. Stoloff, and K. Rajan: *Mater. Sci. Engr.*, A155 (1992), 85.
2 N.S. Stoloff and D.E. Alman: in Intermetallic Matrix Composites, eds., D.L. Anton, et. al., MRS vol. 194, MRS, Pittsburgh, PA, 1990, p. 31.
3 D.A. Hardwick and R.C. Cordi: in Intermetallic Matrix Composites, eds., D.L. Anton, et. a.l, MRS vol. 194, MRS, Pittsburgh, PA, 1990, p. 65.
4 R.G. Rowe and D.W. Skelly: in Intermetallic Matrix Composites II, eds., D.B. Miracle, et. a.l, MRS vol. 273, MRS, Pittsburgh, PA, 1992, p. 411.
5 D.E. Alman, J.A. Hawk, A.V. Petty. Jr., and J.C. Rawers: *JOM*, 46 (1994) (3), 31.
6 D.E. Alman, J.C. Rawers, J.A. Hawk: *Metall. Mater. Trans. A*, 26A (1995), 589.
7 D.E. Alman, C.P. Dogan, J. A. Hawk, and J.C. Rawers: *Mater. Sci.Engr.* ,A192/193 (1995), 624.
8 D.E. Alman and C.P. Dogan: *Metall. Mater. Trans. A,* 26A (1995), 2759.
9 D.E. Alman and J.A. Hawk: in Light Weight Alloys for Alloys for Aerospace Applications- III, eds., E.W. Lee, et. Al., TMS, Warrendale, PA, 1995, p. 531.
10 E.G. Colgan: *Mater. Sci. Rept.* 5 (1990) 1.
11 R.W. Hertzberg, F.D. Lemkey and J.A. Ford: *Trans. AIME*, 233 (1965) 342.
12 G.E. Hoover and R.W. Hertzberg, *Trans. ASM*, 61 (1968) 769.
13 A.S. Yue, F.W. Crossman, A.E. Vodoz and M.I. Jacobson: *Trans. AIME*, 242 (1968) 2441.
14 G.E. Mauer, D.J. Duquette and N.S. Stoloff: *Metall. Trans. A*, 7A (1976) 703.
15 S. Tao and H.D. Embury: *Metall. Trans. A,* 24A (1993) 713.
16 H.E. Boyer and T.L. Gall: Metals Handbook-Desk Top Edition, ASM, Materials Park, OH, 1985, Chapter 6.

TIME DEPENDENT STRESS FIELDS AHEAD OF THE INTERFACE CRACKS IN CREEP REGIME

S.B. BINER
Ames Laboratory, Iowa State University, Ames, IA 50011 U.S.A

ABSTRACT

In this study the behavior of stationary interface cracks in layered materials at creep regime in plane-strain condition and pure opening dominated mode-I load state was studied numerically. The results indicate that the introduction of a transitional layer to the interface of the elastic-creeping bimaterials significantly elevates the stress values ahead of the interface cracks under identically applied load levels at creep regime.

INTRODUCTION

Layered/graded materials hold great promise for high temperature structural applications because they permit components to be designed with tailored properties which reduce both the processing and operationally induced residual stresses to acceptable minimum levels[1]. In recent years substantial progress has been made on the mechanics of interface fracture; excellent summaries can be found [2,3]. These analyses are confined mostly to the assessment of fracture behavior at room temperature; the behavior of interface cracks at creep regime and the role of material parameters on the evolution of time dependent stress-strain fields have not been explored in detail. However, a good understanding of the mechanics of interface cracks for both temperature ranges could be most valuable in the design of these structural components through intelligent manipulation of the interfacial behavior.

In homogenous isotropic materials, the initial response of a cracked body upon application of a load at creep regime, and for some time thereafter, is essentially elastic. During this initial period, creep strains are negligible (i.e., less than the elastic strains) everywhere except in the immediate vicinity of the crack tip. This is, of course, analogous to the small-scale yielding situation at low temperatures; the elastic stress intensity factor, K, provides an adequate description of the crack tip stress-strain field. Once a steady state has been achieved, the stress-strain fields near the crack tip have a *HRR* singularity [4,5] with the amplitude given by a path independent integral C^*, the rate dependent J-integral. Between the two limiting situations of K and C^* control lies a transition period in which the creep strains are higher than the elastic strains over an extensive region ahead of the crack tip. During this period, the integral which defines C^* is path dependent; therefore, neither of these fracture parameters can characterize the stress field during the transition period.

The problem of an interface crack in elastic-plastic materials was originally investigated by Shih and Asaro [6,8]. They presented full numerical solutions for a crack which lies along the interface of elastic-plastic bimaterials. Nearly separable singular fields have been found in the angular zone based on the numerical solutions. It has also been shown that the crack face opens smoothly without rapid oscillations and has *HRR* type singularity in the plastic angular zone.

In this study, the evolution of the time-dependent stress fields ahead of stationary cracks lying at layered interfaces at creep regime and the role of the crack location on the evolution of the stress fields are elucidated. The results obtained are compared with the results for interface cracks in bimaterial systems at creep regime.

Mat. Res. Soc. Symp. Proc. Vol. 434

DETAILS OF THE FEM ANALYSIS

The cases investigated in this study are schematically summarized in Fig. 1a. First, the characteristics of the interface crack between elastic and creeping bimaterials (Fig. 1a top) is investigated. Then, in the following simulations the effects of the transitional layer and the role of the location of the interface crack are studied. In these analysis, first the crack is located at the interface of the creeping sector and the transitional layer (Fig. 1a middle). In the second simulation, the crack is located at the interface of the transitional layer and the elastic sector (Fig. 1a bottom). To simulate the interface cracks a single edge notch (SEN) fracture specimen geometry subject to a remote tension was utilized. Half of the finite element model used in the analyses is shown in Fig. 1b. The smallest element size at the crack tip was 3×10^{-5} of the crack length as indicated in Fig. 1c. To accommodate this large difference in element size and to preserve the aspect ratio of the elements, the finite element discretization elements were scaled exponentially in the radial direction to the crack tip. The analyses were carried out without imposing any symmetry conditions; there were a total of 5037 nodes and 1608 elements in the model.

The stress-strain response of the creeping sector and the transitional layer at creep regime was assumed to obey the power law constitutive relationship. The parameters for the creeping sector, selected as $n= 3$ and $\varepsilon_o \doteq 10^{-12}\,(1/(\text{Nmm}^{-2})^3\ \text{hr})$. Furthermore, Young's modulus relating the elastic strain increments was taken as 10^5 Nmm^{-2} and Poisson's ratio was chosen as $\nu = 0.3$. The Young's modulus of the elastic sector was taken as three times that of the creeping sector. The layered region is assumed to be a transition region between these two extreme regions. Therefore, a composite behavior was assumed for this transition layer with a Young's modulus of 1.5 times that of the creeping sector. For both the elastic sector and the transition layer Poisson's ratio was also taken as $\nu = 0.3$. In many composite systems, (e.g., metal matrix composites reinforced with elastic reinforcements and ceramic composites reinforced with ductile reinforcements), although the creep rate of the composites is much slower than that of the matrix material, the creep exponents (n values) of the composite are usually much higher than the one seen for the matrix material [9-10]. Therefore, the creep properties of this transitional layer were chosen as $n = 5$ and $\varepsilon_o = 10^{-20}\,(1/(\text{Nmm}^{-2})^3\ \text{hr})$. Furthermore, the properties of this layer were also assumed to be constant from one interface to the other, in order to significantly simplify the analyses.

To simulate the constant-load creep test, normal traction was applied to the end of the SEN specimen at the beginning of the solution. All the analyses presented in this study were carried out under identical applied load level.

RESULTS AND DISCUSSION

Resulting C^* values and the time to reach the steady-state for these cases are summarized below:

Crack cases	C* N/mm hour	Transition time (hours)
Fig. 1a - top	0.0195	1873
Fig. 1a-middle	0.0191	2530
Fig. 1a-bottom	0.00383	5773

As can be seen, although the applied loads and the crack lengths were the same, the location of the interface crack in a layered structure has a significant influence on the resulting C^* values at steady state and also on the time required to achieve the steady-state. For the crack at the interface of the

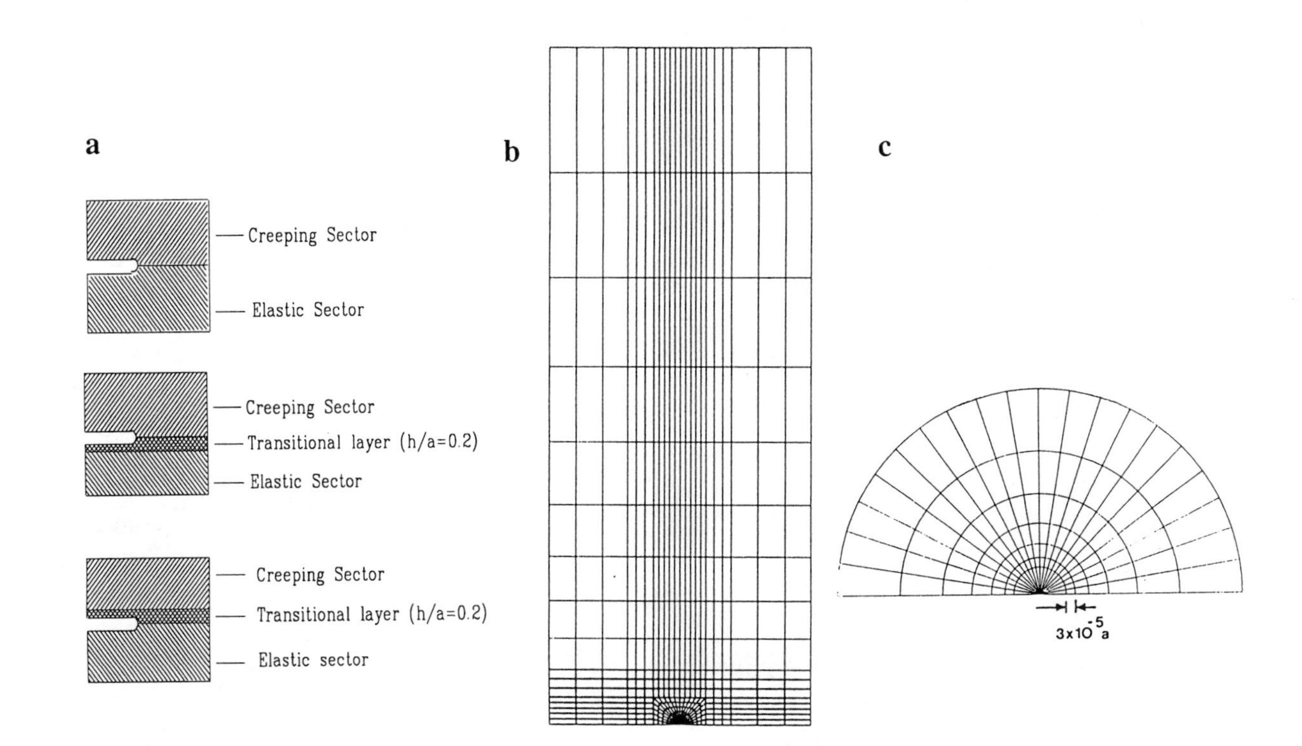

Fig. 1 a-)Schematic representation of the interface crack cases investigated, b-) FEM mesh used in the analysis and c-) the details of the crack tip region.

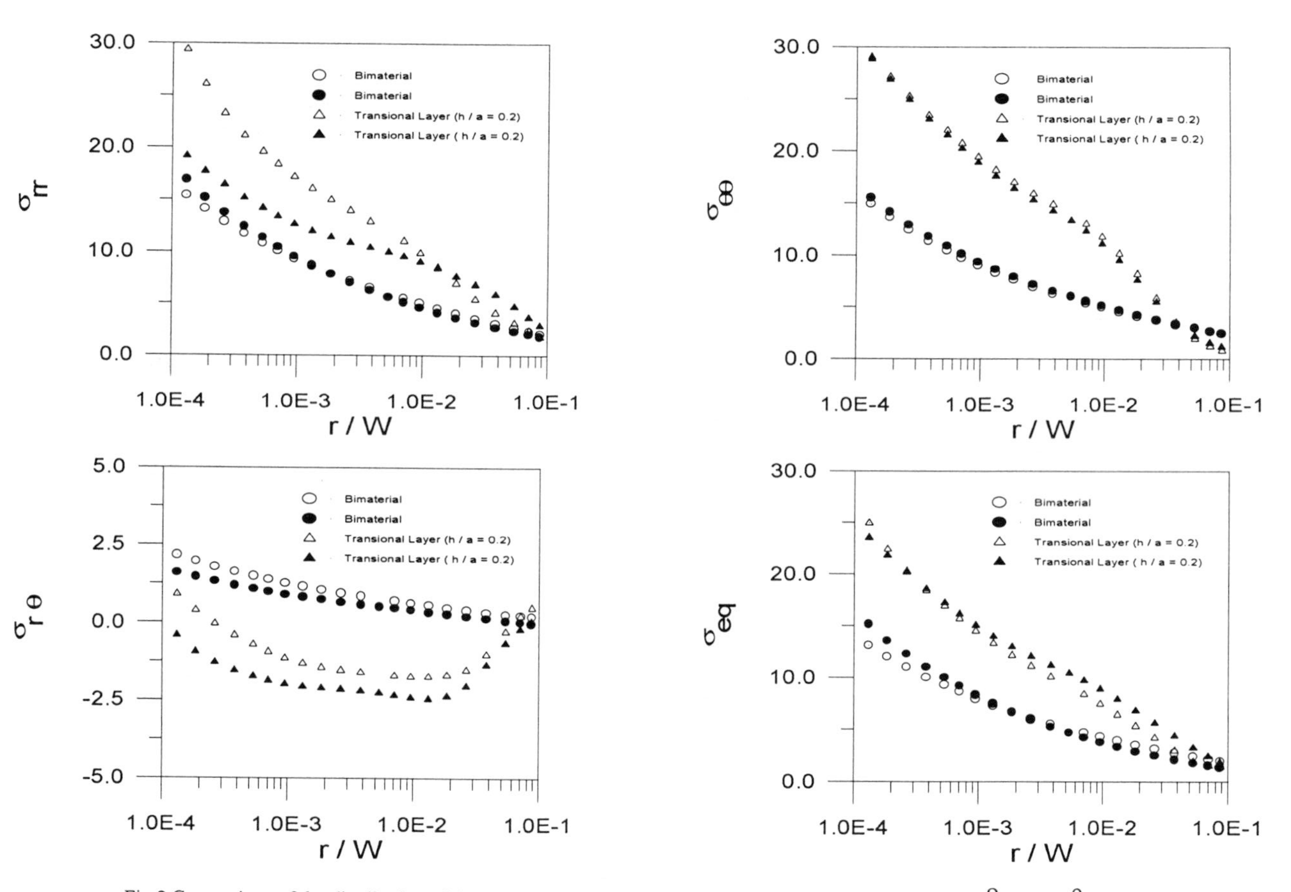

Fig.2 Comparison of the distribution of the stress components at steady-state with radial distance (θ=-7.5^{o} and 7.5^{o}) ahead of the interface cracks in bimaterial and layered material under identical applied loading conditions. In the layered material the interface crack is located at the interface of the elastic sector and the transional layer.

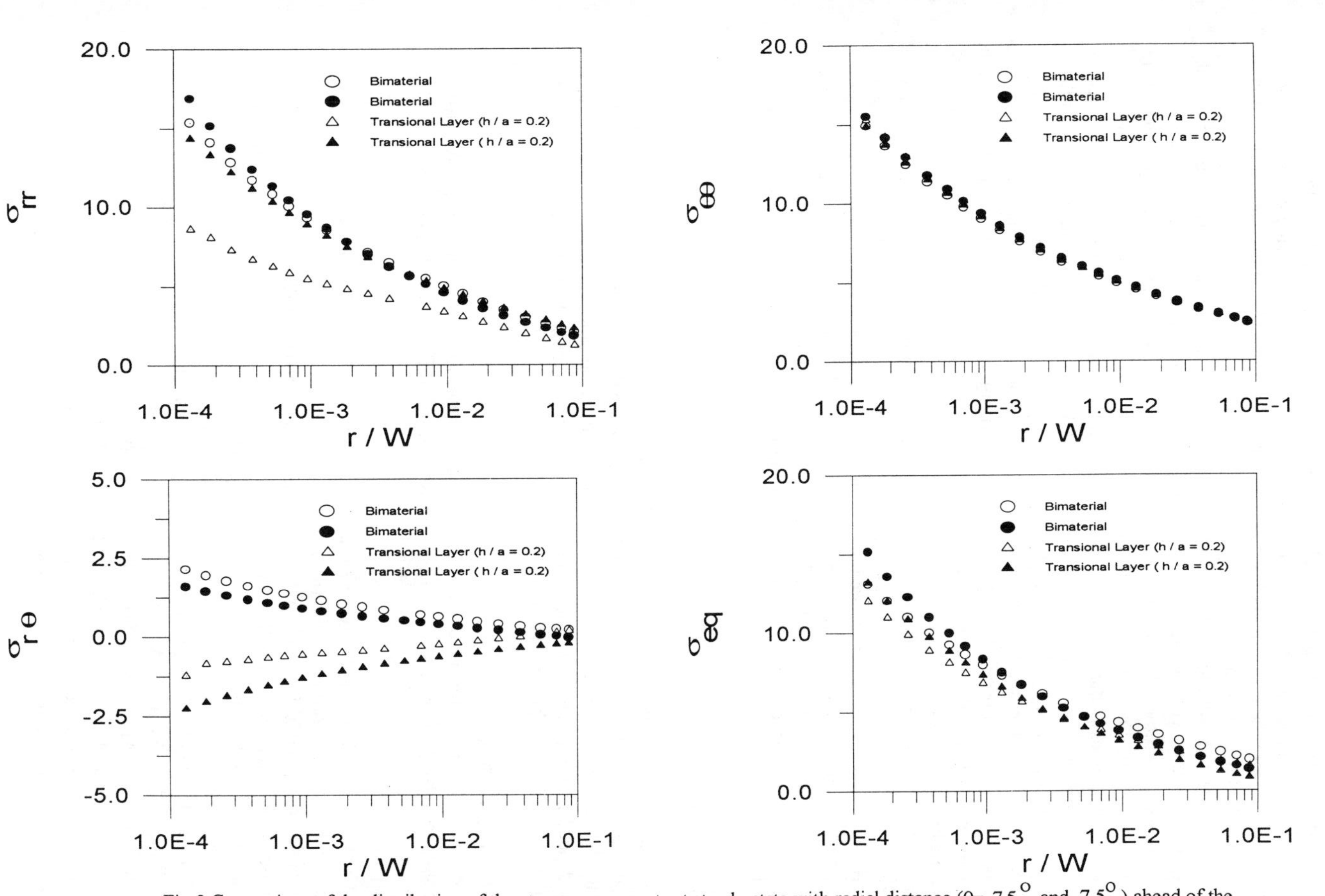

Fig.3 Comparison of the distribution of the stress components at steady-state with radial distance (θ=–7.5^{O} and 7.5^{O}) ahead of the interface cracks in bimaterial and layered material under identical applied loading conditions. In the layered material the interface crack is located at the interface of the creeping sector and the transional layer.

creeping sector and the transitional layer, although the resulting C* was similar to that of the bimaterial case, it required longer time to reach steady-state. For the crack lying on the interface transitional layer, both the magnitude of C* and the time periods were significantly different in comparison tothe bimaterial case.

In Fig. 2, the radial distribution of the stress ahead of the crack at the elastic-creeping bimaterial interface are compared with the stress distribution seen for the interface crack at the transitional layer and the elastic sectors. In this figure, the stress values obtained at the integration points were normalized by the nominal stress acting on the uncracked ligament. In both cases, the solid symbols are the stresses in the elastic sector, and open symbols are the stress in the creeping sector and in the transitional layer, respectively. As can be seen from the figure, the introduction of the transitional layer significantly elevated all the stress components at significant distances ahead of the crack tip. In Fig. 3, the same comparison is made for the interface crack lying between the creep sector and the transitional layer. In contrast to that seen in Fig. 2, in this case the introduction of the transitional layer between the elastic and creeping sectors did not alter the stress distribution that is seen for the bimaterial case.

The layered/graded materials hold great promise to reduce the processing or operationally induced residual stress levels; however, it appears that at creep regime, the resulting stress levels ahead of the interface cracks in such materials are either similar in magnitude or much larger than the stress levels that can occur ahead of the interface cracks in bimaterials. Since the stress magnitudes are widely different, the creep growth rates of such cracks will be significantly different depending on the location of the interface cracks under identically applied loading for a given layered/graded material.

CONCLUSIONS

1. The location of the interface cracks strongly influence the amplitude of the stress field (C^*) and time required to achieve steady-state under identically applied load level.
2. With the introduction of a transitional layer, the resulting stress levels ahead of the interface cracks, depending on the crack location, are either similar in magnitude or much larger than the stress levels for cracks in bimaterial interfaces.

ACKNOWLEDGMENT

This work was performed for the United States Department of Energy under contract W-7405-Eng-82 and supported by the Director of Energy Research, Office of Basic Energy Sciences.

REFERENCES

1. Int. Conf. on "Mechanics and physics of layered and graded material" Eds. S. Suresh and A. Needleman, Davos, Switzerland, 1995 (in-press).
2. J.W. Hutchinson and Z. Suo, Advances in Applied Mech. **29**, 63, (1991).
3. Metal-ceramic interfaces; Acta-Scr. Metall., Proc. Series 4, eds. M. Ruhle, A.G. Evans, M.F. Ashby and J.P. Hirth, Pergamon Press, New York, (1990).
4. J.W. Hutchinson, J. Mech. Phys. Solids. **16**, 13, (1968).
5. J.R. Rice and and G.F. Rosengreen, J. Mech. Phys. Solids. **16**, 1, (1968).
6. C.F. Shih and R.J. Asaro, J. App. Mech. **55**, 299, (1988).
7. C.F. Shih and R.J. Asaro, Int. J. Fract. **42**, 101, (1990).
8. C.F. Shih and R.J. Asaro, J. App. Mech. **56**, 763, (1989).
9. S.B. Biner and W.A. Spitzig, Mat. Sci. Eng. **A150**, 213, (1992).
10. T.G. Nieh, Metall. Trans. **15A**, 139, (1984).

MECHANICAL BEHAVIOR AND CONSTITUTIVE MODELING DURING HIGH TEMPERATURE DEFORMATION OF AL LAMINATED METAL COMPOSITES

R.B. GRISHABER, R.S. MISHRA AND A.K. MUKHERJEE
Department of Chemical Engineering and Material Science, University of California, Davis, California 95616, U.S.A., rbgrishaber@ucdavis.edu

ABSTRACT

A constitutive model for deformation of a novel laminated metal composite (LMC) which is comprised of 21 alternating layers of Al 5182 alloy and Al 6090/SiC/25p metal matrix composite (MMC) has been proposed. The LMC as well as the constituent or neat structures have been deformed in uniaxial tension within a broad range of strain-rates (*i.e.* 10^{-5} to 10^{0} s^{-1}) and homologous temperatures (*i.e.* 0.8 ≥ 0.95 T_m). The results of these experiments have led to a thorough characterization of the mechanical behavior and a subsequent semi-empirical constitutive rate equation for both the Al 5182 and Al 6090/SiC/25p when tested monolithically. These predictive relations have been coupled with a proposed model which takes into account the dynamic load sharing between the elastically stiffer and softer layers when loaded axially during isostrain deformation of the LMC. This model has led to the development of a constitutive relationship between flow stress and applied strain-rate for the laminated structure.

INTRODUCTION

Multi-layer laminate metal composites (LMCs), prepared by hot pressing alternate layers of the unreinforced and reinforced component materials and subsequent hot-rolling, can provide very attractive fracture toughness, damping capacity and also strength properties [1]. The influence of laminate architecture (layer thickness and component volume fraction) as well as component material properties and microstructural residual stress on toughening mechanisms and the resulting crack growth resistance are being actively studied [2, 3]. Interfaces in LMCs were observed to delaminate during the crack propagation leading to blunting [4, 5] or deflection [4, 6] of the crack.

One of the two monolithic (*i.e.* prior to lamination and hereafter referred to as *neat*) layer materials of the LMC studied in this work is a discontinuously reinforced aluminum (DRA) alloy whereas the other is a more conventional solid-solution strengthened aluminum (Al) alloy. Such DRA composites are typically reinforced by particles, whiskers or short fibers. They merit attention because of their desirable properties including low density, high specific stiffness, reduced coefficient of thermal expansion and increased fatigue resistance [7]. In many stiffness-, strength-, and weight-critical applications they can offer higher performance than traditional aluminum alloys at potentially lower cost than organic matrix composites [8]. DRA composites fabricated using elevated-temperature powder metallurgy based Al alloys are also attractive candidates for replacing higher cost titanium alloys in some applications [7]. Thus, the merit of this work is to characterize the mechanical behavior of the layer materials as a function of temperature and loading conditions and then to generate a predictive rate equation for correlating with the observed experimental data.

Mat. Res. Soc. Symp. Proc. Vol. 434

EXPERIMENT

The LMC is comprised of 21 alternating diffusion-bonded Al layers of a 5182 (*i.e.* Al-5Mg, 11 layers) alloy and a 6090 matrix with 25vol.% silicon carbide particulate (*i.e.* Al MMC, 10 layers) ceramic-reinforced alloy as represented in figure 1. From this figure, the average linear intercept grain size within the Al-5Mg and Al MMC layers of the LMC was determined to be 37 and 6 μm, respectably. The average linear intercept grain size of the unlaminated Al-5Mg neat material was determined to be 28 μm. For the corresponding Al

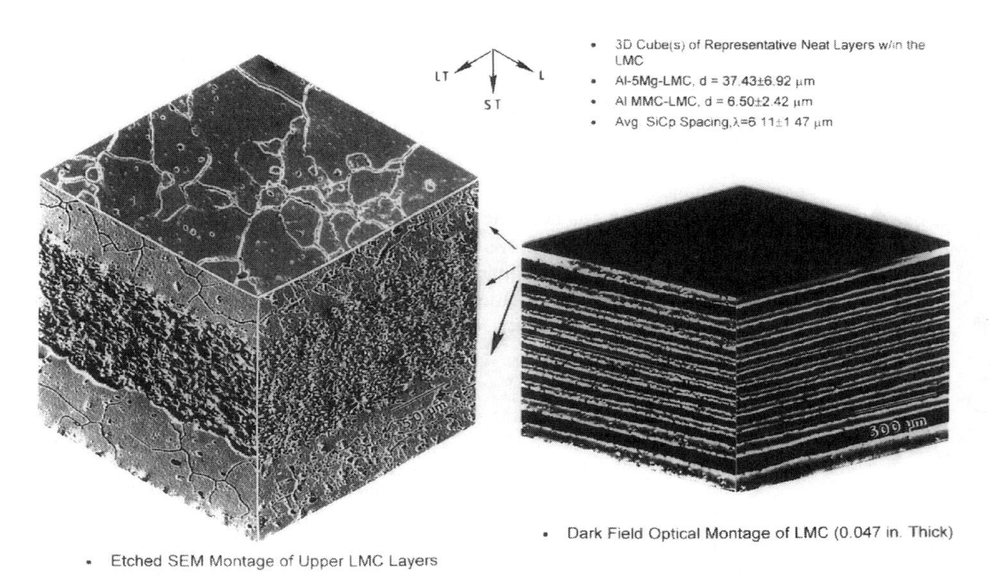

Figure 1. Representation of LMC Microstructure.

MMC neat material, the average linear intercept grain size was observed to be 6 μm. Also, the typical SiCp particulate diameter within the previous matrix was 7 μm. The sheet thickness of the LMC was approximately 1.2 mm with an average layer thickness of 62 μm. Specific details of the processing parameters can be found elsewhere [1, 2]. Uniaxial specimens were electrodischarge machined from the LMC sheet with the rolling direction corresponding to the specimen's tensile axis. Uniaxial constant strain-rate (CSR) tests were conducted within a broad range of strain-rates (*i.e.* 10^{-5} to 10^{0} s^{-1}) and homologous temperatures (*i.e.* $0.8 \geq 0.95$ T_m where T_m is the melting point temperature) in flowing argon (containing an average of 10-15 ppm O_2) at a rate of 170 cc min^{-1} and total pressure of 111.7 kPa.

RESULTS

Al-5Mg Neat Layer Characterization

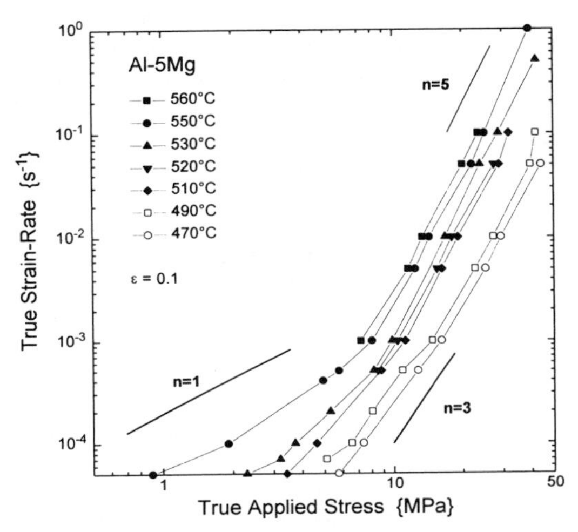

Figure 2. Strain-rate vs. Stress Response of Al-5Mg. (Flow Stress Reported at Strain, ε = 0.1)

The Al-5Mg alloy shows two distinct stress-strain-rate regimes as shown in figure 2. At low CSRs (*i.e.* $\leq 10^{-3}$ s^{-1}), a transition occurs from a solute-dislocation interaction (*i.e.* viscous creep) mechanism with a stress sensitivity exponent of 3 at the lower temperatures (*i.e.* around 470°C, 0.88 T_m) to an apparent Newtonian viscous rate controlling mechanism at a temperature (*i.e.* 550°C, 0.98 T_m) much closer to the melting point of the matrix. The latter mechanism with a stress sensitivity value of 1 can be attributed to either diffusional creep [9-11], Harper-Dorn

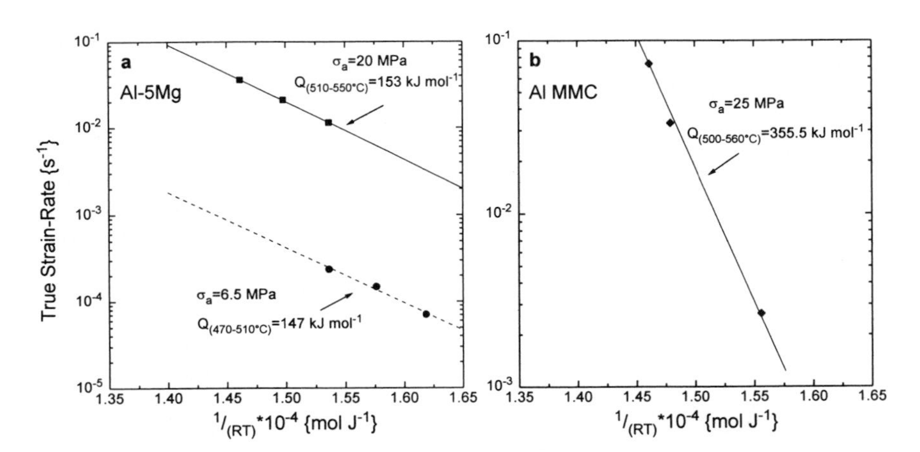

Figure 3. Arrhenius Plots to Determine Apparent Activation Energies of the (a.) Al-5Mg and (b.) Al MMC Neat Materials.

(HD) creep [12-14] or by deformation of a viscous interphase at the grain boundaries [15]. Further study will be required to positively isolate and determine the rate controlling deformation mechanism within this higher homologous temperature regime. However, for purposes of modeling in this effort an appropriate structure dependent constant was selected, assuming a linear relationship between stress and strain-rate (*i.e.* n = 1).

The second regime occurs during transition from intermediate to high (*i.e.* $\geq 10^{-3}$ s^{-1}) applied CSRs, the rate controlling deformation mechanism of viscous creep changes to one of dislocation climb controlled creep at temperatures of 0.75 T_m and above. Correspondingly, the stress sensitivity exponent increases from a value of around 3 to 5. Similar investigations into this material system conducted recently by Lee *et al.* [16] and previously by Mills *et al.* [17] support these findings. In figure 3a, an Arrhenius plot of true strain-rate vs. the inverse of temperature shows that the experimentally determined activation energies of deformation for the low, 147 kJ mol^{-1}, and high, 153 kJ mol^{-1}, stress regions of Al-5Mg are both energetically similar to the value for Al lattice self-diffusivity, 142 kJ mol^{-1}. This equivalency is expected if viscous creep and dislocation climb controlled creep are rate-controlling in the intermediate and high stress regions, respectively [18] (The more exact Darken diffusivity for viscous creep in Mg-atom locked dislocation glide is not available.).

Al 6090/SiC/25p Neat Layer Characterization

The apparent stress exponent, n, of the MMC neat material can be characterized by a transition from a value of 7 to 5 with increasing test temperature (*i.e.* 500°C, 0.88 T_m to 560°C, 0.95 T_m) within a strain-rate range of 10^{-2} to 10^{-1} s^{-1} as seen in figure 4. Furthermore, an apparent approach to a threshold stress dominated deformation regime was observed at higher test temperatures and relatively low applied CSRs (*i.e.* $\leq 10^{-2}$ s^{-1}). Also, in figure 3b, the apparent activation energy for deformation was determined from a typical Arrhenius plot with a value of approximately 356 kJ mol^{-1}. This value is simply too high for any diffusion-related correlation in an aluminum matrix. Thus, in order to model the mechanical behavior of the MMC, the true mechanistic information, from a microscopic viewpoint, must be isolated from the apparent stress sensitivity exponents and also the apparent activation energy of deformation by removing the effects related to the threshold stress state. The numerical and experimental procedures of such a process have been addressed by Pandey *et al.* [19].

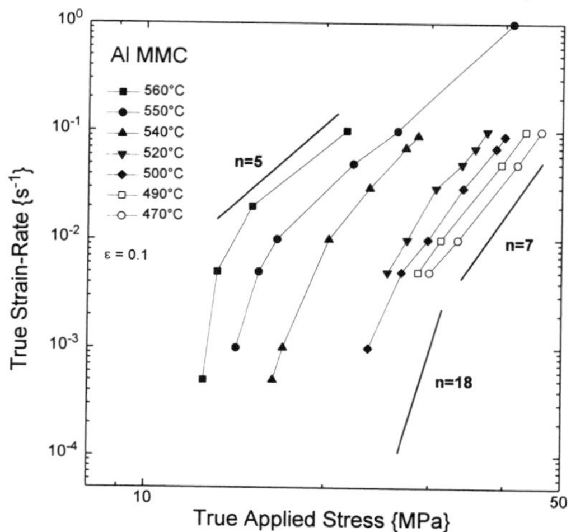

Figure 4. Strain-rate vs. Stress Response of Al MMC. (Flow Stress Reported at Strain, ε = 0.1)

Using this methodology, the true stress sensitivity exponent was determined to have a magnitude of 3 indicative of a rate controlling viscous creep deformation mechanism. This conclusion was substantiated with TEM results where significant intragranular dislocation activity was observed and there was no evidence of subgrains.

Moreover, the true activation energy for deformation, which was obtained by reanalyzing the primary data in terms of an effective stress (*i.e.* the applied stress minus the threshold stress), was determined to be equal to 156 kJ mol^{-1} which compares favorably with that reported in the previous section for Al self-diffusivity. This conclusion also supports viscous creep as being the operative rate controlling mechanism within the Al MMC neat layers.

Modeling High Temperature Deformation

This section will briefly discuss the necessary constitutive equations which govern the deformation behavior of the neat materials as well as that of the laminated structure. At relatively high homologous temperatures, typically above 0.5 T_m, the rate of deformation as a function of applied stress, σ_a, is related by a diffusionally accommodated power-law proportionality. Mukherjee *et al.* [20] used this relationship to develop an accepted semi-empirical constitutive equation for high temperature power-law deformation. The validity of this effort has been addressed elsewhere [21-23]. The constitutive equation to be used in this work is, as follows:

$$\dot{\varepsilon} = A\left(\frac{GbD_o}{kT}\right)\left(\frac{\sigma_a - \sigma_o}{G}\right)^n \left(\frac{b}{d}\right)^p \exp\left(\frac{-Q}{RT}\right) \qquad (1)$$

where σ_o is the threshold stress, n is the stress sensitivity exponent, G is the temperature sensitive shear modulus, b is the Burger's vector, D_o is the pre-exponential diffusivity term, k is the Boltzmann's constant, T is the absolute temperature, d is the linear grain size, p is the grain size exponent, Q is the apparent activation energy for deformation, R is the universal gas constant and A is the mechanism dependent structure constant. It is commonly accepted that the magnitude of the apparent activation energy for deformation, Q, will be equivalent to the activation energy for either self-diffusivity of atoms along the grain boundaries or through the lattice depending on the specific operative and rate controlling micromechanism. Also, for single phase and non-reinforced alloys, the threshold stress value of eqn. 1 is considered negligible in this work.

This constitutive relationship has been shown to reasonably predict the mechanical behavior of dispersion-strengthened materials [19, 24] within the power-law creep controlled deformation region. The method of using an effective stress (*i.e.* σ_a - σ_o) for DRAs rather than the applied stress aides in determining realistic activation energies and stress exponent values [25]. The recent works of Bieler *et al.* [26] and an overview by Mishra *et al.* [27] have shed significantly more information on this particular area.

Constitutive Modeling of Laminated Metal Composites

Using the results and constitutive relations previously discussed, the primary focus of this section will be to present a theoretical model for predicting the micromechanical response to an applied stress by the laminated structure. First, the following assumptions will be considered:

a) the laminated structure will be loaded such that the all of the individual layers are acted upon by an applied force in a parallel configuration,
b) isostrain and correspondingly isostrain-rate conditions will apply continuously throughout the entire deformation process,

c) load sharing to maintain the previous assumption will occur within the LMC where the elastically stiffer layers will accumulate a larger portion of the total applied load in order to maintain strain compatibility,

d) the initial volume fractions of each neat phase within the deforming gage volume will not vary as a function of strain accumulation.

Thus, given the above set of assumptions, it would be reasonable to expect that a LMC would exhibit a rule-of-mixtures response attributed to the mechanical behavior associated with each neat layer under similar deformation conditions. This can be visualized by considering the issue of strain compatibility where the elastically stiffer layers will accumulate a greater share of the overall applied load such that isostrain conditions are maintained.

The individual layers which make up the LMC are all loaded axially. It is expected that the operative rate controlling mechanism within each of the laminate's neat layers will be equivalent to that determined for its monolithic component deformed under identical conditions. Therefore, using principles discussed in a review by Lilholt [28] for load sharing and correspondingly stress sharing under elastically loaded isostrain conditions, the stress of the composite, σ_{LMC}, can be written as:

$$\sigma_{LMC} = (V \cdot \sigma)\big|_{alloy} + (V \cdot \sigma)\big|_{MMC} \qquad (2).$$

Schematically, this model for isostrain deformation has been illustrated in figure 5.

The approach to modeling the deformation behavior of the LMC is similar to that originally presented by McDanels [29] for creep of an unidirectional continuous-fiber composite and later advanced by Kelly and Street [30, 31] and more recently by Goto and McLean [32-34]. The composite model in this treatment considers the condition where both the reinforced fibers and the matrix phase deform by power-law creep. We consider the

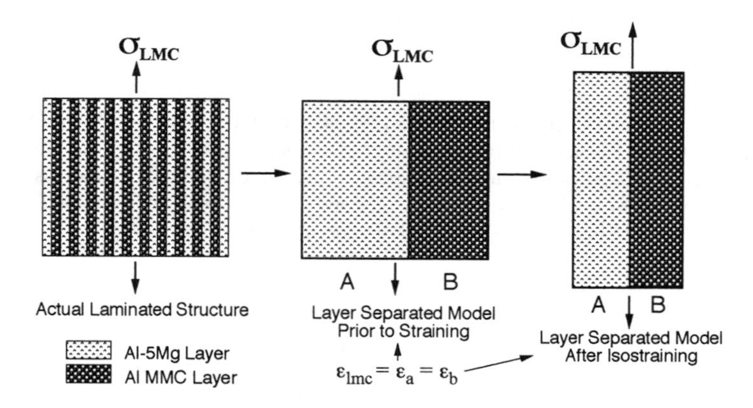

Figure 5. Schematic Model of Isostrain Loading Configuration for the Laminate.

reinforcement fibers in the previous models to be equivalent to the MMC layers within the LMC. As shown in figure 6, the response of the laminated structure to an applied constant strain-rate can be modeled with eqn. 2 using the constitutive predictions generated by eqn. 1 for the alloy and MMC neat structures deformed monolithically under identical conditions.

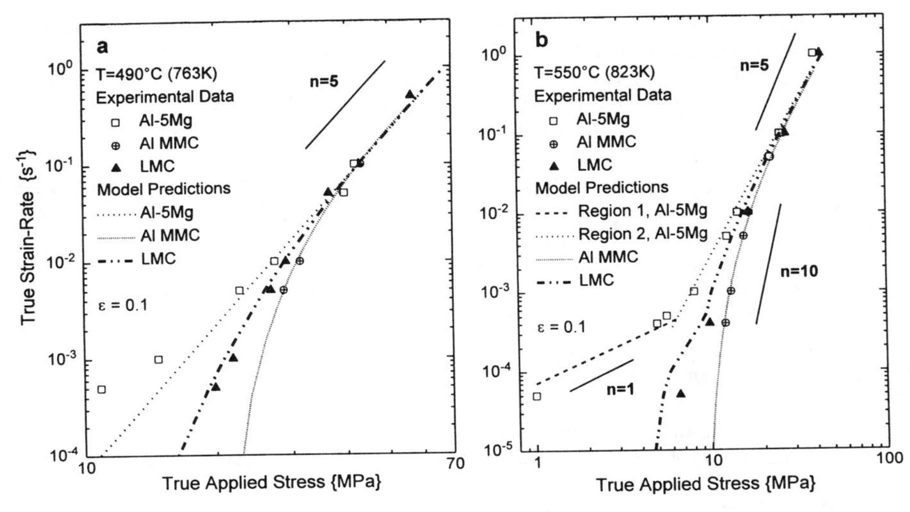

Figure 6. Deformation Modeling of the LMC at (a.) 490°C and (b.) 550°C.

The constitutive predictions for the deformation response of the LMC and the two neat materials at both 490°C and 550°C are presented in this figure with experimental results. From this figure, it is clearly evident that the isostrain model presented in this work can be used to reasonably predict the high temperature mechanical behavior of Al-matrix laminated metal composites.

CONCLUSIONS

By *a priori* characterization of the high temperature mechanical behavior of the individual layer materials which comprise the laminated structure of interest, a model which accounts for load sharing between the elastically stiffer and softer layers during isostrain deformation has been presented and experimentally verified.

ACKNOWLEDGMENTS

This work is supported by grant no. F49620-93-1-0466 from the Department of Defense (DOD) Air Force Office of Scientific Research (AFOSR) under the auspices of Dr. Charles H. Ward, Capt. USAF, Bolling AFB, DC. The authors would also like to especially acknowledge Dr. Donald R. Lesuer and Dr. Chol K. Syn of Lawrence Livermore National Laboratory in California for providing the material used in this work.

REFERENCES

1. C. K. Syn, D. R. Lesuer and O. D. Sherby in Light Materials for Transportation Systems, ed. N. J. Kim, (Center for Advanced Aerospace Materials, Kyongju, Korea, 1993), p. 763.

2. D. Lesuer, C. Syn, R. Riddle, O. Sherby, to be published in Symposium on Intrinsic and Extrinsic Fracture Mechanisms in Discontinuously Reinforced Inorganic Composite Systems (TMS Meeting, Las Vegas, NV, 1995).
3. M. Manoharan, L. Ellis, J. J. Lewandowski, Scripta Met. et Mater., **24**, 1515 (1990).
4. R. O. Ritchi, W. Yu, and R. J. Bucci, Eng. Fract. Mechanics, **32**, 362 (1989).
5. C. K. Syn, S. Stoner, D. R. Lesuer, O. D. Sherby, to be published in Symposium on High Performance Ceramic & Metal Matrix Composites (TMS Meeting, San Francisco, CA, 1994).
6. T. M. Osman, M. S. Thesis, Case Western Reserve University, 1993.
7. W. H. Hunt, T. M. Osman and J. J. Lewandowski, J.O.M., **45**, 30 (1993).
8. M. B. McKimpson, E. L. Pohlenz and S. R. Thompson, J.O.M., **45**, 26 (1993).
9. F. R. N. Nabarro, Rep. Conf. Strength Solids, 1948, 75-90.
10. C. Herring, Applied Physics, **21**, 437 (1950).
11. R. L. Coble, Applied Physics, **34**, 1679 (1963).
12. J. Harper and J. E. Dorn, Acta Metall., **5**, 654 (1957).
13. J. Fiala and T. G. Langdon, Mater. Sci. Engr., **A151**, 147 (1992).
14. O. A. Ruano, J. Wolfenstine, J. Wadsworth and O. D. Sherby, Acta Metall., **39**, 661 (1991).
15. B. Baudelet, M. C. Dang, F. Bourdeaux, Scripta Metall. Mater., **26**, 573 (1992).
16. D. H. Lee, K. T. Hong, D. H. Shin and S. W. Nam, Mater. Sci. Engr., **A156**, 43 (1992).
17. M. J. Mills, J. C. Gibeling and W. D. Nix, Acta Metall., **33**, 1503 (1985).
18. A. K. Mukherjee in Materials Science and Technology, eds. R. W. Cahn, P. Haasen and E. J. Kramer, vol. ed. H. Mughrabi, (VCH, **6**, 1993) pp. 407-460.
19. A. B. Pandey, R. S. Mishra and Y. R. Mahajan, Scripta Metall., **24**, 1565 (1990).
20. A. K. Mukherjee, J. E. Bird and J. E. Dorn, Trans. ASM, **62**, 155 (1969).
21. A. K. Mukherjee in Treatise on Materials Science and Technology, ed. R. J. Arsenault, Vol. 6, 163 (1975).
22. O. D. Sherby and J. Wadsworth in Progress in Materials Science, Vol. 33, 169 (1989).
23. H. J. Frost and M. F. Ashby, Deformation-Mechanism Maps (Pergamon Press, Oxford, 1982), p. 12.
24. R. S. Mishra, A. G. Paradkar and K. N. Rao, Acta Metall., **41**, 2243 (1993).
25. R. S. Mishra and A. K. Mukherjee, Scripta Metall., **25**, 271 (1991).
26. T. R. Bieler, R. S. Mishra and A. K. Mukherjee in Materials Science Forums, Vols. 170-172, 65 (1994).
27. R. S. Mishra, T. R. Bieler and A. K. Mukherjee, Acta Metall. Mater., **43**, 877 (1995).
28. H. Lilholt in Mechanical Properties of Metallic Composites, ed. S. Ochiai (Marcel Dekker, Inc., NY, 1994), pp. 389-471.
29. D. McDanels, R. A. Signorelli and J. W. Weeton, NASA Report TN-4173 (1967).
30. A. Kelly and K. N. Street, Proc. R. Soc., **A328**, 267 (1972a).
31. A. Kelly and K. N. Street, Proc. R. Soc., **A328**, 283 (1972b).
32. S. Goto and M. McLean, Scripta Metall., **23**, 2073 (1989).
33. S. Goto and M. McLean, Acta Metall., **39**, 153 (1991).
34. S. Goto and M. McLean, Acta Metall., **39**, 165 (1991).

Mechanical Behavior of a Ni/TiC Microlaminate Under Static and Fatigue Loading

Y.C. Her, P.C. Wang, J.-M. Yang, and R.F. Bunshah
Department of Materials Science and Engineering, University of California, Los Angeles, CA 90095

ABSTRACT

The mechanical behavior and damage mechanisms of the Ni/TiC microlaminate composites under static and cyclic loading were investigated. The relationship between the ultimate tensile strength and the layer thickness at both room temperature and 600°C was studied. The fatigue life and the evolution of the stiffness reduction under various maximum applied stress levels were determined. The results revealed that the ultimate tensile strength linearly increased as the laminate layer thickness decreased. Also, the microlaminate exhibited a non-progressive fatigue behavior.

INTRODUCTION

Microlaminated composites consist of a number of alternating layers of two different materials, such as metal/metal, metal/ceramic or ceramic/ceramic systems. As a potential high strength, high toughness material, microlaminated composites have a significant advantage over fibrous composites because of their in-plane isotropic properties. The properties of the microlaminated composites depend upon the structure and properties of each of the components, the respective volume fraction, the interlaminar spacing, their mutual solubility and the possible formation of a brittle intermetallic compound between them. Thus, the structure and physico-mechanical properties of these microlaminated composite materials can be varied and controlled within wide ranges. This will open up the possibility of tailoring materials with unique combination of properties to be used either as self-supporting shapes or in the form of coatings for various applications. Several microlaminate systems have been investigated, which include Ni/Cu [1-3], Cu/Fe [3], Au/Ni [4], Mo/W [5], TiC/TiB_2 [6], Ni/TiC [7], Nb_3Al/Nb [8], and Cr_2Nb/Nb [8].

This work was conducted to study the mechanical behavior of a Ni/TiC microlaminated composite. The mechanical behavior and damage mechanisms under both static and cyclic loading were investigated.

MATERIALS AND EXPERIMENTAL PROCEDURES

The deposition of the Ni/TiC microlaminates was carried out by alternating direct evaporation for the pure nickel layers and activated reactive evaporation (ARE) for the TiC layers. A water-cooled shield was placed between the TiC and Ni evaporation sources to prevent the mixture of the vapor streams prior to deposition onto a rotating substrate. The substrate was located above the sources and heated by a radiant heater from the back side. The schematic diagram of the apparatus is shown in Fig.1 [3]. The volume fraction of each material and the individual layer thickness was controlled by adjusting the deposition rate of the respective material and the rotation speed of the substrate. The temperature of the deposition was 620 °C. The diameter of the condensates with the thickness varying along the radial axes was about 40 cm.

Mat. Res. Soc. Symp. Proc. Vol. 434 © 1996 Materials Research Society

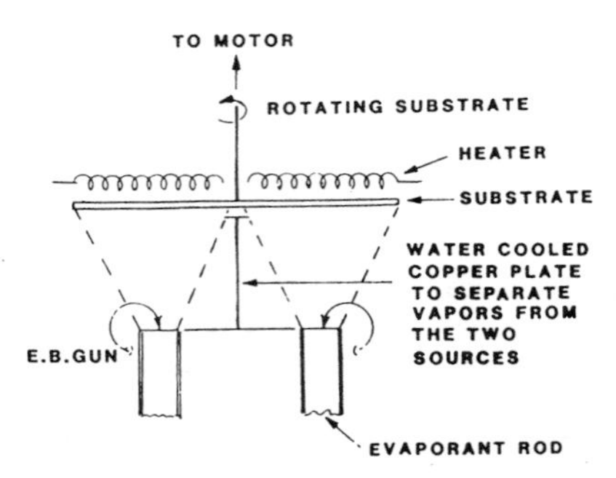

Fig.1 Schematic diagram of the vacuum evaporation apparatus with two sources for depositing laminate composites on a rotating substrate

Microstructure of the resulting Ni/TiC composite was analyzed by X-ray diffraction and SEM with energy-dispersive spectroscopy (EDS).

Tensile tests were conducted at room temperature in air and at 600°C in an argon atmosphere on an Instron testing machine. The cross-head speed was 0.5mm min^{-1}. Straight-edge specimens with dimensions of 6.25mm (0.25in) x 100mm (4in) were cut from a single deposit in the circumferential direction. A thin foil strain gauge with a gauge length of 3.2mm was mounted on the specimen to measure the corresponding strain. Low cycle fatigue tests were conducted on a closed-loop, servo-hydraulic Instron testing machine. Fatigue specimens were cut to the dimensions of 6.25mm (0.25in) x 76mm (3in). Specimen edges were polished by SiC sand papers to 600 grid in order to reduce the damage done during machining. All specimens were tested in a load-controlled mode at a frequency of 15 Hz and R=0.1. The maximum applied stresses varied from 250 to 550 MPa. Thin foil strain gauges were epoxy bonded on each specimen in order to monitor the evolution of the stiffness degradation. The fracture surfaces of specimens and the polished cross sections near the fracture surface after tensile and fatigue testing were examined by a scanning electron microscope.

RESULTS AND DISCUSSIONS

1. Microstructure and Composition

The cross section of the Ni/TiC microlaminate, as shown in Fig.2, clearly shows discrete alternating layers of Ni and TiC in the composite. X-ray diffraction studies confirmed that all the laminates were composed of Ni and TiC. No interdiffusion between the Ni and TiC was found in the as-deposited state from the micrographs. By utilizing direct evaporation for the nickel layers and ARE for the TiC layers, a Ni/TiC microlaminate composite with Ni:TiC ratio of 5.5:1 was obtained. The volume fraction of the Ni and TiC layer was measured to be 85 and 15 %, respectively.

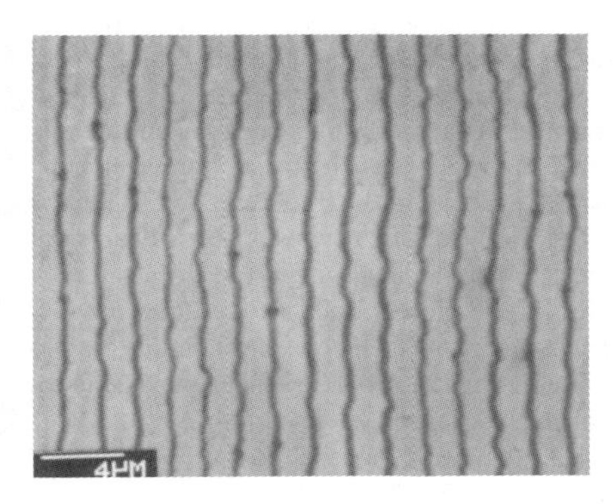

Fig.2 SEM micrograph of the Ni/TiC microlaminate

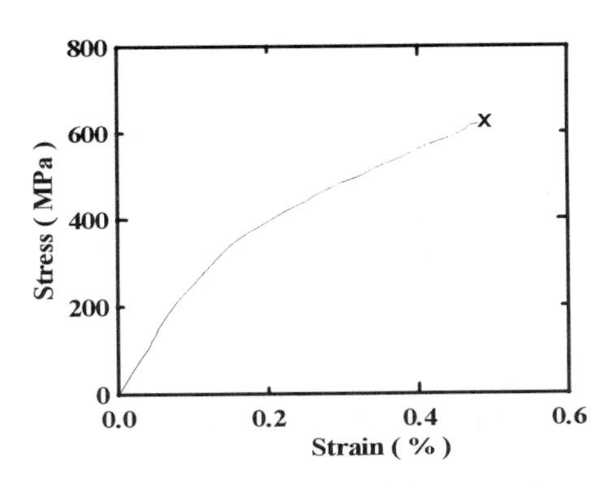

Fig.3a Typical stress-strain curve of the Ni/TiC microlaminate at room temperature

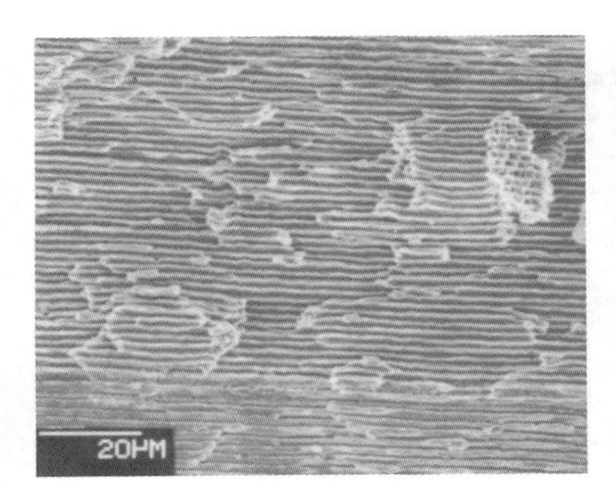

Fig.3b Tensile fracture surface of the Ni/TiC microlaminate

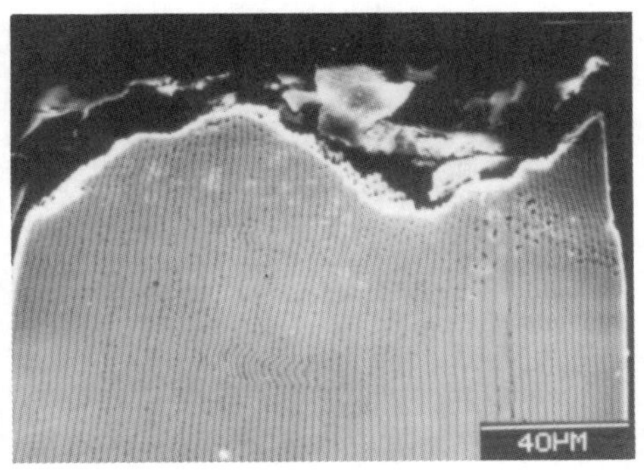

Fig.3c Cross-sectional area near the tensile fracture surface

2. Tensile behavior

A typical stress-strain curve of the Ni/TiC microlaminate composite with layer thickness of 1.74 μm tested at room temperature is shown in Fig.3a. The microlaminate exhibits a short linear stress-strain behavior, followed by a non-linear behavior prior to final fracture. The initial Young's modulus was measured to be 262 GPa, which is close to that predicted by the rule of mixture. The ultimate tensile strength and the strain-to-failure were measured to be 620 MPa and 0.48 %, respectively. The UTS of a vapor-deposited Ni was measured to be 276 and 73 MPa at room temperature and 600°C, respectively [7]. It is obvious that the UTS of a Ni/TiC microlaminate is superior to that of the pure Ni. Fig.3b shows the fracture surface of the Ni/TiC microlaminate. The nickel layers with knife edge fracture profile failed by ductile shear fracture and the TiC layers failed in a brittle manner. Fig.3c is the cross-sectional area near the fracture surface. Fragmentation of the brittle TiC layers and localized yielding of the ductile Ni layers adjacent to the fractured TiC layer were observed near the fracture surface. Based upon the stress-strain response and fracture characteristics, the damage sequences of the Ni/TiC microlaminate under tensile loading was proposed as follows. As the applied tensile load increased, the weakest TiC layer failed first and the load was transferred to the neighboring Ni layers and the surviving TiC layers. This load redistribution resulted in the microyielding of the Ni layers, multiple cracking of the same TiC layer, and the rupture of the other TiC layers. As the applied stress increased, the accumulated damage resulted in the reduction of the load carrying capacity of the microlaminate. Once the applied load exceeds the residual load carrying capacity of this material, catastrophic failure of the Ni/TiC microlaminate occurred. It was found that the zone of the fragmentation of the brittle TiC layers and the localized yielding of the ductile Ni layers were constrained within 100 μm which was much smaller than the gauge length (3.2mm). Therefore, the strain-to-failure is small. The non-linear stress-strain behavior is due to the yielding of Ni layers and the progressive rupture of the TiC layers.

Fig.4 represents the effect of the layer thickness on the ultimate tensile strength (UTS) of the microlaminate composite at room temperature and at 600 °C, respectively. It clearly indicates that at both room and elevated temperatures, the UTS increased linearly as the layer thickness decreased.

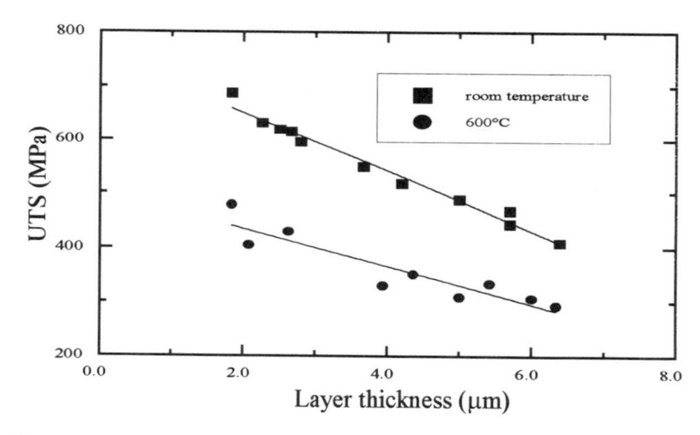

Fig.4 UTS vs. laminate layer thickness (one Ni layer plus one TiC layer) at room temperature and 600°C

3. Low cycle fatigue behavior

The maximum applied stress as a function of fatigue life of the Ni/TiC microlaminates is plotted in Fig.5a. The microlaminate exhibited a wide range of scattering in fatigue life at maximum applied stresses between 350-550 MPa. At σ_{max} = 250 MPa, the microlaminate did not fail after being loaded to 10^6 cycles. The stiffness reduction as a function of fatigue cycles at various stress levels is shown in Fig.5b. It indicates that no significant variation in the stiffness was observed during fatigue testing until the specimen failed catastrophically. The above results strongly suggest that the Ni/TiC microlaminate exhibits a non-progressive fatigue behavior in which neither stiffness reduction nor microstructural damage was detected before catastrophic failure. A typical fracture surface of the Ni/TiC microlaminates after fatigue testing at different stress levels is shown in Fig.6. Processing defects such as flake (Fig.6a) and spit (Fig.6b) defects were found on the fracture surface of each specimen. The spit defect is produced by a small droplet ejected from the molten pool, which lands on the substrate and is incorporated into the coating. The flake defect is produced by accelerated coating deposition on a foreign particle. Since the bonding between the processing defect and the deposit is usually poor, decohesion at the interface between the defect and the deposit may trigger the rapid rupture of TiC layers. This would lead to the catastrophic failure of the laminate. Therefore, the fatigue life of the Ni/TiC microlaminate is controlled by the size, distribution and bonding conditions of growth defects.

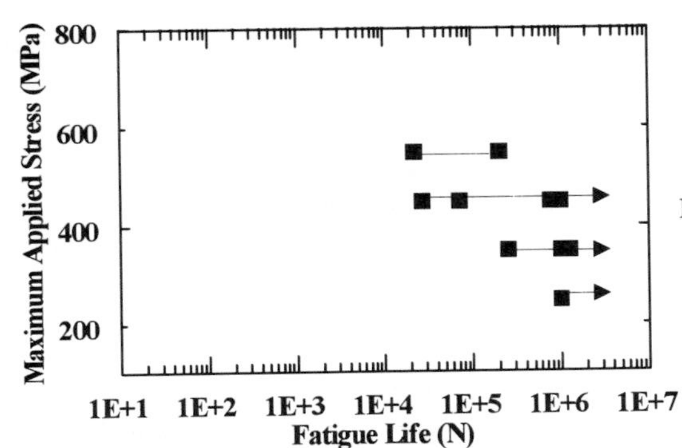

Fig.5a The S-N curve of the Ni/TiC microlaminate

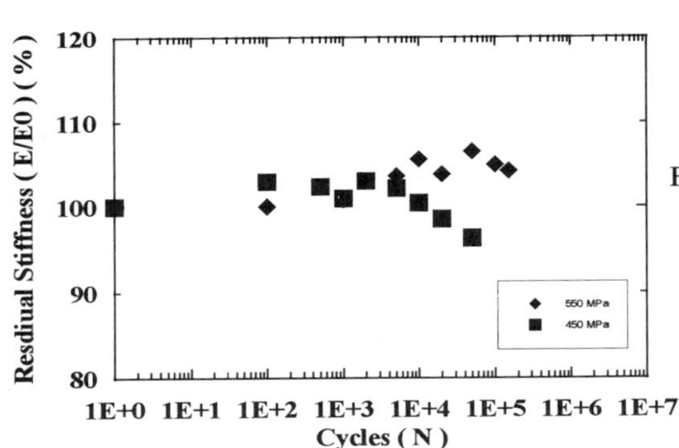

Fig.5b The stiffness reduction curve of the Ni/TiC microlaminate

Fig.6a The fatigue fracture surface of the Ni/TiC microlaminate with a flake defect

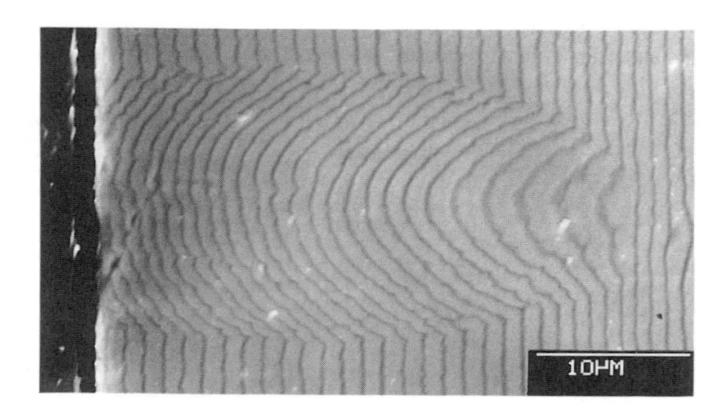

Fig.6b A spit defect in the Ni/TiC microlaminate

CONCLUSIONS

1. Ni/TiC microlaminate composites have been successfully produced by direct evaporation for the nickel layers and ARE for the TiC layers. The typical stress-strain curve of the Ni/TiC microlaminate composite exhibited an initial linear elastic region followed by a non-linear stress-strain behavior. The UTS of the Ni/TiC microlaminate composites increased linearly as the layer thickness decreased at both room temperature and 600 °C.
2. The Ni/TiC microlaminated composites exhibit a non-progressive fatigue crack propagation behavior. The fatigue life of the microlaminates depended strongly on the size, distribution, and bonding conditions of the processing defects.

ACKNOWLEDGMENTS

This work was partially supported by the National Science Foundation (DDM9057030).

REFERENCES

1. T.Tsakalakos and J.E.Hilliard, J. Appl. Phys. 54,734(1983)
2. R.C.Cammarata, T.E.Schlesinger, K.Kim, S.B.Qadri, and A.S.Edelstein, Appl. Phys. Lett. 56,1862(1990)
3. R.F.Bunshah, R.Nimmagadda, and H.J.Doerr, The Solid Films, 72(1980) 261-275
4. W.M.C.Yang, T.Tsakalakos and J.E.Hilliard, J. Appl. Phys. 48,876(1977)
5. M.Vill, D.P.Adams, S.M.Yalisove, and J.C.Bilello, Acta. Metall. Mater. Vol.43, 4. No.2,427(1995)
6. B.A.Movchan, A.V.Demchishin, G.F.Badilenko, R.F.Bunshah, C.Sans, C.Deshpandey and H.J.Doerr, Thin solid films, 97(1982) 215-219
7. R.F.Bunshah, C.Sans, C.Deshpandey, H.J.Doerr, B.A.Movchan, A.V.Demchishin, G.F.Badilenko, and Thin solid films, 96(1982) 59-66
8. R.G.Rowe, D.W.Skelly, M.Larsen, J.Heathcote, G.Lucas, and G.R.Odette, Mater. Res. Soc. Symp. Proc. vol.323, Boston. p.461, 1933
9. R.F.Bunshah, Handbook of deposition technologies for films and coatings, p.211 (1994)

FATIGUE CRACK GROWTH IN ALUMINUM LAMINATE COMPOSITES

P. B. HOFFMAN*, R. D. CARPENTER** AND J. C. GIBELING**
* Technology Center, Columbian Chemicals Company, P.O. Box 96, Swartz, LA 71281
** Division of Materials Science and Engineering, University of California, Davis, CA 95616

ABSTRACT

Fatigue crack growth has been measured in a laminated metal composite (LMC) consisting of alternating layers of AA6090/SiC/25p metal matrix composite (MMC) and AA5182 alloy. This material was tested in both as-pressed (F temper) and aged (T6 temper) conditions. Corresponding crack growth measurements were made in self-laminates of both the MMC and AA5182 materials to examine the role of the interfaces.

The LMC-T6 material has a significantly higher fracture toughness than its MMC constituent but exhibits only modest improvements in nominal fatigue crack growth resistance and a lower nominal threshold. The LMC-T6 shows high levels of crack closure which reveal poor intrinsic crack growth resistance. Self-laminated AA6090/SiC/25p MMC in the F temper has lower microhardness, lower strength and superior intrinsic crack growth resistance compared to this material in the T6 condition. The fatigue fracture surface of T6 temper MMC remains macroscopically smooth despite the development of a large amount of crack closure. The self-laminated AA5182 exhibits a significant improvement in fatigue crack growth resistance after heat treatment and a corresponding change in crack morphology from smooth to very rough. This change leads to a large increase in roughness-induced closure. The effects of heat treatment are attributed to differences in interfacial strength.

INTRODUCTION

Discontinuously reinforced aluminum (DRA) metal matrix composites (MMCs) have received widespread attention as candidate materials for applications in which high specific strength and stiffness are requirements. While a variety of material systems have been produced to meet these primary goals, the relatively low fracture toughness of these composites compared to conventional aluminum alloys has remained of concern. For this reason, several investigators have devised extrinsic methods to improve the fracture toughness of MMCs [1-5]. One approach that has shown particular promise involves lamination of the DRA with a more ductile alloy. This approach has been shown to lead to substantial increases in fracture toughness in model alloys created by epoxy bonding as well as demonstration alloys prepared by diffusion bonding [3-5].

Many of the applications for which high fracture toughness is desired, especially in aerospace structures, also require that materials exhibit good fatigue crack propagation resistance. Unfortunately, relatively little is known about the mechanisms of fatigue crack propagation in laminated structures. Our previous work on aluminum laminates composed of alternating layers of AA6090/SiC/25p and AA5182 demonstrated that although the fracture toughness of the laminate in the nominal T6 temper is superior to that of its MMC constituent, it exhibits only modestly superior fatigue crack growth resistance and a lower nominal threshold [6]. The AA5182 layers in the laminate exhibited substantial amounts of shear failure and fracture surface roughness, suggesting that both plasticity and roughness-induced crack closure provide some

Mat. Res. Soc. Symp. Proc. Vol. 434

extrinsic toughening, at least at intermediate growth rates. Other work by Chawla and Liaw on roll bonded pure aluminum and 304 stainless steel revealed that crack growth rates in the laminate were lower than in either component [7]. These limited studies suggest that the fatigue crack growth behavior of composite laminates may depend on the properties of the individual laminates as well as the characteristics of the interfaces, which are both affected by heat treatment.

Because of the limitations of our current knowledge, the goal of the present study is to expand on our previous efforts to understand fatigue crack propagation in laminated aluminum composites. We seek to understand the influence of heat treatment on the behavior of composite laminates as well as self-laminates made of each of the two constituents. Of particular concern is the influence of matrix strength and interface debonding on crack growth rates in the near-threshold regime.

EXPERIMENTAL

Laminated metal composites (LMCs) were produced at Lawrence Livermore National Laboratory in the same manner as previously reported [4-6]. Each LMC consists of twenty-one alternating layers produced by stacking eleven layers of AA5182 with ten layers of AA6090/SiC/25p, each layer measuring 51 mm by 51 mm square and 2.5 mm thick. Prior to stacking, the layers were chemically cleaned to remove the surface oxide scale. The stacked layers were then hot pressed at 450°C in an argon gas atmosphere to approximately 1/3 of their initial thickness. Due to friction between the press platens and the outer layers, these layers were under greater lateral constraint than the inner layers, resulting in their being thicker than the inner layers. In addition, self-laminates of both AA5182 and AA6090/SiC/25p were produced by a similar method in order to study separately the effects of composite lamination and layering. All laminates were finished by heating to 450°C and warm rolling in the same direction as the rolling direction of the as-received sheets. The hot pressed and warm rolled laminates were cut into the required compact type (C(T)) specimens in the L-T orientation with respect to the rolling direction of the original sheet material by wire electrostatic-discharge-machining. To produce a T6 condition, specimens were then heated in air at 560°C for one hour, quenched in agitated warm water and aged at 160°C for sixteen hours followed by air cooling. This heat treatment represents the nominal treatment to develop peak strength in the AA6090 alloy, and is not expected to alter the microstructure or properties of the solid solution strengthened AA5182.

Fatigue crack propagation experiments were conducted in accordance with ASTM standard E-647-88a using C(T) specimens machined in the crack divider orientation in which a crack grows simultaneously in all layers [8]. The thickness of the composite laminate specimens was 10.0 mm, while the AA5182 self-laminate specimens were 10.9 mm thick and the AA6090/SiC/25p self-laminates were 11.0 mm thick. All specimens had dimensions of width 44.45 mm and height 53.34 mm. Testing was performed in ambient air with a sinusoidal wave form at 20 Hz and a load ratio $R = 0.1$. Low and near-threshold regions of crack growth were investigated with K-decreasing tests using a K-controlled step method with a stress gradient $c = -0.08\ mm^{-1}$.

The servohydraulic testing system and associated data acquisition and control software were the same as those used in the previous experiments on metal composites [6,9]. Crack length measurements were made in real time using a COD-based compliance method [10]. In addition, crack closure detection was based on calculation of the maximum correlation coefficient in a least squares regression fit of load *vs.* COD data. Closure was defined as the point at which the correlation coefficient, excluding the load reversal region at the peak of the cycle, begins to decrease. This condition reflects a deviation of the load *vs.* COD curve from linearity. The load at this point is defined as the closure load and the corresponding stress intensity at closure is K_{cl}.

RESULTS

The nominal fatigue crack propagation rates in the laminate for the T6 and F tempers are compared to the behavior of the monolithic MMC in Figure 1. These data reveal that in the F temper, the LMC exhibits significantly higher crack growth rates than the MMC. However, the behavior of the LMC is modestly superior to that of the MMC when both are tested in the T6 condition, although the laminate appears to exhibit a slightly lower threshold. Based on the complex structure of these materials, it is reasonable to expect relatively high levels of crack closure. This expectation is confirmed by the measured K_{cl} results illustrated in Figure 2, showing a dramatic rise in closure levels from the start of the test (high ΔK) to the end of the test (low ΔK). The LMC in the T6 condition exhibits the highest closure levels, confirming that the fatigue crack growth resistance of this material can be attributed to extrinsic mechanisms.

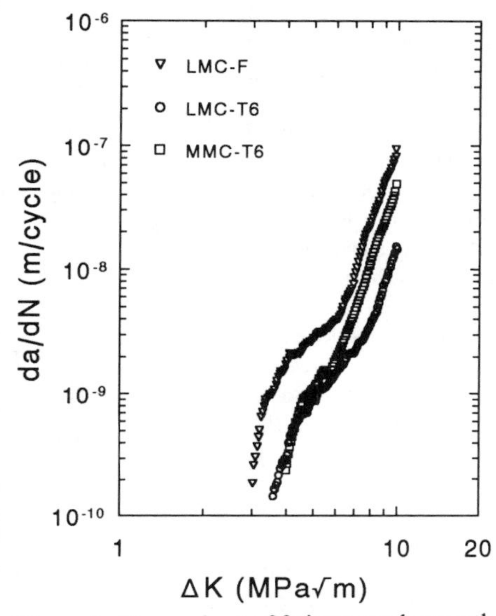

Figure 1. Comparison of fatigue crack growth behavior in the LMC-F, LMC-T6 and MMC-T6.

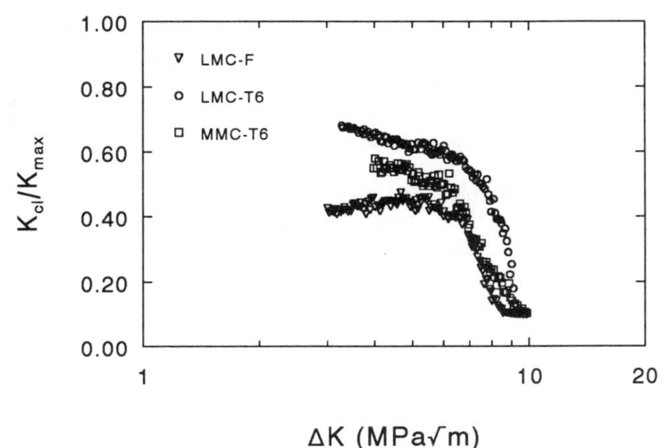

Figure 2. Relative crack closure levels in the LMC-F, LMC-T6 and MMC-T6.

In a previous paper, we noted that the crack tip remains macroscopically straight in these materials with slight tunneling in the MMC layers [6]. Figure 3 shows the crack surface farther back from the crack tip. In this illustration, the darker layers are the AA5182 while the lighter layers are the AA6090/SiC/25p. In the F temper (Figure 3(a)), the interfaces between the layers are fairly diffuse and there is no evidence of interfacial debonding. In contrast, the specimen tested in the T6 temper reveals much more distinct interfaces that also exhibit some debonding that is characteristic of this temper. This observation suggests that the difference in behavior between the two materials may be partly related to different degrees of constraint of the deformation in the more ductile AA5182. In addition, observation of the fracture surfaces reveals that the AA5182 does not fail in a ductile manner, but rather that crack propagation is accompanied by extensive shear deformation and tearing in these layers.

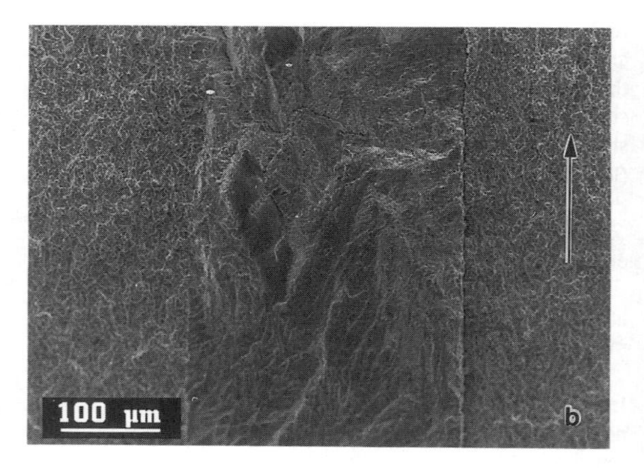

Figure 3. Representative SEM micrographs of fatigue crack surface at high ΔK for (a) LMC-F and (b) LMC-T6. (Darker center layers are AA5182. Arrows indicate crack growth direction.)

In order to explore more fully the contributions of the individual materials, self-laminates of each were tested in both the F and T6 tempers. Again, we note that the T6 designation used here simply connotes the fact that the materials were subjected to a certain heat treatment that nominally produces peak strength in the AA6090 matrix. The AA5182 does not exhibit precipitation hardening as is revealed through microhardness measurements, which show no change before and after the heat treatment. The crack growth rates for the LMC-F fall between those of the two constituents in the near threshold regime, as shown in Figure 4(a). At higher growth rates, the laminated composite and AA5182 self-laminate exhibit similar behaviors. Another important observation is that the AA5182 self-laminate exhibits very poor fatigue crack growth resistance in the as-pressed condition, suggesting that is may be a poor choice for lamination when the design goal is to improve fatigue crack growth resistance. When these self-laminates are tested in the T6 temper, the situation is altered considerably (Figure 4(b)). Although the data exhibit considerable scatter due to large crack deflection effects and the concomitant difficulty of making reliable measurements of crack length using the COD-based compliance method, the results reveal several interesting trends. Again, the data for the composite laminate fall between the results for the constituent self-laminates. However, the relative positions of the AA5182 and AA6090/SiC/25p are reversed. In particular, the fatigue crack growth resistance of the AA5182 is improved significantly, even though no change in microstructure, precipitate structure or strength is

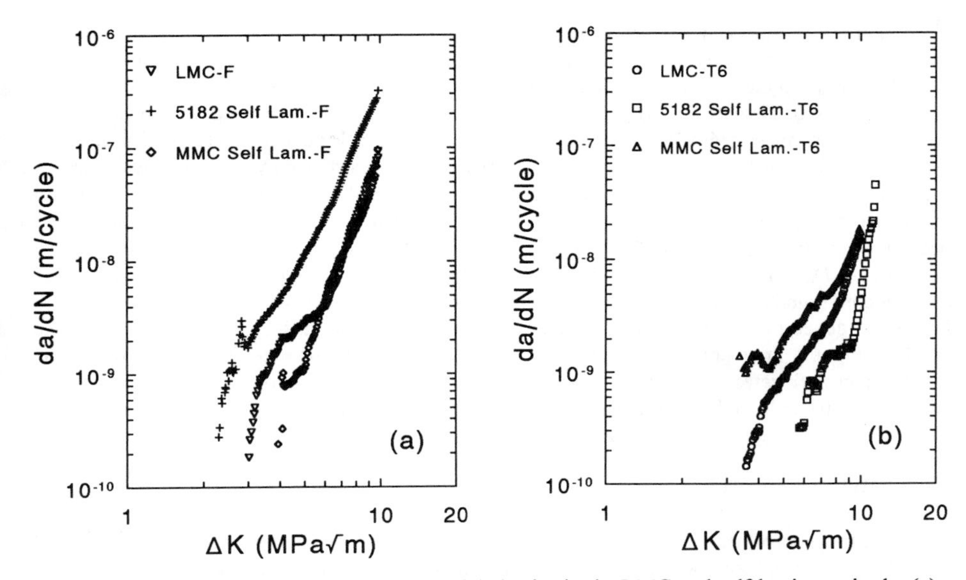

Figure 4. Comparison of fatigue crack growth behavior in the LMC and self-laminates in the (a) as-pressed (F) condition and (b) heat treated (T6) condition.

expected. In contrast, the fatigue crack growth behaviors of the other two materials are not significantly different from the F temper. The good apparent fatigue crack growth behavior of the AA5182 is again attributed to extrinsic factors, as revealed through a plot of the crack closure stress intensity factors (Figure 5). The closure levels in the AA5182 alloy after heat treatment are especially high, and can be attributed the very rough fracture surface observed in this material. The heat treated AA5182 represents an appropriate choice for lamination with AA6090/SiC/25p in order to improve the fatigue crack growth resistance of the laminate. Finally, we note that the relative crack growth rates of the laminate and its constituents as well as the high closure levels are similar to observations reported for ductile phase toughened intermetallic compounds [11,12].

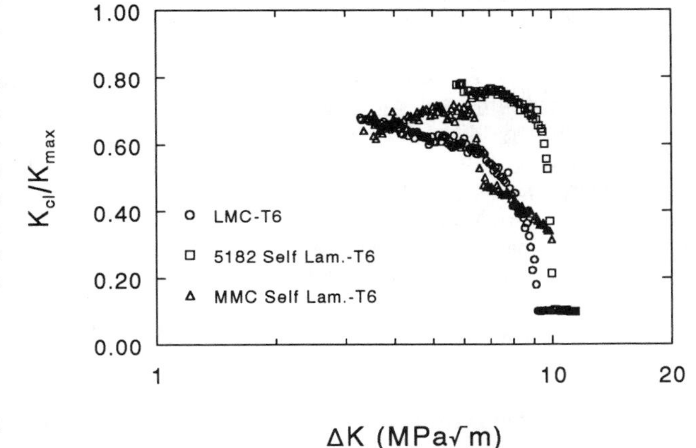

Figure 5. Relative crack closure levels in the LMC and the self-laminates in the heat treated (T6) condition.

CONCLUSIONS

The results of the present investigation reveal that lamination of a discontinuously reinforced aluminum metal matrix composite (AA6090/SiC/25p) with a ductile solid solution strengthened AA5182 alloy leads to improved fatigue crack growth resistance in the near-threshold regime. Fatigue crack growth rates in the laminated composite lie between those observed in self-laminates of the two constituents in both the as-pressed and heat treated conditions. Nonplanar crack growth in the AA5182 layers of the composite results in crack deflection and shear failure, enhancing fatigue crack growth resistance by extrinsic mechanisms. Heat treatment from the as-pressed to the nominal age hardened condition significantly influences interfacial strength as well as the strength of the DRA constituent. In particular, the nominal T6 treatment results in low interfacial strength, thereby reducing the degree of constraint exerted by the stronger DRA layers on the AA5182 layers. This condition leads to greater crack deflection and roughness-induced crack closure, which enhances the fatigue crack growth resistance. Further evidence of the importance of interfacial strength is found in the observation that, while the T6 treatment does not alter the strength of the AA5182 self-laminate, it significantly affects the crack growth rates. Together, these observations indicate the importance of interfacial and constituent strength, and suggest that additional study of interfacial properties is needed to fully characterize the fatigue crack growth resistance of composite laminates.

ACKNOWLEDGMENTS

The authors wish to thank Dr. C. K. Syn and Dr. D. R. Lesuer of Lawrence Livermore National Laboratory for supplying the laminate material and providing partial research support.

REFERENCES

1. V. C. Nardone, in Intrinsic and Extrinsic Fracture Mechanisms in Inorganic Composite Systems, edited by John J. Lewandowski and Warren H. Hunt, Jr. (The Minerals, Metals and Materials Society, Warrendale, PA, 1995), pp. 85-92.
2. L. Yost Ellis and J. J. Lewandowski, Mater. Sci. Engr. **A183**, 59 (1994)
3. T. M. Osman, J. J. Lewandowski, W. H. Hunt, D. R. Lesuer and R. Riddle, in Intrinsic and Extrinsic Fracture Mechanisms in Inorganic Composite Systems, edited by John J. Lewandowski and Warren H. Hunt, Jr. (The Minerals, Metals and Materials Society, Warrendale, PA, 1995), pp. 103-111.
4. C. K. Syn, D. R. Lesuer and O. D. Sherby, in Proc. Intl. Conf. Advanced Synthesis of Engineered Structural Materials, edited by J. J. Moore, *et al.*, (ASM Intl., Materials Park, OH, 1992), p. 149.
5. D. Lesuer, C. Syn, R. Riddle and O. Sherby, in Intrinsic and Extrinsic Fracture Mechanisms in Inorganic Composite Systems, edited by John J. Lewandowski and Warren H. Hunt, Jr. (The Minerals, Metals and Materials Society, Warrendale, PA, 1995), pp. 93-102.
6. P. B. Hoffman and J. C. Gibeling, Scripta Metall. Mater. **32**, 901 (1995)
7. K. K. Chawla and P. K. Liaw, J. Mater. Sci., **14**, 2143 (1979).
8. ASTM Standard E647-88a, **3.01**, 646 (1989).
9. T. J. Sutherland, P. B. Hoffman and J. C. Gibeling, Metall. and Mater. Trans., **25A**, 2453 (1994).
10. A. Saxena and S. J. Hudak, Jr., Int. J. Fracture, **14**, 453 (1978).
11. K. T. Venkateswara Rao, G. R. Odette and R. O. Ritchie, Acta Metall. Mater. **42**, 893 (1994).
12. L. Murugesh, K. T. Venkateswara Rao and R. O. Ritchie, Scripta Metall. Mater. **29**, 1107 (1993).

FATIGUE BEHAVIOR OF SCS-6/TITANIUM/TITANIUM ALUMINIDE HYBRID LAMINATED COMPOSITE

P.C. WANG, Y.C. Her, and J.-M. YANG
Department of Materials Science and Engineering, University of California, Los Angeles, CA 90095

ABSTRACT

The fatigue behavior of the SCS-6 silicon carbide fiber-reinforced Ti-6Al-4V/Ti-25Al-10Nb hybrid laminated composite was investigated at room temperature. The accumulation of fatigue damage in the form of matrix cracking was measured as a function of loading cycles and applied stress levels. The residual stiffness and residual tensile strength of the post-fatigued specimens were determined. The comparison of the crack growth behavior of the hybrid composite with both the SCS-6/Ti-6-4 and SCS-6/Ti-25-10 composites will also be discussed.

INTRODUCTION

Continuous silicon carbide fiber-reinforced titanium alloy matrix composites have been evaluated as possible structural materials for gas turbine engines. More recently, composites based upon the ordered titanium-aluminum intermetallic alloys such as Ti_3Al and TiAl have also been under intensive development for potential applications [1]. These intermetallic alloys have lower density and better environmental stability than typical disordered titanium alloys. However, these intermetallic alloys also suffer from inherent problems such as low ductility and fracture resistance at ambient temperature. Therefore, efforts have been made to further improve the damage tolerance, mechanical and environmental reliability, and elevated-temperature properties of the titanium based composites to meet the design requirements. Recently, the hybridization of titanium alloys and titanium aluminide intermetallic alloys has been demonstrated as a viable approach to tailor the properties of the composites [2]. The hybrid laminated composite is expected to be used at temperatures beyond those attainable in conventional titanium matrix composites and also to improve the damage tolerance of the titanium aluminide matrix composites. Matrix hybridization has been demonstrated as an effective approach in fiber-reinforced polymer/metal matrix [3-5], and metal/metal matrix [6-8] composites for years. The objective of this study is to investigate the fatigue damage evolution and mechanical property degradation of a hybrid composite namely, SCS-6/Ti-6Al-4V/Ti-25Al-10Nb. Comparison of fatigue resistance between the hybrid and monolithic composites is also provided.

MATERIALS AND EXPERIMENTAL PROCEDURES

The material used in this study was an unidirectional SCS-6 fiber-reinforced Ti - 6at% Al - 4at% V (Ti-6Al-4V) and Ti - 25at% Al- 10at% Nb - 3at% V - 1at% Mo (Ti-25-10) hybrid composite. The microstructure of the resulting composite was shown in Fig. 1. The properties of the constituents and the composite are listed in Table 1. The Ti-25-10 laminae were placed between the SCS-6/Ti-6-4 lamina, and the lay-up was consolidated by hot isostatic pressing. The volume fraction of the SCS-6 fiber, Ti-25-10 and Ti-6-4 was measured to be 20%, 35%, and 45%, respectively. Unnotched straight-edged specimens with dimensions of 76.2 mm x 6.35

Mat. Res. Soc. Symp. Proc. Vol. 434

mm x 3.07 mm were cut from the panel and, subsequently, epoxy bonded with metallic end tabs on both ends. A clip-on extensometer with gage length of 12.7 mm was attached to one side of the sample with rubber bands to measure the corresponding strain during testing. All the fatigue testings were performed at room temperature with a stress ratio, R = 0.1, frequency, f = 10 Hz, while the maximum applied stresses ranged from 400 to 600 MPa. Matrix crack length was measured by using a surface replication technique under an optical microscope. Some of the fatigue tests were interrupted, and the samples were pulled to fracture quasi-statically in order to measure the post-fatigued tensile strength. For comparison, the crack growth behavior, and the matrix crack growth rates of the SCS-6/Ti-6-4 (V_f = 35%) and SCS-6/Ti-25-10 (V_f = 35%) were also investigated at σ_{max} = 600 MPa.

Table 1: Material properties of the hybrid composite and its constituents

	SCS-6/Ti-6-4/Ti-25-10	Ti-6-4	Ti-25-10
fiber modulus (GPa)	400	NA	NA
modulus (GPa)	210	110 [9]	125 [11]
tensile strength (MPa)	1138	920 [10]	853 [12]
B	-	4×10^{-11} [9]	2.43×10^{-11} [12]
n	-	3 [9]	3.9 [12]

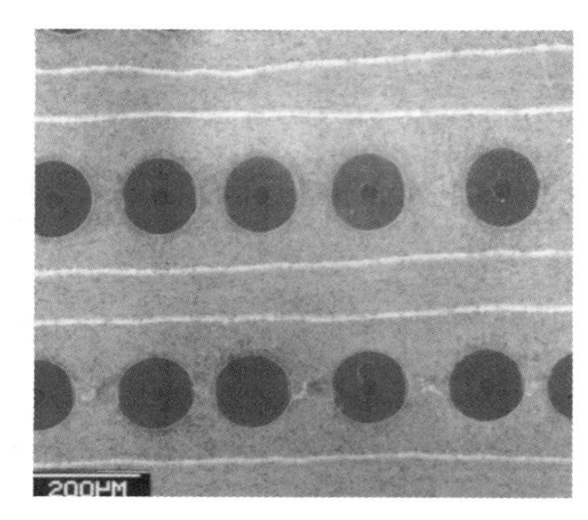

Fig. 1: The microstructure of the SCS-6/Ti-6-4/Ti-25-10 composite.

RESULTS

The S-N curve and the damage of the hybrid composite accumulated under low cycle fatigue loading have been reported by Jeng et al. [2]. Typically, the matrix cracks initiated from the fractured fiber or the damaged fiber-matrix interfacial reaction layer near the specimen edge, and propagated perpendicular to the fiber direction. Minimal fiber breakage and limited interfacial debonding were found in the wake of the matrix crack tip. Most of the fibers ruptured near the fracture surface resulting in minimal fiber pull-out. A close observation of the ruptured matrix revealed that there are two distinct regions: a fatigue induced brittle fracture region and an overload induced ductile fracture region. However, in the overload region, quasi-ductile tearing was found in the Ti-25-10 layer, while the Ti-6-4 layer exhibited classical dimple structure, as shown in Fig. 2. SEM observation showed that the fatigue crack front (indicated by the star curve) of this composite was periodically wavy, as shown in Fig. 3.

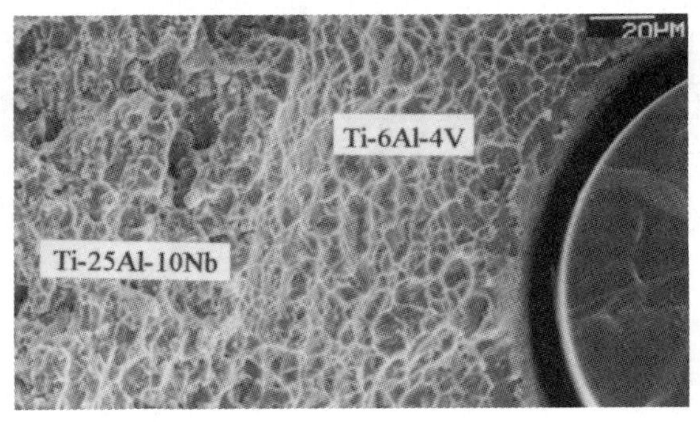

Fig. 2: Matrix fracture morphology in the overload region.

The leading crack front was always found in the brittle Ti-25-10 layer, while the lagged crack front was usually located on the centerline of fiber array. Also, the amplitude of the wavy crack front was about 200 µm. It clearly indicates that the SCS-6/Ti-6-4 layers successfully retarded the rapid crack growth in the brittle Ti-25-10 layers.

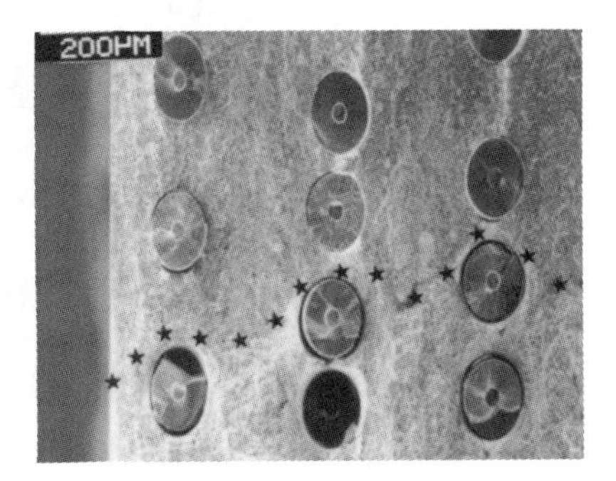

Fig. 3: The wavy fatigue crack front in the hybrid composite.

Fig. 4 shows the stiffness reduction and the total matrix crack length as a function of fatigue cycles for the hybrid composite tested at 500 MPa. The stiffness reduction curve can be classified into two regimes: a regime without any stiffness reduction followed by a rapid stiffness reduction regime which corresponds to the fatigue crack propagation. The stiffness reduction occurred when the matrix crack was initiated at N = 2.5×10^5 cycles under σ_{max} = 500 MPa. There was no plateau in the stiffness reduction curve as found in the SCS-6/Ti-15-3 or SCS-6/Ti-22-23 composites [12,13]. Also, quasi-steady state crack propagation behavior was found, in which the crack length increased linearly with the fatigue cycles. Microstructural observation revealed that most of the fibers remained intact in the wake of crack tip. Therefore, it is evident that matrix cracking primarily accounted for the stiffness reduction. The average quasi-steady-state crack growth rates, residual stiffness, and fatigue lives for tests performed at both 500 and 600 MPa are listed in Table 2. The average crack growth rates of both the SCS-6/Ti-6-4 and SCS-6/Ti-25-10 composites are listed in Table 3 for comparison. It was found that the crack growth rates in the hybrid composite were sensitive to the applied stresses. As a result, when the maximum applied stress was increased from 500 to 600 MPa the crack growth rate increased by 100%. In addition, no fatigue damage was detected after cyclically loading the hybrid composite for 10^6 cycles when σ_{max} was reduced to 400 MPa.

Table 2: Fatigue Performance of SCS-6/Ti-6-4/Ti-25-10 hybrid laminated composite

maximum applied stress (MPa)	average crack growth rate $x10^{-9}$ (m/cycle)	residual stiffness (%)	fatigue life (cycles)
600	15.6	92.6	$2.23x10^5$
500	6.8	94.1	$4.1x10^5$

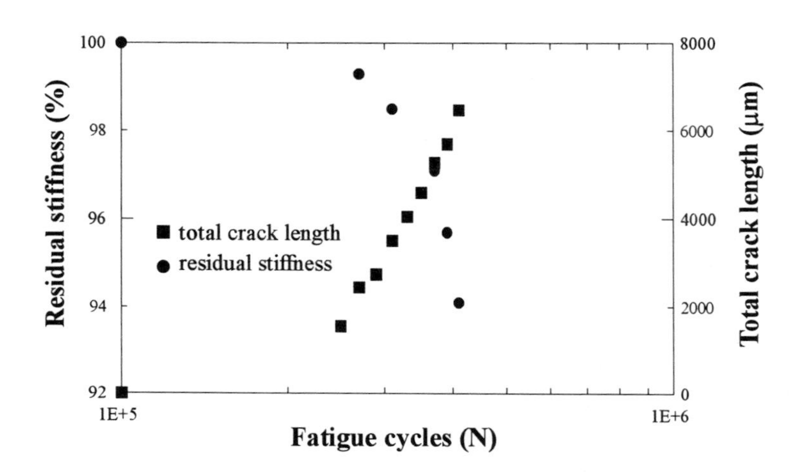

Fig. 4: The stiffness reduction and the total crack length curves as a function of fatigue cycles.

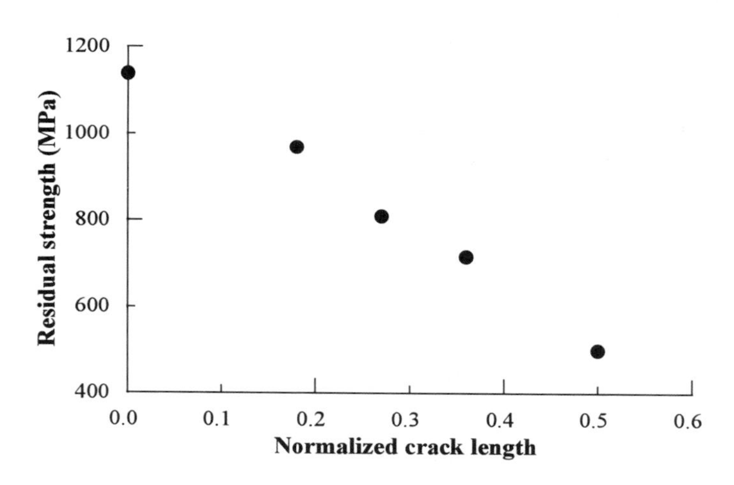

Fig. 5: The post-fatigued tensile strength as a function of the normalized crack length (a/w).

The post-fatigued tensile strength as a function of the normalized matrix crack length is shown in Fig. 5. It clearly reveals that the tensile strength decreased significantly as the matrix crack propagated. As the normalized crack length (a/w) increased to 0.5, the tensile strength dropped to 500 MPa, which corresponds to a 56% reduction in strength.

Table 3: Experimental and predicted crack growth rates
for composites with various fiber volume fraction

crack growth rate x10^{-9} (m/cycle)	$V_f = 35\%$ (measured)	$V_f = 35\%$ (predicted)	$V_f = 31\%$ (predicted)
SCS-6/Ti-6-4	4.3	4.3	6.1
SCS-6/Ti-25-10	14.4	16.2	24.8

DISCUSSION

The quasi-steady-state crack growth rate of a fiber bridged matrix crack has been modeled by Cox using the J-integral methodology [15]. Also, the model has been successfully extended for predicting the quasi-steady-state crack growth rate of the SCS-6/Ti-22Al-23Nb composite by Wang et al. [16]. By using the methodology developed by Wang et al. and assuming that the average interfacial frictional stress (τ) was equal to 35 MPa [13], the quasi-steady state crack growth rates of both SCS-6/Ti-6-4 and SCS-6/Ti-25-10 composites under σ_{max} = 600 MPa were predicted, as listed in Table 3. The parameters of Paris' law, $da/dN = B(\Delta K_{tip})^n$, required in the calculation of both Ti-6-4 and Ti-25-10 alloys are also listed in Table 1. It was found that the predicted results for composites with 35% fiber content agreed with the experimental data quite well, suggesting that Cox's model was applicable for both composites. Additionally, it reveals that the crack growth rate of the ductile matrix SCS-6/Ti-6-4 composite is much slower than that of the brittle matrix SCS-6/Ti-25-10 composite. For the hybrid composite, the fiber volume fraction of the SCS-6/Ti-6-4 monolayer was found to be 31%. Therefore, due to the existence of the brittle Ti-25-10 laminae, the crack growth rate of the hybrid composite is expected to be faster than that of monolithic SCS-6/Ti-6-4 composite with V_f = 31%. Crack growth rates of both SCS-6/Ti-6-4 and SCS-6/Ti-25-10 composites were predicted by using Cox's model based upon V_f = 31%, as shown in Table 3. It clearly indicates that the crack growth rate of the hybrid composite is, as expected, greater than that of SCS-6/Ti-6-4 composite. However, it is much smaller than that of the SCS-6/Ti-25-10 composite. It demonstrates that even though the overall fiber volume fraction of the hybrid composite is only 20%, the measured crack growth rate of the hybrid composite is comparable to that of the composites with V_f = 31%. This is due to its unique microstructural design. Therefore, hybridization provides an effective approach for tailoring the fatigue properties of the fiber-reinforced titanium-based composites.

CONCLUSION

1. The unnotched hybrid composites exhibited a quasi-steady-state crack propagation behavior under cyclic loading. Similar phenomena have been observed in other SCS-6/Ti matrix composites.

2. The residual tensile strength decreased significantly as the matrix crack length increased. When the normalized crack length was 0.5, the strength reduction was found to be as

great as 56%. It is believed that the low residual strength is due to fiber breakage that occurred during crack propagation.

3. Matrix hybridization has proven to be a feasible approach to improve the fatigue resistance of the brittle matrix and to reduce the manufacturing cost.

ACKNOWLEDGMENTS

This work is supported by the Air Force Office of Scientific Research (F49620-93-1-0320) Dr. Walter Jones is the program monitor. The authors also thanks Mr. Kirk Cheng for reviewing the manuscript.

REFERENCES

1. H. A. Lipsitt, in High-Temperature Ordered Intermetallic Alloys I, Vol. 39, edited by C.C. Koch, C. T. Liu, and N. S. Stoloff (Material Research Society, 1985), p. 351.
2. S.M. Jeng and J.-M. Yang, J. Mater. Sci. 27, 5357-5364 (1992).
3. R. O. Ritchie, W.K. Yu, and R. J. Bucci, Engg. Fract. Mech. 32 (3), 361-377, (1989).
4. C.T. Lin, P.K. Kuo, and F.S. Yang, Composites 22 (2), 135-141, (1991).
5. G. Freischmidt, R. S. P. Coutts, and M. N. Janardhana, J. Mater. Sci. letts. 13, 1027-1031, (1994).
6. K. K. Chawla and P. K. Liaw, J. Mater. Sci. 14, 2143-2150, (1979).
7. J. Sandovsky, Kovove Materialy 25 (6), 749-756, (1987).
8. M. S. Madhukar, A. Fared, J. Awerbuch, and M. J. Kaczak, in High Temperature/High Performance Composites, edited by F. D. Lemkey, S. G. Fishman, A. G. Evans, and J. R. Strife (Materials Research Society, Vol. 120), p. 121-8, (1988).
9. J. G. Bakuckas and W. S. Johnson, J. Comp. Tech. Res. 15 (3), 242-255, (1993)
10. R. T. Bhatt and H. H. Grimes, Metall. Trans 13A, Nov. 1933-1938, (1982).
11. S.M. Jeng, C.J. Yang, J.-M. Yang, D. G. Rosenthal, and J. Goebel in Intermetallic Matrix Composites, edited by D. L. Anton, P. L. Martin, D. B. Miracle, and R. McMekking, (Materials Research Society Vol. 194), p. 279, (1990).
12. D. P. DeLuca, B. A. Cowles, F. K. Haake, and K. P. Holland, Fatigue and Fracture of Titanium Aluminides, (Materials Laboratory, Wright Research and Development Center, Air Force Systems Command), p. 33, 133, (1990).
13. P.C. Wang, S.M. Jeng, and J.-M. Yang, Mater. Sci. & Eng. A200, 173-180, (1995).
14. P.C. Wang, S.M. Jeng, and J.-M. Yang, Acta Metall. et Mater. 1996 in press.
15. B. N. Cox, Mech. Mater. 15, 87-98, (1993).
16. P.C.Wang and J.-M. Yang, submitted to Metall. Trans. A.

EVALUATION OF IMPACT DAMAGE RESISTANCE IN LAMINATED COMPOSITES USING RESINS WITH DIFFERENT CROSSLINK DENSITIES

A. J. LESSER
Polymer Science and Engineering Department, University of Massachusetts, Amherst, MA 01003

ABSTRACT

It is generally recognized that fiber-reinforced laminated composites are susceptible to damage resulting from low-velocity impacts. Over recent years, many strategies have been devised to increase the fracture toughness of resin matrix materials with the aim of improving the composites overall resistance to impact damage. One popular strategy for enhancing the fracture toughness of thermosets involves increasing its molecular weight between crosslinks which, in turn, enhances the resins ductility. In this paper, we investigate the efficiency of this toughening approach with regard to resisting damage in composite laminates subjected to low-velocity impacts. Generic damage characteristics and mechanisms are reviewed and it is shown that two different events occur during the impact process. First, the laminate experiences a local failure which resembles a Hertzian fracture process followed by subsequent delamination between the plies. Results are presented illustrating the effects that systematically increasing the molecular weight between crosslinks of the resin has on each of these mechanisms. Also, the residual compressive strength (Compression After Impact) of the laminates made with these resins is presented.

INTRODUCTION

A Considerable amount of work has been published in recent years regarding the damage resistance and damage tolerance of laminated composites. This subject has received attention because it turns out that a laminates resistance to impact damage and its residual strength are key issues in the design of composite structures. It has been shown that a low velocity impact by foreign objects can create non-visible damage which can severely degrade the compression strength of the laminate[1]. In 1985, a report was issued by the Office of Technology Assessment[2] which identified impact resistance and delamination as the two key areas for composite research and development.

One of the most commonly used tests for measuring a composites resistance to impacts is referred to as the Compression After Impact (CAI) test[3,4] Although these types of tests are widely used, there is a great deal of controversy with regard to the results obtained from them. One of the inherent sources of confusion in CAI testing is that they actually measure the combined effects of two separate phenomena: damage resistance and damage tolerance[5-7]. Damage resistance is a measure of the composites ability to resist damage during an impact event, and damage tolerance is a measure of the residual performance of a pre-damaged composite. Additional sources of confusion in CAI testing arise from the inability to compare results from different test standards due to different specimen geometries, different support fixtures and different impact energies. Also, subtle variations in the CAI support conditions have shown to cause large changes in the final results[8]. These issues arise because both the amount of damage caused by impact and the residual panel strength are a combined result of the material properties of the composite (resin and fiber) as well as the geometry of the panel together and its associated boundary conditions during the test.

Mat. Res. Soc. Symp. Proc. Vol. 434 © 1996 Materials Research Society

In previous work, we have shown that the damage induced in laminated composites subjected to subperforating impacts is governed by two basic phenomena[9,10]. During an impact event, the damage initiates a Hertzian type cone of microcracks together with pairs of radial cracks within individual plies. Once the cone is developed, delamination occurs between the plies in a quasi-static fashion coincident with the increasing impact load. In this paper we present results from an investigation whose aim was to study these mechanisms with a family of resins whose fracture toughness is systematically increased.

A significant amount of effort was placed in development of high performance resins for composite applications. Some of the more popular approaches include introduction of a soft-particle (rubbery or thermoplastic) second phase into the matrix[11], increasing the molecular weight between crosslinks of the resin[12], or a combination of the two[13]. In this paper we present results from a systematic investigation using the second toughening approach; namely increasing the molecular weight between crosslinks.

EXPERIMENT

The resin formulation used in this study included a bisphenol A epoxy (EPON 825) cured with stoichiometric amounts of ethylenediamine (tetra-functional) and methylethylenediamine (di-functional). The ethylenediamine was used for crosslinking the network and the methylethylenediamine was used for chain extension. Resin plaques and composite panels were fabricated with by combining stoichiometric ratios of these curing agents with the epoxy to achieve desired crosslink densities.

Neat resin plaques of 3.2 mm (1/8-inch) nominal thicknesses were made from each resin system. Measurements of the resin fracture toughness (K_q) were conducted in accordance with the ASTM D5045-91 on compact tension specimens. Pre-cracking of the test specimens was accomplished with the aid of a razor blade and small mallet on the specimens after they were conditioned in liquid nitrogen.

Unidirectional prepregs were made for each resin with AS4 un-sized fibers. Afterward, 32 to 40 ply $[45/90/-45/0]_{ns}$ quasi-isotropic panels and test specimens were cut in accordance with the Boeing BSS 7260 specification. Next, the quasi-isotropic panels were impacted in accordance with the Boeing with an impact energy equal to 6.675 kJ/m (1500 in-lb/in) of the panel's thickness. The impacts were conducted on a Dynatup Model 8250 Instrumented Impact Test Machine at an ambient temperature of 23°C. Traces of the impact load history and impact energy (calculated from the load) were stored digitally.

After impacting, all panels were subjected to an ultrasonic investigation and either compression tested in accordance with the Boeing Test method or sectioned and polished for fractographic studies. The ultrasonic method adopted for this study incorporated the use of a Panametrics pulse-echo C-scan system using a 15 MHz 1/4-inch focused transducer.

RESULTS AND DISCUSSION

Damage Mechanisms

Previous investigations on impacted panels[9,10] have shown that the damage characteristics generally observed in these panels are a combined result of two separate events. The first event is a local failure in the panel which produces central micro-cracks and radial cracks within each ply of the laminate. Following this initial local failure subsequent growth of delamination proceeds between the plies as the panel accommodates additional load. It was shown earlier[9,10] that the central micro-cracks and larger pairs of radial cracks resemble those reported in Hertzian type failures characterized in other materials[14,15]. These types of failures are usually reported in brittle materials and generally contain two types of crack patterns: a cone crack commonly referred to as a *Hertzian Cone* and radial cracks. The cone crack consists of a conical shaped crack pattern completely encompassing the region under the contact location and the radial cracks initiate from the cone and propagate outward in a radially symmetric fashion. A comparison of this crack pattern is shown together with that produced in an isotropic material in Figure 1.

The development of the Hertzian cone occurs once the impact load reaches a critical threshold level during the impact event. This is illustrated on a typical impact load history in Figure 2 (note that the load in Figure 2 is normalized by the panel thickness for later comparisons). If a panel is impact loaded (or quasi-statically loaded) below this threshold level no significant damage is detected. However, once this threshold level is reached, a sudden unloading is detected in the impact load history resulting from the development of the cone and radial cracks.

Once the impact load surpasses the threshold level at which the Hertzian cone and radial cracks are produced, delamination occurs between the plies with the size of the delaminations increasing with increasing load (see Figure 3). The delaminations between individual plies is shown to be *butterfly* shaped with its major axis coincident with the fiber direction of the lower ply. This effect is attributed to the bending stiffness mismatch between individual plies producing interfacial stresses which mimic the delamination pattern[16]. The size of the delaminations increase coincident with the increase in load. This regime in the load history is shown in Figure 2.

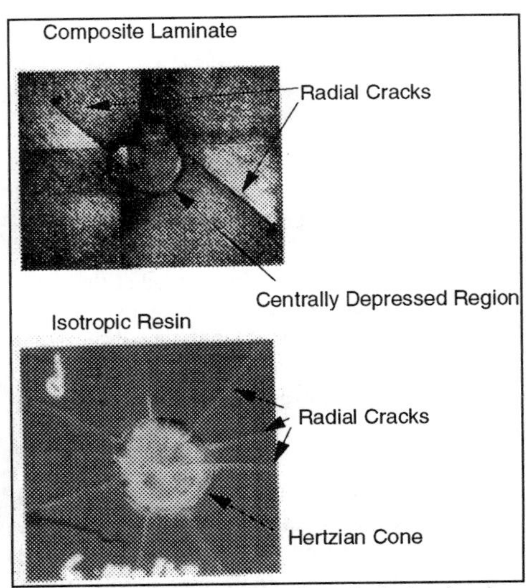

Figure 1: Hertzian crack pattern in isotropic glass (below) and fiber reinforced laminate (above).

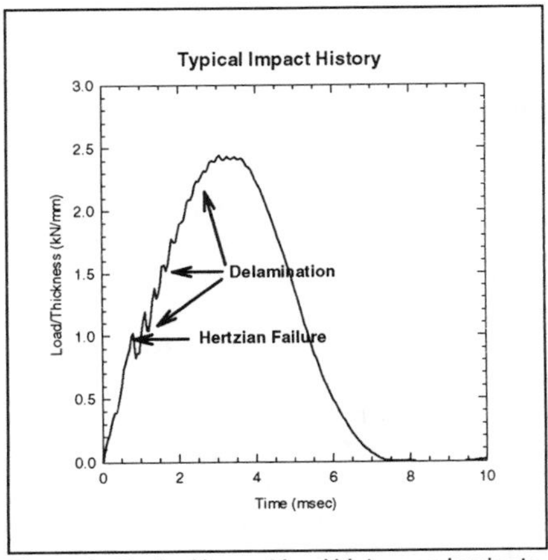

Figure 2: Typical impact load history on laminate.

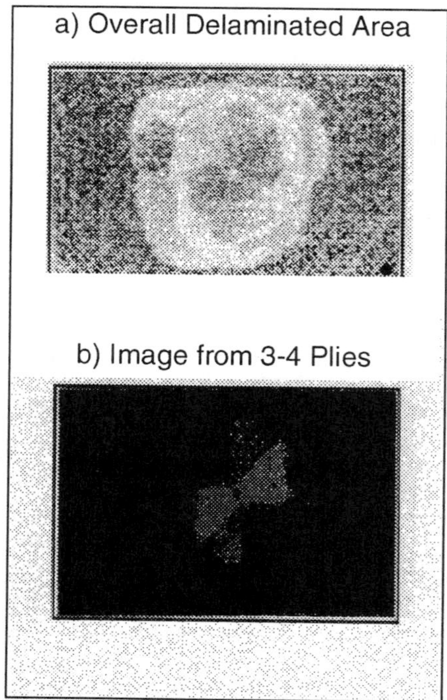

Figure 3: Ultrasonic image of delaminated regions; a) overall, b) 3 and 4 plies.

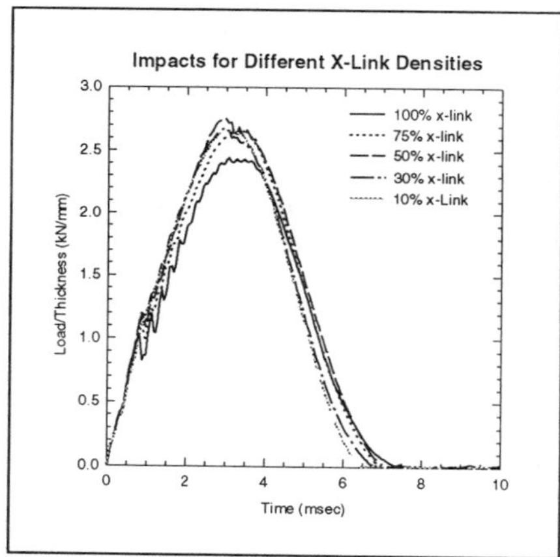

Figure 4: Impact load histories for panels fabricated with different crosslink densities.

Effect of Crosslink Density

Impact tests were conducted on quasi-isotropic panels fabricated with resins of varying crosslink densities ranging from 100% (i.e., fully crosslinked) to 10%. Figure 4 illustrates typical impact curves (normalized by the panel thickness) for five of these crosslink densities.

Careful inspection of these curves (c.f. Figure 4) show that all the curves load at the same rate prior to reaching the threshold level. However, the threshold level itself changes as the crosslink density of the resin is changed. The resin formulation with the highest threshold level was achieved at a 50% crosslink density with a 15% increase in the threshold level. However, beyond the 50% crosslink density level no further increase in the threshold level is realized and even decreases at very low crosslink densities.

Similar behavior is seen in the maximum impact load (see Fig. 4) with the highest load occurring in the 50% crosslinked system. Again the difference in the maximum load in the 50% crosslinked system is approximately 15% higher than that measured on the fully crosslinked system. These results qualitatively agree with the maximum tensile strength measured in these resin systems in that the strongest resin formulation was measured at 50% crosslink density.

Panels from each resin system were ultrasonically scanned after the impact tests to characterize the extent of damage produced during the impacts. Measurements of the overall delaminated area were recorded for each crosslink density and are presented in Figure 5. These results show that the overall delaminated area decreases from 47 cm^2 to 8 cm^2 with a corresponding decrease in crosslink density from 100% to 10%.

Compression After Impact tests were conducted on each of these panels in accordance with the Boeing BSS7260 CAI standard to assess their residual strengths. Figure 6 illustrates the results from these tests over the crosslink densities. These results show that residual strength increases consistent with decreases in the crosslink density.

CONCLUSIONS

The results from this investigation show that the overall resistance to damage and corresponding residual compressive strength of the composite continually increase as the molecular weight between crosslinks increases. This suggests that the highest resistance to overall impact damage is achieved through the highest molecular weight between crosslinks. This higher damage resistance, in turn, translates to a higher residual strength of the composite laminate. This implies that a thermoplastic resin matrix material would provide the highest resistance to impact damage which is consistent with our expectations.

However, our results also show that the highest threshold level before damage is initiated occurs at approximately 50% crosslink density for the resin and curing agent studied herein. This suggests that an optimum crosslink density is desired in applications where the emphasis would be to produce a laminate with the highest resistance to damage initiation.

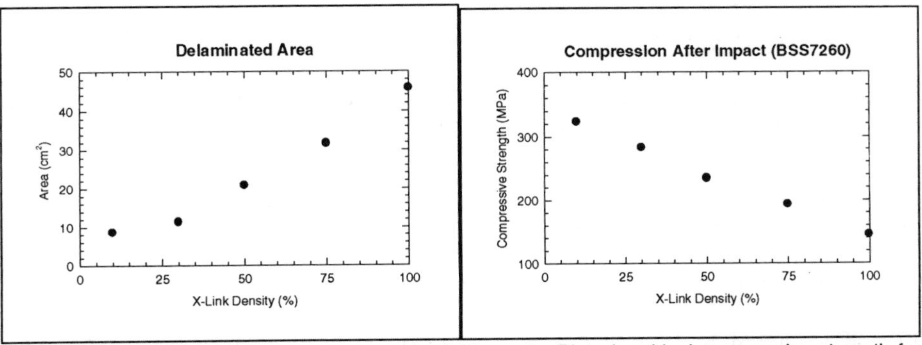

Figure 5: Overall delaminated area measured by C-scan on impacted panels at different xlink densities.

Figure 6: Plot of residual compressive strength for laminates with different xlink densities.

REFERENCES

1 Reifsnider, K.L., (1989) *Handbook of Composites*, C. T. Herakovich et al., eds., Elsivier Science Publishers, Amsterdam

2 Office of Technology Assessment, (1985) *New Structural Materials Technologies Opportunities for the Use in Advanced Ceramics and Composites.* , US Congress Tech. Mem., US Gov. Printing Office, Washington, D.C.

3 Boeing Specification Support Standard, (1982). *BSS 7260*, Code Ident. No. 81205, Boeing Aerospace, Los Angeles, CA.

4 NASA Reference Publication 1092, (1983). Compiled by ACEE Composites Project Office, Langley Research Center, Hampton, Virginia

5 Lee, S. M., Zahuta, P. (1991) *J. of Composite Materials* ,25, pp 204-222.

6 Cairns, D. S., Lagace, P. A. (1992) *J. of Reinforced Plastics and Composites*, 11, pp 395-412.

7 Hull, D., Shi, Y. B. (1993) *Composite Structures*, 23, pp 99-120.

8 Falabella, R. Olesen, K. A. Boyle, M. A. (1990) *Proc. of 35th Int. SAMPE Tech.Conf.*, pp 1454 -1465.

9 Lesser, A.J., Filippov, A.G. (1991) *Proc. of 36th International SAMPE Tech. Conf*, 36, pp 886-900

10 Lesser, A.J., Filippov, A.G. (1994) *Int. J. of Damage Mechanics*, 3, pp 408 - 432

11 Bucknell, C. B. (1977) *Toughened Plastics*, Applied Science, N.Y.

12 Bravenec, L. D., Dewhirst, K. C., Filippov, A.,G., (1988) "Lightly Crosslinked Thermosets", *33rd Int. SAMPE Symposium,* March 7-10, pp 1377-1384.

13 Lesser, A.J., Puccini, P.M., Modic, M.J., (1992) "Mechanical Performance of Single and Multiphase Resin Systems under Constrained Stress States", *Society of the Plastics Industry,* ERF, March 9-11, pp 120- 127

14 Acton, M. R., Wilks, J. (1989) *J. of Material Science* , 24, pp 4229-4238.

15 Chaudhri, M. M., Liangyi, C. (1989). *J. of Material Science* ,24, pp 3441-3448

16 Liu, D. (1988) *J. of Composite Materials* 22, pp 674 - 692.

INHERENTLY SMART LAMINATES OF CARBON FIBERS IN A POLYMER MATRIX

XIAOJUN WANG, D.D.L. CHUNG
Composite Materials Research Laboratory, Furnas Hall, State University of New York at Buffalo, Buffalo, NY 14260-4400

ABSTRACT

Real-time monitoring of fatigue damage and dynamic strain in inherently smart and continuous unidirectional (0° and 90°) and bidirectional (0°/90°) carbon fiber epoxy-matrix composites by electrical resistance (R) measurement was achieved. Upon cyclic tension (0°) of 0° or 0°/90° composites, R (0°) decreased reversibly, while R perpendicular to the fiber layers increased reversibly, though R in both directions changed irreversibly by a small amount after the first cycle. Upon fatigue testing of the 0° composite at a maximum stress of 57-58% of the fracture stress, the peak R (0°) in a cycle irreversibly increased both in spurts and continuously, due to fiber breakage, which started at 30% of the fatigue life. For the 90° composite, R(0°) increased reversibly upon tension (0°) and decreased reversibly upon compression (0°).

INTRODUCTION

Fatigue failure is the cause of numerous disasters. In order to prevent such disasters, fatigue damage is monitored prior to fatigue failure. Another method is to use past experience to predict the lifetime of the structure under consideration. The former method is more effective for fatigue failure prevention than the latter, but requires more effort. Fatigue monitoring is conventionally performed by monitoring at a frequency of at most once a loading cycle because the monitoring is restricted to damage and reversible strain cannot be monitored. Because fatigue monitoring conventionally occurs without dynamic strain monitoring, the loading cycle at which damage occurs cannot be determined unless the loading is periodic.

A method of fatigue monitoring is acoustic emission [1,2], which suffers from inability to monitor dynamic strain. Less common is the method involving measurement of the structure's electrical resistance, which increases due to damage. In other words, the structure is inherently smart. Previous work using resistance to monitor fatigue was carried out on a CaF_2-matrix SiC-whisker composite [3], but dynamic strain monitoring was not performed. In order for dynamic strain and damage to be simultaneously monitored with a single method, that method must involve a measurand which changes in value reversibly during reversible straining and changes irreversibly during damage. In this work, this was achieved for continuous carbon fiber polymer-matrix by using the resistance as the measurand.

Continuous carbon fiber polymer-matrix composites are attractive in that they combine high strength, high modulus and low density. Previous work on a polymer-matrix composite containing a combination of continuous glass fibers and continuous carbon fibers has shown that the resistance of this composite increases irreversibly upon damage (due to the fracture of the carbon fibers) [4], but fatigue monitoring and reversible resistivity changes (dynamic strain

Mat. Res. Soc. Symp. Proc. Vol. 434 © 1996 Materials Research Society

monitoring) were not explored. Our previous work on unidirectional (0°) continuous carbon fiber polymer-matrix composite [5] has shown that the 0° and transverse (perpendicular to fiber layers) resistivities of the composite reversibly and respectively decrease and increase upon dynamic 0° straining, due to the reversible change in the degree of fiber alignment. This paper is an extension of Ref. 5 to demonstrate fatigue monitoring.

EXPERIMENTAL METHODS

Composite samples were constructed from individual layers cut from a 12 in. wide unidirectional carbon fiber prepreg tape manufactured by ICI Fiberite (Tempe, AZ). The product used was Hy-E 1076E, which consisted of a 976 epoxy matrix and 10E carbon fibers.

The composite laminates were laid up in a 4 x 7 in. (102 x 178 mm) platen compression mold with laminate configuration $[0]_8$ and $[90]_{32}$ for unidirectional composites and $[0_6/90_6]$ for bidirectional composites. The individual 4 x 7 in. fiber layers were cut from the prepreg tape. The layers were stacked in the mold with a mold release film on the top and bottom of the layup. No liquid mold release was necessary. The density of the laminate was 1.52 ± 0.01, 1.52 ± 0.03 and 1.50 ± 0.03 g/cm^3 for 0°, 90° and bidirectional composites respectively. The thickness of the laminate was 1.0, 4.5 and 1.5 mm for 0°, 90° and bidirectional composites respectively. The volume fraction of carbon fibers in the composite was 58%, 57% and 52% for 0°, 90° and bidirectional composites respectively. The laminates were cured using a cycle based on the ICI Fiberite C-5 cure cycle. The curing occurred at 179 ± 6°C (355 ± 10°F) and 0.61 MPa (89 psi) for 120 min. Afterwards, they were cut to pieces of size 160 x 14 mm, 153 x 12 mm and 178 x 7 mm for 0°, 90° and bidirectional composites respectively. Glass fiber reinforced epoxy end tabs were applied to both ends on both sides of each piece, such that each tab was 30 mm long and the inner edges of the end tabs on the same side were 100 mm apart. The tensile strength was 1280 ± 109, 62 ± 4.9 and 695 ± 18 MPa for 0°, 90° and bidirectional composites respectively. The tensile ductility was 1.07 ± 0.12, 0.78 and 1.06 ± 0.18 % for 0°, 90° and bidirectional composites respectively. The resistivity (0°) was 3.3 x 10^{-2}, 57.4 and 8.84 x 10^{-2} Ω.cm for 0°, 90° and bidirectional composites respectively; that perpendicular to the fiber layers was 21.35 and 45.7 Ω.cm for unidirectional and bidirectional composites respectively.

The resistance R was measured in the 0° direction and the direction perpendicular to the fiber layers using the four-probe method, while either cyclic tension (0°) was applied. Silver paint was used for all electrical contacts. The four probes consisted of two outer current probes and two inner voltage probes. R refers to the sample resistance between the inner probes. For measuring R (0°), the four electrical contacts were around the whole perimeter of the sample in four parallel planes that were perpendicular to the 0° axis, such that the inner probes were 60 mm apart and the outer probes were 78 mm apart. For measuring R perpendicular to the fiber layers, the current contacts were centered on the largest opposite faces parallel to the 0° axis and in the form of open rectangles of length 70 mm in the 0° direction, while each of the two voltage contacts was in the form of a solid rectangle (length 20 mm in the 0° direction) surrounded by a current contact (open rectangle). Thus, each face had a current contact surrounding a voltage contact. A strain gage was attached to the center of one of the largest opposite faces parallel to the 0° axis for samples for measuring R (0°) as well as samples for measuring R perpendicular to the fiber layers. A Keithley 2001 multimeter was used at 0.5-1.0 mA and 1.5-2.0 V. The displacement rate was 1.0 mm/min for the 0° and bidirectional composites, and 0.5 mm/min for the 90° composite. A hydraulic mechanical testing system (MTS 810) was used for tension-tension cyclic loading (0°) at 1 s per cycle, with stress ratio (minimum to maximum stress) 0.05 and maximum stress 740 MPa (58% of fracture stress, strain = 0.56%) for the 0° composite. Multiple samples were used for each test.

RESULTS AND DISCUSSION

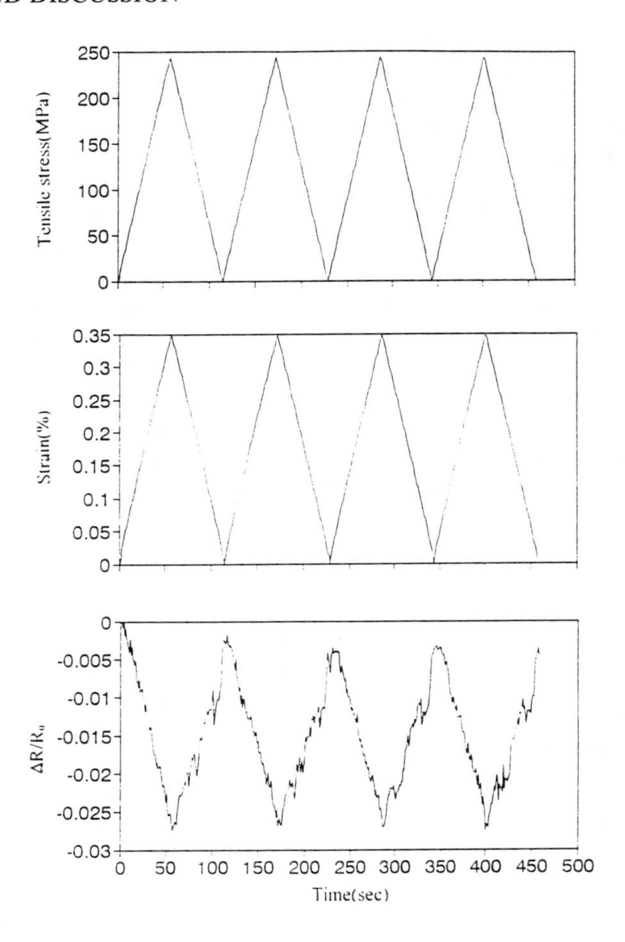

Fig. 1 Variation of $\Delta R/R_o$ (0° direction), tensile stress and tensile strain with cycle number during cyclic tension to a stress amplitude equal to 35% of the fracture stress for bidirectional composite.

Fig. 1 shows the tensile stress, strain and $\Delta R/R_o$ (0°) obtained simultaneously for the bidirectional composite during cyclic tension to a stress amplitude equal to 35% of the breaking stress. The strain returned to zero at the end of each cycle. Because of the small strains involved, $\Delta R/R_o$ is essentially equal to the fractional increase in resistivity. The $\Delta R/R_o$ decreased upon loading and increased upon unloading in every cycle, such that R irreversibly decreased slightly after the first cycle. The irreversible decrease is due to the irreversible decrease in the degree of waviness of the 0° fibers after the first cycle.

A length increase without any resistivity change would have caused R to increase during tensile loading. In contrast, R was observed to decrease. Furthermore, the observed magnitude of $\Delta R/R_o$ was about 20 times that of $\Delta R/R_o$ calculated by assuming that $\Delta R/R_o$ was only due to length increase and not due to any resistivity change. Hence, the contribution of $\Delta R/R_o$ from the length increase is negligible compared to that from the resistivity change.

$\Delta R/R_o$ perpendicular to the fiber layers increased reversibly upon loading (0°). The small value of $\Delta R/R_o$ for the bidirectional composite compared to the unidirectional composite [5] is due to the 90° fibers in the former and that these fibers make the increase in the degree of 0° fiber alignment only slightly affect the chance that adjacent fiber layers touch one another.

Both dynamic tensile and compressive loadings were conducted on the 90° composite. Stress (0°) was applied while R (0°) was measured and the fibers were in the 90° direction. R increased upon tension and decreased upon compression. The effect was not totally reversible due to plastic deformation, which is attributed to the polymer matrix, as the polymer matrix dominated the mechanical properties perpendicular to the fibers. The reversible resistance changes under tension and compression are due to piezoresistivity, which refers to the phenomenon in which the resistivity increases upon increase of the distance between filler units during tension and decreases upon decrease of this distance during compression.

The strain sensitivity or gage factor is defined as the reversible part of $\Delta R/R_o$ divided by the strain amplitude. Its values (0°) are shown in Table 1 for the 0° and 90° composites as well

Table 1 Strain sensitivity (or gage factor) in the 0° direction.

Composite type	Maximum stress / Fracture stress	Strain sensitivity
0° unidirectional	32%	-35.7*
	36%	-37.6*
Bidirectional	19.4%	-5.7*
	35%	-7.1*
90° unidirectional	/	2.2*
	/	-2.1^{+}

*Tension
$^{+}$Compression

as the bidirectional composite. The magnitude of the strain sensitivity is much higher for the 0° composite than the bidirectional composite and is smallest for the 90° composite. This means that the contribution of the 90° fibers (i.e., piezoresistivity) to the resistance change observed in the bidirectional composite is small. The strain sensitivity for the bidirectional composite is not equal to the average of the values for the 0° and 90° composites, due to the interaction between the 0° and 90° fiber layers in the bidirectional composite.

$\Delta R/R_o$, tensile stress and tensile strain (all 0°) were simultaneously obtained during cyclic tension-tension loading of the 0° composite. As cycling progressed beyond 218,277 cycles (55% of fatigue life), the peak R (at the end of a cycle) significantly but gradually increased, such that the increase did not occur in every cycle, but occurred in spurts. Beyond 353,200 cycles (89% of fatigue life), the increase of the peak R occurred continuously from cycle to cycle rather than in spurts. At 396,457 cycles (99.9% of fatigue life), the increase became more severe, such that spurts of increase occurred on top of the continuous increase. The severity kept increasing until failure at 396,854 cycles, at which R abruptly increased. The last spurt before the final abrupt increase occurred at 396,842 cycles (99.997% of fatigue life).

The early period in which the peak R increased discontinuously in spurts is attributed to minor damage in the form of fiber breakage which did not occur in every cycle. The discontinuous nature of the increase implies that the increase is not due to plastic deformation, but is due to fiber breakage. The subsequent period in which the peak R increased continuously but gradually is attributed to fiber breakage which occurred in every cycle. The subsequent period in which the peak R increased rapidly, both in spurts and continuously, is attributed to more extensive fiber breakage, which occurred in the final period before failure. Fiber breakage was visually observed along the long edges of the composite sample during fatigue testing. Thus, by following the increase in the peak R, the degree of damage can be monitored progressively in real time.

Assuming that the irreversible resistance increase during fatigue testing is due to fiber breakage and that the resistivity of the undamaged portion of the composite does not change during testing, the fraction of fibers broken is equal to the fractional decrease in the effective cross-sectional area of the unidirectional composite. Fig. 2 shows a plot of this fraction as a function of the percentage of fatigue life. Fiber breakage started to occur at 30% of the life, though appreciable growth of the fraction of fibers broken did not start till 45% of the life. Fiber breakage occurred in spurts from 45% to 89% of the life, due to its not occurring in every cycle. The smallest spurt involved 0.6% of the fibers breaking. This corresponds to 1020 fibers breaking. Thus, each spurt involved the breaking of multiple fibers. This is reasonable since the fibers were in bundles of 6000 fibers. The smallest spurt involved the breaking of a fraction of a fiber bundle. At 89% of the life, fiber breakage started to occur continuously rather than in spurts. Immediate failure occurred when 70% of the fibers were broken.

ACKNOWLEDGEMENT

This work was supported in part by the Center for Electronic and Electro-Optic Materials of the State University of New York at Buffalo.

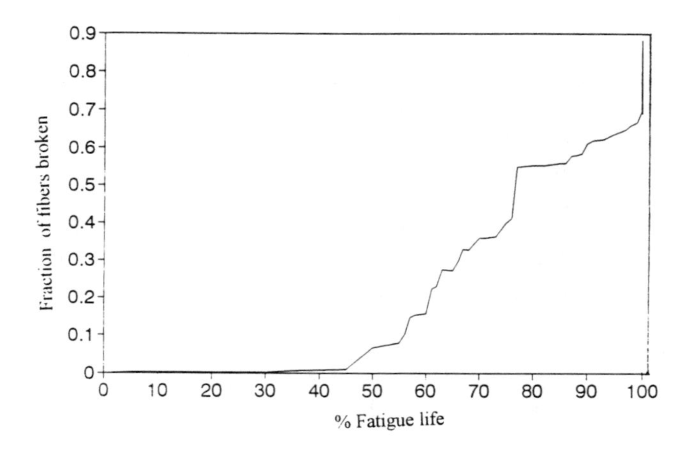

Fig. 2 Variation of the fraction of fibers broken with the percentage of fatigue life during tension-tension fatigue testing up to failure for 0° unidirectional composite.

REFERENCES

1. Krzysztof J. Konsztowicz and Denise Fontaine, J. Am. Ceramic Soc. 73 (10), 2809-2814 (1990).

2. E. Santos-Leal and R.J. Lopez, Measurement Science & Technology 6(2), 188-195 (1995).

3. Atsumu Ishida, Masaru Miyayama and Hiroaki Yanagida, J. Am. Ceram. Soc. 77 (4), 1057-1061 (1994).

4. Norio Muto, Hiroaki Yanagida, Masaru Miyayama, Teruyuki Nakatsuji, Minoru Sugita and Yasushi Ohtsuka, J. Ceramic Soc. Japan 100 (4), 585-588 (1992).

5. Xiaojun Wang and D.D.L. Chung, Smart Mater. Struct., in press.

BRICK STRUCTURE IMPROVED BY USING CEMENT MORTAR CONTAINING SHORT CARBON FIBERS

MINGGUANG ZHU* and D.D.L. CHUNG*
*Composite Materials Research Laboratory, Furnas Hall, State University of New York at Buffalo, Buffalo, NY 14260-4400, U.S.A.

ABSTRACT

The addition of short carbon fibers in the optimum amount of 0.5% of the cement weight to mortar increased the brick-to-mortar bond strength by 150% under tension and 110% under shear when the gap between the adjoining bricks was fixed, and by 50% under tension and 44% under shear when the gap between the adjoining bricks was allowed to freely decrease due to the weight of the brick above the joint. This effect is attributed to the decrease of the drying shrinkage by the fiber addition. The drying shrinkage decrease was particularly large at 2 to 24 h of curing. At 24 h, the shrinkage was decreased by 50% by the addition of fibers in the amount of 0.5% of the cement weight. Fibers in excess of the optimum amount gave less bond strengthening due to increased porosity in the mortar.

INTRODUCTION

The durability of brick constructions is limited by the bonding strength between brick and mortar [1]. The bonding strength can be affected by the mortar strength, mortar shrinkage, brick strength, joint thickness, interface morphology and chemical bond [1-4]. In particular, the drying shrinkage of the mortar greatly weakens the brick-to-mortar bond because of the fact that the mortar shrinks as it cures in a brick-to-brick joint, while the adjoining bricks do not shrink. This situation is even more serious in the head joints (i.e., vertical joints) of the masonry wall and in the tuck pointing of old brick constructions, since the gaps between the bricks are fixed.

Carbon fiber reinforced mortar provides an avenue for increasing the brick-mortar bonding strength, for the carbon fibers can increase the mortar strength and decrease the drying shrinkage as well [5]. It was reported that, by adding 0.35 vol.% short carbon fibers to new mortar, the shear bond strength of the old-new mortar joint was increased by up to 89% [6]. It is important to investigate the usefulness of carbon fiber reinforced mortar in the masonry field for the purpose of obtaining brick-mortar bonds of improved strength.

In this work, it was found that the addition of carbon fibers in the optimum amount of 0.5% by weight of cement greatly increased the brick-to-mortar bond strength. The fractional increase was 150% under tension and 110% under shear when the gap between the adjoining bricks was fixed, and was 50% under tension and 44% under shear when the gap between the adjoining bricks was allowed to freely decrease due to the weight of the brick above the joint. Carbon fiber contents beyond the optimum gave lower strengths due to increased porosity.

EXPERIMENT

Materials and sample preparation

The carbon fibers used were 5 mm long, 10 μm in diameter, unsized and based on isotropic pitch. They were provided as Carboflex by Ashland Petroleum Co. (Ashland, Kentucky) and were used in amounts of 0, 0.25, 0.5, 1 and 2% of the cement weight (equivalent

Mat. Res. Soc. Symp. Proc. Vol. 434

to 0, 0.175, 0.35, 0.70 and 1.4 vol.% respectively). The cement was Portland cement, Type I, from Lafarge Corporation (Southfield, MI). The sand was natural sand (100% passing 2.36 mm sieve, 99.91% SiO_2). Silica fume was used in the amount of 15% of the weight of the cement. Methylcellulose was used in the amount of 0.4% of the cement weight. The defoamer (Colloids 1010) always used along with methylcellulose was in the amount of 0.13 vol.%. The water reducing agent powder was TAMOL SN (Rohm and Haas), which contained 93-96% sodium salt of a condensed naphthalenesulfonic acid. The water-reducing-agent/cement ratio was 2% for fiber contents of 0, 0.25 and 0.5% of the weight of the cement, and was 3% when the fiber content was 1% or 2% of the weight of the cement. In general, the slump of carbon fiber reinforced cement tends to decrease with increasing carbon fiber content. The water/cement ratio was 0.475 at all fiber contents.

The bricks used were of two types, namely red 3-hole bricks (202 x 90 x 56 mm) and red paving bricks (197 x 89 x 57 mm). They were extruded, wire cut, clay units that comply with the requirements of ASTM C216 and ASTM C902. The 3-hole bricks were used for tensile debonding testing, such that the joining surface was the smooth edge surface of the brick. The paving bricks (with no hole) were used for shear debonding testing, such that the joining surface was the rough large surface of the brick.

Methylcellulose was dissolved in water and then the fibers and defoamer were added and stirred by hand for about 2 min. Then sand, cement, water and water reducing agent were successively added and mixed in a Hobart mixer with a flat beater for 5 min.

The cement in the mortar requires water to initiate and continue the hydration for setting and hardening. Since the bricks may absorb water from the mortar, thereby decreasing the bonding strength [7], all the bricks in this work were immersed in water for about 3 h. Then they were wiped dry and the mortar was applied on them. By water immersion, the possibility of water absorption by the bricks was reduced.

The gap between the adjoining brick surfaces were either kept fixed during curing (by placing two parallel 11-mm diameter steel bars between the bricks and near the edges of the bricks and using a C-clamp to keep the gap between the bricks constant, as shown in Fig. 1) or allowed to decrease freely during curing due to the weight of the brick above the joint (by not having any steel bar between the bricks). In the first case, the steel bars were placed on the

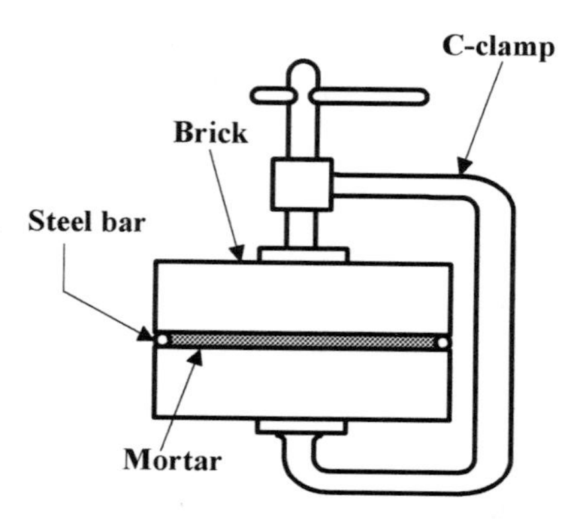

Fig. 1 Joint configuration for the case of the gap being fixed.

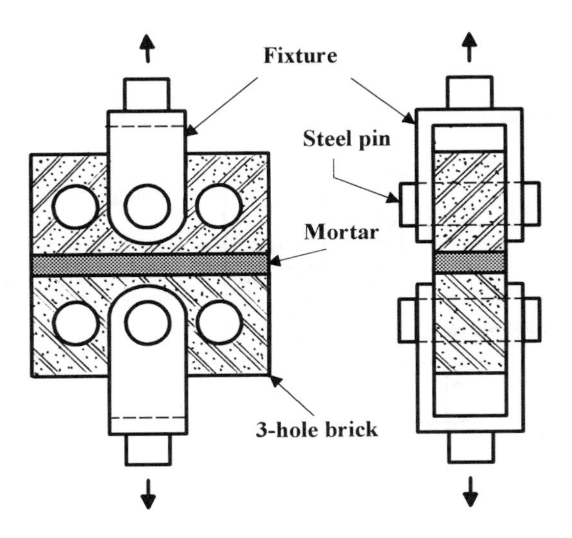

Fig. 2 Sample configuration for tensile debonding testing.

lower brick and the mortar was placed to fill the space between the bars. Then the upper brick was placed on top of the bars and a C-clamp was used to hold everything together. Thus, the gap between the bricks was kept constant during the curing of the mortar. The excess mortar was squeezed out from the gap and was scraped off. Special fixtures were attached to the steel bars during the sample preparation in order to make sure that the bars were in the right position and were parallel to the edge of the brick. In the latter case, the same procedure was followed, except that the C-clamp was not used and the two steel bars were carefully withdrawn from the joint prior to curing; the pressure due to the brick above the joint was 2.1 kPa for the tensile debonding test and was 1.5 kPa for the shear debonding test. The former case is relevant to the joining of the vertical edges of side-by-side bricks in a brick wall and to the tuck pointing (repair) of brick structures. The latter case is relevant to the joining of the horizontal surfaces of bricks on top of one another in a brick wall.

No vibrator was used on the mortar. Curing of the mortar took place in air, with the temperature at 20-25°C and the relative humidity at 55-65%.

Testing

Debonding tests were conducted under tension (Fig. 2) and shear (Fig. 3), using a Sintech 2/D screw-type mechanical testing system at 7 days of curing. Under tension, the force was applied in a direction perpendicular to the joint. Under shear, the force was applied in a direction parallel to the joint. Six samples of each fiber content were tested for each configuration. In both cases, debonding occurred mostly at the brick-mortar interface. Table 1 gives the joint strengths obtained for various fiber contents. At any fiber content, the joint strength, whether under tension or shear, was much higher when the gap was not fixed than when the gap was fixed, as expected. Whether under tension or shear, and whether the gap was fixed or not, the

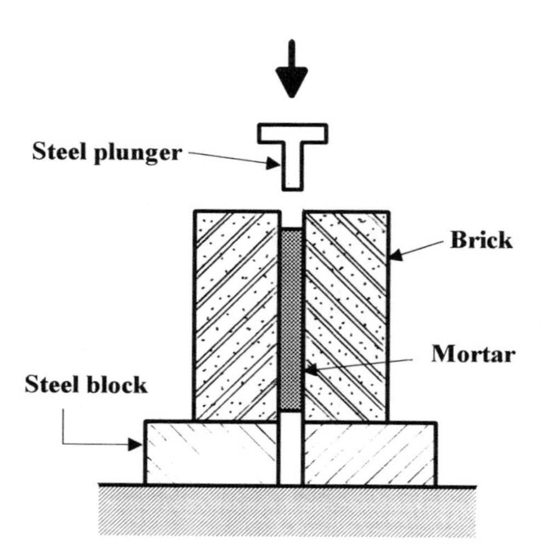

Fig. 3 Sample configuration for shear debonding testing.

joint strength increased with increasing fiber content up to 0.5% of the cement weight and then decreased with further increase in the fiber content.

The ASTM method C29/C29M-91a was used to measure the porosity and density of mortars with different fiber contents. As shown in Table 2, the porosity increased with increasing fiber content. Since the fiber addition increased the viscosity of the mortar mixture and no vibrator was used, the chance that the air voids generated during the mixing period would remain in the mortar after curing increased with increasing fiber content. This is believed to be the main reason why the mortar porosity increased with the fiber content.

Optical microscopy of the cross section of the brick-mortar interface for the case without fibers shows that, in the presence of pressure during curing (i.e., the case of the gap being not

Table 1. Brick/Brick joint strength (kPa)

Fiber/cement ratio (%)	Tensile*		Shear*	
	Gap not fixed	Gap fixed	Gap not fixed	Gap fixed
0	348.5(±5%)	39.4(±5%)	437.3(±4%)	126.5(±5%)
0.25	406.1(±4%)	61.4(±5%)	506.6(±4%)	178.4(±6%)
0.5	524.4(±3%)	97.3(±4%)	631.7(±5%)	267.5(±3%)
1.0	285.6(±6%)	49.7(±6%)	429.1(±5%)	224.0(±6%)
2.0	174.1(±5%)	35.8(±8%)	454.3(±6%)	217.7(±7%)

*Average of 6 tests

Table 2. Porosity and density of mortars

Fiber/cement ratio (%)	Porosity (%)	Density (g/cm^3)
0	1.5	1.937
0.25	5.8	1.824
0.5	10.1	1.741
1.0	24.8	1.458
2.0	29.5	1.371

fixed), the mortar filled the irregularities in the brick surface, but, in the absence of pressure (i.e., the case of the gap being fixed), the mortar did not, thus leaving voids at these irregularities. This is mainly why the joint strength (Table 1) is higher for the case of the gap being not fixed. Another reason is that the shrinkage stress at the brick-mortar interface is less when the gap is not fixed. Similar microscopy for the case of the gap being not fixed shows that the pores became more abundant as the fiber content increased and some of the pores even occupied the irregularities in the brick surface. That the joint strength decreases with increasing porosity (air content) has been previously reported [7].

The drying shrinkage strain during curing was measured for curing ages from 2 to 24 h using a Perkin-Elmer thermal mechanical analyzer (TMA7) operated at room temperature as a

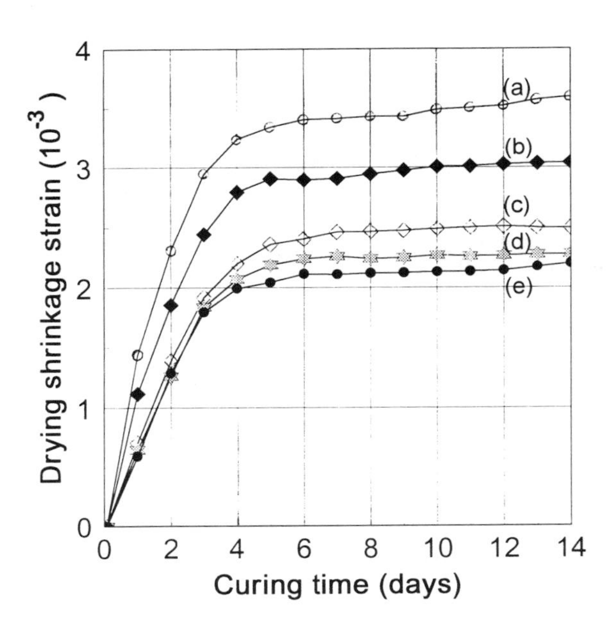

Fig. 4 Drying shrinkage strain vs. curing time from 2 h to 14 days. Fibers in amounts of (a) 0%, (b) 0.25%, (c) 0.5%, (d) 1.0% and (e) 2.0% of the cement weight.

function of time and for curing ages from 1 to 14 days in accordance with ASTM C490-83a. In the former method, the 5-mm diameter cylindrical specimen was 12 mm long along the cylindrical axis (direction of length measurement) and the length change measurement accuracy was ± 0.00001 mm (± 0.01 μm). In the latter method, the 25.4 mm x 25.4 mm rectangular specimen was 286 mm long in the direction of length measurement and the length change measurement accuracy was ±0.0025 mm. Fig. 4 shows the complete picture from 2 h to 14 days. The fiber addition reduced the drying shrinkage most severely at low fiber contents, i.e., 0.25 and 0.5% of the cement weight. Further increase in the fiber content gave only incremental effects.

DISCUSSION

This work shows that the addition of carbon fibers to mortar in the amount of only 0.5% of the cement weight greatly increases the brick-to-mortar bond strength. The fractional strength (tension or shear) increase is larger when the gap between the adjoining bricks is fixed than when the gap is not fixed. The fractional strength increase is similar under tension and shear. This bond strengthening is practically useful. When considering the high labor cost of building or repairing a brick structure, the increased mortar cost due to the fiber addition is incremental in the overall cost. When considering the mortar price alone, the fiber addition in the amount of 0.5% of the cement weight (together with the addition of methylcellulose, silica fume and defoamer) results in a mortar price increase of approximately 30%.

The fiber addition decreases the drying shrinkage and increases the porosity of the mortar. The decrease in the drying shrinkage helps to increase the bond strength, whereas the increase in the porosity decreases the bond strength. Thus the overall effect is that the highest bond strength occurs at an intermediate fiber content of 0.5% of the cement weight.

REFERENCES

1. Edgar Jung, in Brick and Block Masonry, edited by de Courcy, John W., Elsevier Applied Science, 1988, pp. 182-193.

2. Caspar J.W.P. Groot, in Brick and Block Masonry, edited by de Courcy, John W., Elsevier Applied Science, 1988, pp. 175-181.

3. Arnold W. Hendry, Structural Brickwork, Wiley, 1981, pp. 29-31.

4. S.J. Lawrence, and H.T. Cao, in Brick and Block Masonry, edited by de Courcy, John W., Elsevier Applied Science, 1988, pp. 194-204.

5. Pu-Woei Chen and D.D.L. Chung, Composites, V. 24, No. 1, 1993, pp. 33-52.

6. Pu-Woei Chen, Xuli Fu and D.D.L. Chung, Cement Concr. Res. 25(3), 491-496 (1995).

7. Bruce T. Wright, Rick D. Wilkins, and George W. John, Masonry: Design and Construction, Problems and Repair, ASTM STP 1180, edited by Melander, John M., and Lauersdorf, Lynn R., American Society for Testing and Material, Philadelphia, 1993, pp. 197-210.

AUTHOR INDEX

SUBJECT INDEX